Advances in MATHEMATICAL ECONOMICS

Aims and Scope. The project is to publish *Advances in Mathematical Economics* once a year under the auspices of the Research Center for Mathematical Economics. It is designed to bring together those mathematicians who are seriously interested in obtaining new challenging stimuli from economic theories and those economists who are seeking effective mathematical tools for their research.

The scope of *Advances in Mathematical Economics* includes, but is not limited to, the following fields:

- Economic theories in various fields based on rigorous mathematical reasoning.
- Mathematical methods (e.g., analysis, algebra, geometry, probability) motivated by economic theories.
- Mathematical results of potential relevance to economic theory.
- Historical study of mathematical economics.

Authors are asked to develop their original results as fully as possible and also to give a clear-cut expository overview of the problem under discussion. Consequently, we will also invite articles which might be considered too long for publication in journals.

More information about this series at http://www.springer.com/series/4129

Toru Maruyama
Editor

Advances in Mathematical Economics

Volume 23

Springer

Editor
Toru Maruyama
Professor Emeritus
Keio University
Tokyo, Japan

ISSN 1866-2226 ISSN 1866-2234 (electronic)
Advances in Mathematical Economics
ISBN 978-981-15-0715-1 ISBN 978-981-15-0713-7 (eBook)
https://doi.org/10.1007/978-981-15-0713-7

This Springer imprint is published by the registered company Springer Nature Singapore Pte Ltd.
The registered company address is: 152 Beach Road, #21-01/04 Gateway East, Singapore 189721, Singapore

Editor's Preface

The first issue of *Advances in Mathematical Economics* was published in 1999. It was basically a collection of articles presented at the Symposium on Mathematical Analysis in Economic Theory, which had been held in Tokyo in the autumn 1997, on the occasion of Professor Gérard Debreu's visit to Japan. Looking at the pictures in my photo album, memories of various scenes of the conference come up to my mind.

Since then, we have been continuing to publish this series annually for 20 years in order to bring together those mathematicians who are seriously interested in obtaining new stimuli from economic theories and those economists who are seeking effective mathematical tools for their research. I really miss several leading figures in our project who passed away during these 20 years, including Professors Debreu, Kiyosi Itô, Leonid Hurwicz, Marcel K. Richter, Michihiro Ohyama, Yoichiro Takahashi and Masaya Yamaguti.

I must confess that the editorial board has been suffering from the difficulty in collecting abundant contributions to our series. It is, in part, due to the downsizing of the population in the mathematical economics world. However, the problem is not all that simple. Taking account of the advice of our publisher, we reflected well upon the policy of our editorial work. And finally, we arrived at the decision to close this series and to grope for certain new style of publication more effective to promote scientific communications between mathematicians and economists. We are still discussing various possibilities together with our publisher.

This 23rd issue is projected as a special volume to celebrate the completion of the series.

On behalf of the editorial board of the series, I would like to devote my sincere thanks to many authors and readers who have been supporting our venture. It is also impossible for me to exaggerate my deep gratitude to the members of the board for their warmest cooperation. In particular, Professor Shigeo Kusuoka's contribution as my partner deserves a special mention. I have to acknowledge the generous support of the Oak Society Inc., which provided an office space as well as the secretarial service for our editorial work.

We sincerely hope for the nice rebirth of our publication.

Tokyo, Japan
August 9, 2019

Toru Maruyama

Contents

Evolution Problems with Time-Dependent Subdifferential Operators

Charles Castaing, Manuel D. P. Monteiro Marques, and Soumia Saïdi

Abstract In this paper, we study the existence of solutions for evolution inclusions governed by a time-dependent subdifferential operator with multivalued perturbations in a separable Hilbert space. Several applications are presented.

Keywords Fractional differential · Skorokhod · Subdifferential · Sweeping process · Time-dependent evolution · Variational

Article type: Research Article
Received: June 1, 2019
Revised: July 3, 2019

1 Introduction

Let H be a separable Hilbert space and $[0, T]$ $(T > 0)$ be an interval of $\mathscr{R}$. Consider the subdifferential of a time-dependent proper lower semicontinuous convex function $\varphi(t, \cdot)$ of H into $[0, +\infty]$, denoted $\partial\varphi(t, \cdot)$ and the effective domain

JEL Classification: C62, C65
Mathematics Subject Classification (2010): 34A60, 49J52, 47J22, 34G25, 49J53

C. Castaing
IMAG, University of Montpellier, CNRS, Montpellier, France

M. D. P. Monteiro Marques (✉)
CMAFcIO, Departamento de Matemática, Faculdade de Ciências da Universidade de Lisboa, Lisboa, Portugal
e-mail: mdmarques@fc.ul.pt

S. Saïdi
LMPA Laboratory, Department of Mathematics, Mohammed Seddik Ben Yahia University, Jijel, Algeria

T. Maruyama (ed.), *Advances in Mathematical Economics*, Advances in Mathematical Economics 23, https://doi.org/10.1007/978-981-15-0713-7_1

of the function $\varphi(t,\cdot)$ denoted dom $\varphi(t,\cdot)$, for each $t \in [0,T]$. In [23], Peralba has established existence and uniqueness results for absolutely continuous solutions to the initial value problems

$$\begin{cases} -\dot{u}(t) \in \partial\varphi(t,u(t)), \text{ a.e. } t \in [0,T] \\ u(0) = u_0 \in \text{dom } \varphi(0,\cdot). \end{cases} \tag{P_1}$$

Firstly, we study the existence of absolutely continuous solutions to evolution inclusions of the form

$$\begin{cases} -\dot{u}(t) \in \partial\varphi(t,u(t)) + F(t,u(t)), \text{ a.e. } t \in [0,T] \\ u(0) = u_0 \in \text{dom } \varphi(0,\cdot), \end{cases} \tag{P_2}$$

where $F : [0,T] \times H \to H$ is a convex weakly compact valued mapping satisfying the condition $F(t,x) \subset \beta(t)(1 + ||x||)\overline{B}_H$, for all $t \in [0,T]$ and $x \in H$, for some non-negative function $\beta(\cdot) \in L^2_{\mathscr{R}}([0,T])$. Problem (P_2) includes as special cases several problems in applied mathematics, e.g. in Control Theory and in Mathematical Economics. Noteworthy subclasses of problems are those of evolution inclusions governed by a sweeping process (or Moreau process [20]), associated with a closed convex valued and absolutely continuous mapping $C : [0,T] \to H$

$$-\dot{u}(t) \in N_{C(t)}(u(t)) + F(t,u(t)), \text{ a.e. } t \in [0,T] \tag{P_3}$$

and parabolic variational inequalities of the form

$$\begin{cases} -\dot{u}(t) \in \partial\varphi(t,u(t)) + N_{C(t,u(t))}(u(t)), \text{ a.e. } t \in [0,T] \\ u(0) = u_0 \in C(0,u_0), \end{cases} \tag{P_4}$$

where $N_{C(t,x)}(x)$ denotes the normal cone of the closed convex moving set $C(t,x)$ at x.

Existence and relaxation of solutions of problems with subdifferential operators have been studied in the articles [4, 14, 15, 24–26, 28, 29].

Secondly, we state the existence and uniqueness of an absolutely continuous solution to the evolution inclusion of the form

$$f(t) - Au(t) \in \partial\varphi(t, \frac{du}{dt}(t)), \text{ a.e. } t \in [0,T] \tag{P_5}$$

where $f : [0,T] \to H$ is a bounded continuous mapping, φ is a normal (lower semicontinuous) convex integrand and A is a linear continuous coercive operator in a finite-dimensional space H. We also study a second variant dealing with absolutely continuous solutions to evolution inclusions governed by a convex sweeping process

$$f(t) - Au(t) \in N_{C(t)}(\frac{du}{dt}(t)), \text{ a.e. } t \in [0,T] \tag{P_6}$$

where $C : [0, T] \to H$ is a bounded closed convex moving set having a continuous variation. Problems (P_5) and (P_6) are related to a class of parabolic variational inequalities in applied mathematics (cf. Barbu [3], section 5.2) and to the evolution inclusions studied by Adly et al. [1] ((1.5), (1.6), (5.19)), namely

$$-A_1\frac{du}{dt}(t) - A_0u(t) + f(t) \in N_{C(t)}(\frac{du}{dt}(t)), \text{ a.e. } t \in [0, T], \tag{P_7}$$

where A_0, A_1 are two bounded linear symmetric positive operators in H, $f : [0, T] \to H$ is a bounded continuous mapping and $C : [0, T] \to H$ is a bounded closed convex moving set having a continuous variation.

The paper is organized as follows. Section 2 is devoted to notation and definitions. In particular, we recall and develop in Sects. 3 and 4 several regularity properties of the maximal monotone operator $A(t) := \partial\varphi(t, \cdot)$ ($t \in [0, T]$) in H and its extension $\mathscr{A}$ to $L^2_H([0, T])$. In the last sections, we state the main results of the paper and their applications to Skorohod problems, to a fractional differential inclusion and to a second order evolution inclusion.

2 Notation and Preliminaries

In the following, we present some basic definitions, and notations which will be used throughout this paper. Let $[0, T]$ ($T > 0$) be an interval of $\mathscr{R}$, H a separable Hilbert space, $\mathscr{L}([0, T])$ the sigma-algebra of Lebesgue measurable sets in $[0, T]$ and $\mathscr{B}(H)$ the sigma-algebra of Borel sets in H. We denote by $\overline{B}_H(x, r)$ the closed ball of center x and radius r and $\overline{B}_H$ the closed unit ball of H. The space $\mathscr{C}_H([0, T])$ of all continuous functions $x : [0, T] \to H$ is endowed with the usual sup–norm. For $p \in [1, +\infty[$ (resp. $p = +\infty$), we denote by $L^p_H(I)$ (resp. $L^\infty_H(I)$) the space of measurable functions $x : [0, T] \to H$ such that $\int_0^T ||x(t)||^p dt < +\infty$ (resp. which are essentially bounded) endowed with the usual norm $||x||_{L^p_H([0,T])} = (\int_0^T ||x(t)||^p dt)^{\frac{1}{p}}$, $1 \le p < +\infty$ (resp. endowed with the usual essential supremum norm $||x||_\infty$). We recall that the topological dual of $L^1_H([0, T])$ is $L^\infty_H([0, T])$. Let φ be a lower semicontinuous convex function from H into $\mathscr{R} \cup \{+\infty\}$ which is proper in the sense that its effective domain dom φ defined by

$$\text{dom } \varphi := \{x \in H : \varphi(x) < +\infty\}$$

is nonempty and, as usual, its Fenchel conjugate is defined by

$$\varphi^*(v) := \sup_{x \in H}[\langle v, x\rangle - \varphi(x)].$$

It is often useful to regularize φ via its Moreau envelope

$$\varphi_\lambda(x) := \inf_{y\in H}[\varphi(y) + \frac{1}{2\lambda}||x - y||^2] \text{ for } \lambda > 0.$$

The subdifferential $\partial\varphi(x)$ of φ at $x \in \text{dom } \varphi$ is

$$\partial\varphi(x) = \{v \in H : \varphi(y) \geq \langle v, y - x\rangle + \varphi(x) \ \forall y \in \text{dom } \varphi\}$$

and its effective domain is $\text{D}(\partial\varphi) = \{x \in H : \partial\varphi(x) \neq \emptyset\}$.

It is well known that if φ is a proper lower semicontinuous convex function, then its subdifferential operator $\partial\varphi$ is a maximal monotone operator.

If $A : H \to H$ is a maximal monotone operator, then its graph is sequentially strongly-weakly closed, that is, if $x = \lim_{n\to\infty} x_n$ strongly in H and $y = \lim_{n\to\infty} y_n$ weakly in H, where $x_n \in \text{D}(A)$ and $y_n \in Ax_n$, then $x \in \text{D}(A)$ and $y \in Ax$.

For any set-valued operator $A : H \to H$, the range of A is

$$R(A) := \bigcup_{x\in H} Ax.$$

Denote by I_H the identity operator on H and by A^{-1} the inverse operator of A. For any subset S of H, $y \mapsto \delta^*(y, S)$ is the support function of S. A set-valued mapping F is scalarly upper semicontinuous if, for any $y \in H$, the real-valued function $x \mapsto \delta^*(y, F(x))$ is upper semicontinuous. We refer to [8], for details concerning convex analysis and measurable set-valued mappings.

3 Results with Single-Valued Time-Dependent Perturbation

We recall here an important result due to Peralba (see [23]).

Theorem 1 *Let $\varphi : [0, T] \times H \to [0, +\infty]$ be such that*

(H_1) *for each $t \in [0, T]$, the function $x \longmapsto \varphi(t, x)$ is proper lower semicontinuous and convex,*

(H_2) *there exist a ρ-Lipschitz function $k : H \longrightarrow \mathscr{R}^+$ and an absolutely continuous function $a : [0, T] \to \mathscr{R}$, with a non-negative derivative $\dot{a} \in L^2_{\mathscr{R}}([0, T])$, such that for every $(t, s, x) \in [0, T] \times [0, T] \times H$*

$$\varphi^*(t, x) \leq \varphi^*(s, x) + k(x)|a(t) - a(s)|.$$

Let $u_0 \in \operatorname{dom}\varphi(0,\cdot)$. *Then, the differential inclusion*

$$\begin{cases} -\dot{u}(t) \in \partial\varphi(t, u(t)), \text{ a.e. } t \in [0, T] \\ u(0) = u_0 \in \operatorname{dom}\varphi(0,\cdot) \end{cases} \tag{P_1}$$

has a unique absolutely continuous solution $u(\cdot)$ *on* $[0, T]$. *Moreover, for all* t, $u(t) \in \operatorname{dom}\varphi(t,\cdot)$ *and the function* $t \mapsto \varphi(t, u(t))$ *is absolutely continuous on* $[0, T]$.

The following proposition contains crucial estimates in the study of evolution problems with perturbation $-\dot{u}(t) \in \partial\varphi(t, u(t)) + h(t)$ [24, Propositions 3.3 and 3.4].

Proposition 1

(*a*) *The unique solution* $u(\cdot)$ *of* (P_1) *satisfies*

$$||\dot{u}||_{L^2_H([0,T])} \le [\sqrt{T}k(0)||\dot{a}||_{L^2_{\mathscr{R}}([0,T])} + \frac{\rho^2}{4}||\dot{a}||^2_{L^2_{\mathscr{R}}([0,T])} + \varphi(0, u_0) - \varphi(T, u(T))]^{\frac{1}{2}}$$
$$+ \frac{\rho}{2}||\dot{a}||_{L^2_{\mathscr{R}}([0,T])}.$$

(*b*) *If* $h \in L^2_H([0, T])$ *and* $u_0 \in \operatorname{dom}\varphi(0,\cdot)$, *then the following problem*

$$\begin{cases} -\dot{u}(t) \in \partial\varphi(t, u(t)) + h(t), \text{ a.e. } t \in [0, T] \\ u(0) = u_0 \in \operatorname{dom}\varphi(0,\cdot), \end{cases}$$

admits a unique absolutely continuous solution $u(\cdot)$ *that satisfies*

$$||\dot{u}|| \le \frac{1}{2}(\rho + 1)||\dot{a} + |h||| + ||h||$$
$$+ [\sqrt{T}k(0)||\dot{a} + |h||| + \frac{(\rho + 1)^2}{4}||\dot{a} + |h|||^2 + \varphi(0, u_0) - \varphi(T, u(T))]^{\frac{1}{2}},$$

where $|h|$ *is the function defined by* $|h| : t \mapsto ||h(t)||$ *for all* $t \in [0, T]$ *and the* $L^2_{\mathscr{R}}([0, T])$ *norm is denoted by* $||\cdot||$, *for shortness.*

4 Extension to $L^2_H([0, T])$ of Subdifferential Operator $\partial\varphi(t,\cdot)$

In order to prove the main result of this section (Proposition 6) we recall some results developed by Peralba [23].

Proposition 2 *Let $\varphi : H \longrightarrow [0, +\infty]$ be proper lower semicontinuous convex, then, for any $z \in H$, the function*

$$x \longmapsto \phi(x) = \frac{1}{2}||x - z||^2 + \varphi(x)$$

has a strict minimum. The unique minimum point, where this minimum value is reached, is called proximal point of z relatively to φ and is denoted by $prox_\varphi\, z$. The mapping $prox_\varphi$ is a contraction[1] *and*

$$\forall x \in H : \ \phi(x) \geq \phi(prox_\varphi\, z).$$

Proposition 3 *Using the notation of Proposition 2, one has*

$$\forall x \in H : \ \phi(x) \geq \phi(prox_\varphi\, z) + \frac{1}{2}||x - prox_\varphi\, z||^2.$$

Proof As $prox_\varphi\, z$ is the point where the function ϕ reaches its minimum, one has

$$0 \in \partial\phi(prox_\varphi\, z);$$

by additivity (see [19]) of the subdifferentials of $x \mapsto \varphi(x)$ and $x \mapsto \frac{1}{2}||x - z||^2$, this may also be expressed as

$$z - prox_\varphi\, z \in \partial\varphi(prox_\varphi\, z), \quad \text{i.e.,}$$

$$\forall x \in H : \ \langle z - prox_\varphi\, z, x - prox_\varphi\, z\rangle + \varphi(prox_\varphi\, z) \leq \varphi(x),$$

which is the required estimate.

Proposition 4 *Let $\varphi : H \longrightarrow [0, +\infty]$ be a proper lower semicontinuous and convex function; then, for all $\lambda > 0$, one has*

$$\partial\varphi_\lambda(x) = prox_{\frac{1}{\lambda}\varphi^*} \frac{x}{\lambda} = \frac{1}{\sqrt{\lambda}} prox_{\varphi^*(\frac{\cdot}{\sqrt{\lambda}})} \frac{x}{\sqrt{\lambda}} = \frac{1}{\lambda} prox_{\lambda\varphi^*(\frac{\cdot}{\lambda})}\, x.$$

Proposition 5 *If φ satisfies conditions (H_1) and (H_2) then, for each $\lambda > 0$, there exists a function M_λ defined from H to $\mathscr{R}^+$, 4ρ-Lipschitz and satisfying, for any s and t in $[0, T]$ and any $z \in H$*

$$||prox_{\lambda\varphi^*(t, \frac{\cdot}{\lambda})}\, z - prox_{\lambda\varphi^*(s, \frac{\cdot}{\lambda})}\, z||^2 \leq M_\lambda(z)|a(t) - a(s)|.$$

[1] For more details on the prox mappings, we refer to Moreau [18].

Proof Having fixed z, set

$$p(t) = prox_{\lambda\varphi^*(t,\frac{\cdot}{\lambda})}\, z \text{ and } \gamma(t,x) = \frac{1}{2}||x - z||^2 + \lambda\varphi^*(t, \frac{x}{\lambda}).$$

Applying Proposition 3 to each convex lower semicontinuous function $\lambda\varphi^*(t, \frac{\cdot}{\lambda})$, one has

$$\forall x \in H : \gamma(t,x) \geq \gamma(t, p(t)) + \frac{1}{2}||x - p(t)||^2.$$

Then, for each $x \in H$ and $s \in [0, T]$, by hypothesis (H_2) and the previous inequality, one has

$$\begin{aligned}\gamma(s,x) &= \frac{1}{2}||x - z||^2 + \lambda\varphi^*(s, \frac{x}{\lambda})\\ &\geq \frac{1}{2}||x - z||^2 + \lambda\varphi^*(t, \frac{x}{\lambda}) - \lambda k(\frac{x}{\lambda})|a(t) - a(s)|\\ &\geq \gamma(t, p(t)) + \frac{1}{2}||x - p(t)||^2 - \lambda k(\frac{x}{\lambda})|a(t) - a(s)|.\end{aligned}$$

We may also write

$$||x - p(t)||^2 \leq 2\lambda k(\frac{x}{\lambda})|a(t) - a(s)| + 2[\gamma(s,x) - \gamma(t, p(t))];$$

if applied to $x = p(s)$, this yields

$$||p(t) - p(s)||^2 \leq 2\lambda[k(\frac{p(t)}{\lambda}) + k(\frac{p(s)}{\lambda})]|a(t) - a(s)|. \tag{1}$$

Indeed, $\gamma(s, p(s)) \leq \gamma(s, p(t))$, because $p(s)$ is the point where $\inf_{x\in H} \gamma(s,x)$ is reached; therefore,

$$\begin{aligned}\gamma(s, p(s)) - \gamma(t, p(t)) &\leq \gamma(s, p(t)) - \gamma(t, p(t))\\ &= \lambda\varphi^*(s, \frac{p(t)}{\lambda}) - \lambda\varphi^*(t, \frac{p(t)}{\lambda})\\ &\leq \lambda k(\frac{p(t)}{\lambda})|a(t) - a(s)|.\end{aligned}$$

Clearly, now the inequality in the statement of the Proposition may be obtained from (1) by taking

$$M_\lambda(z) := 4\lambda \sup_{t\in[0,T]} k(\frac{p(t)}{\lambda}),$$

which is well defined if we show that p and $t \mapsto k(\frac{p(t)}{\lambda})$ are continuous functions on $[0, T]$. Since k is ρ-Lipschitz continuous, then we obtain from the above inequalities

$$||p(t) - p(s)||^2 \leq 2\rho||p(t) - p(s)|||a(t) - a(s)| + 4\lambda k(\frac{p(t)}{\lambda})|a(t) - a(s)|;$$

hence,

$$||p(t) - p(s)|| \leq \rho|a(t) - a(s)| + \left[4\lambda k(\frac{p(t)}{\lambda})|a(t) - a(s)| + \rho^2|a(t) - a(s)|^2\right]^{\frac{1}{2}}$$

and p is continuous on $[0, T]$.

Finally, M_λ is 4ρ-Lipschitz, since

$$M_\lambda(z) - M_\lambda(z') \leq 4\lambda \sup_{t\in[0,T]} |k(\frac{p(t)}{\lambda}) - k(\frac{p'(t)}{\lambda})|,$$

where $p'(t) = prox_{\lambda\varphi^*(t,\frac{\cdot}{\lambda})}\, z'$ and k is ρ-Lipschitz:

$$|k(\frac{p(t)}{\lambda}) - k(\frac{p'(t)}{\lambda})| \leq \rho||\frac{p(t)}{\lambda} - \frac{p'(t)}{\lambda}|| \leq \frac{\rho}{\lambda}||z - z'||.$$

Denote by $A(t) := \partial\varphi(t, \cdot)$ the maximal monotone operator in H associated with $\partial\varphi(t, \cdot)$, $t \in [0, T]$, where φ satisfies conditions (H_1) and (H_2). Let us consider the operator $\mathscr{A} : L^2_H([0, T]) \to L^2_H([0, T])$ defined by

$$\mathscr{A}x = \{y \in L^2_H([0, T]) : y(t) \in A(t)x(t), \text{ a.e. } t \in [0, T]\}.$$

Then, $\mathscr{A}$ is well defined since by Theorem 1 the evolution inclusion

$$-\dot{u}(t) \in A(t)u(t) = \partial\varphi(t, u(t)), \text{ a.e. } t \in [0, T], \ u(0) \in \text{dom } \varphi(0, \cdot)$$

admits a unique absolutely continuous solution.

For the sake of completeness, we reproduce an important result from Peralba [23, Proposition 30].

Proposition 6 *If for any $t \in [0, T]$, $A(t) = \partial\varphi(t, \cdot)$ where φ satisfies conditions (H_1) and (H_2), then $\mathscr{A}$ is a maximal monotone operator.*

Proof It is obvious that $\mathscr{A}$ is monotone. Indeed, let $y_i \in \mathscr{A}x_i$, $i = 1, 2$, then,

$$\langle y_1 - y_2, x_1 - x_2\rangle_{L^2_H([0,T])} = \int_0^T \langle y_1(t) - y_2(t), x_1(t) - x_2(t)\rangle dt$$

with $\langle y_1(t) - y_2(t), x_1(t) - x_2(t)\rangle \geq 0$ for a.e. $t \in [0, T]$.

To prove that it is maximal monotone, it is enough to show that

$$R(I_{L^2_H([0,T])} + \mathscr{A}) = L^2_H([0, T]),$$

i.e. that given any $g \in L^2_H([0, T])$ there is a $v \in L^2_H([0, T])$ such that

$$g \in v + \mathscr{A}v.$$

Remark that for any $t \in [0, T]$, $A(t)$ is maximal monotone; then, if we set

$$v(t) = [I_H + A(t)]^{-1}g(t),$$

the function v satisfies

$$g(t) \in [I_H + A(t)]v(t).$$

It remains to prove that $v \in L^2_H([0, T])$.

One has

$$\begin{aligned} v(t) &= [I_H + A(t)]^{-1}g(t) = J_1(t)g(t) \\ &= g(t) - A_1(t)g(t) \\ &= g(t) - prox_{\varphi^*(t,\cdot)}g(t), \end{aligned}$$

where $J_\lambda = (I_H + \lambda A)^{-1}$, $\lambda > 0$ and $A_\lambda = \lambda^{-1}(I_H - J_\lambda)$ (with the help of Proposition 4). We take $h(t) := prox_{\varphi^*(t,\cdot)}g(t)$ and check that $||h||^2$ is integrable.

First, h is measurable, as it results from [5, Theorem 1, p. 174] applied to the measurable functions $i : [0, T] \to [0, T] : t \mapsto t$ and $g : [0, T] \to H : t \mapsto g(t)$ and to the continuous function $f : [0, T] \times H \to H : (t, x) \mapsto prox_{\varphi^*(t,\cdot)}\, x$.

Notice that the function f is continuous, since the applications $prox$ are contractions and by Proposition 5, one has

$$||f(t, x) - f(t_0, x_0)|| \leq ||x - x_0|| + \sqrt{M_1(x_0)}\sqrt{|a(t) - a(t_0)|}.$$

Let us prove that $||h||^2$ is integrable. One has

$$\begin{aligned} ||h(t)||^2 &= ||prox_{\varphi^*(t,\cdot)}g(t) - prox_{\varphi^*(t,\cdot)}0||^2 + ||prox_{\varphi^*(t,\cdot)}0||^2 \\ &\quad + 2\langle prox_{\varphi^*(t,\cdot)}g(t) - prox_{\varphi^*(t,\cdot)}0, prox_{\varphi^*(t,\cdot)}0\rangle \end{aligned}$$

$$\leq ||g(t)||^2 + ||prox_{\varphi^*(t,\cdot)}0||^2 + 2||g(t)||||prox_{\varphi^*(t,\cdot)}0||$$
$$= [||g(t)|| + ||prox_{\varphi^*(t,\cdot)}0||]^2.$$

Then, the function $||h||^2$ is integrable because the function $t \longmapsto prox_{\varphi^*(t,\cdot)}0$ is continuous and $g \in L^2_H([0,T])$.

Now, we are able to state the following useful proposition.

Proposition 7 *Let $A(t) := \partial\varphi(t,\cdot)$, $\forall t \in [0,T]$ where φ satisfies conditions (H_1) and (H_2). Let $(x_n)_n$ and $(y_n)_n$ be two sequences in $L^2_H([0,T])$ satisfying*

(i) $y_n(t) \in A(t)x_n(t)$, $\forall n \in \mathcal{N}$ *and for a.e.* $t \in [0,T]$,
(ii) $(x_n)_n$ *strongly converges to* x *in* $L^2_H([0,T])$,
(iii) $(y_n)_n$ *weakly converges to* y *in* $L^2_H([0,T])$.

Then, one has $y(t) \in A(t)x(t)$ a.e. $t \in [0,T]$.

Proof By (i), one has $y_n \in \mathscr{A}x_n$ where $\mathscr{A}$ is the maximal monotone operator associated with $A(t)$ given in Proposition 6. Since the graph of $\mathscr{A}$ is sequentially strongly-weakly closed, by $y_n \in \mathscr{A}x_n$ and by (ii) and (iii), one deduces that $y \in \mathscr{A}x$, so that coming back to the definition of $\mathscr{A}$, one gets $y(t) \in A(t)x(t)$ a.e. $t \in [0,T]$.

5 Main Results

5.1 Existence Results for Problem (P_2)

We begin with a compactness result.

Lemma 1 *Assume that φ satisfies conditions (H_1) and (H_2), and that* dom $\varphi(t,\cdot)$ *is ball-compact, for every $t \in [0,T]$. Let $X : [0,T] \to H$ be measurable with convex and weakly compact values, and $X(t) \subset \beta(t)\overline{B}_H, \forall t \in [0,T]$, where $\beta \in L^2_{\mathscr{R}^+}([0,T])$. Let S^2_X denote the set of all $L^2_H([0,T])$-selections of X. Then, the set $\mathscr{X} := \{u_f : f \in S^2_X\}$ of absolutely continuous solutions to the evolution inclusions*

$$\begin{cases} -\dot{u}_f(t) \in \partial\varphi(t, u_f(t)) + f(t), \text{ a.e. } t \in [0,T], \; f \in S^2_X \\ u_f(0) = u_0 \in \text{dom } \varphi(0,\cdot) \end{cases}$$

is a compact subset of $\mathscr{C}_H([0,T])$.

Proof For each $f \in S^2_X$, we have for a.e. $t \in I$, $||f(t)|| \leq \beta(t)$ so that, by the estimate given in Proposition 1, the solution set $\mathscr{X} := \{u_f : f \in S^2_X\}$ is bounded and equicontinuous in $\mathscr{C}_H([0,T])$. As $u_f(t) \in$ dom $\varphi(t,\cdot)$ and dom $\varphi(t,\cdot)$ is ball-compact for each $t \in [0,T]$, we conclude that $\mathscr{X}$ is relatively compact in $\mathscr{C}_H([0,T])$. Finally, we take a sequence (u_{f_n}) in $\mathscr{X}$, converging uniformly to a continuous mapping w. We may assume that (f_n) weakly converges in $L^2_H([0,T])$ to $f \in S^2_X$, since S^2_X is convex weakly compact in $L^2_H([0,T])$, and also that $(\dot{u}_{f_n})$ weakly converges to some $z \in L^2_H([0,T])$. It follows that $w(t) = u_0 + \int_0^t z(s)ds$ and $\dot{w} = z$. Since $u_{f_n} \in \mathscr{X}$, we have

$$-\dot{u}_{f_n}(t) \in \partial\varphi(t, u_{f_n}(t)) + f_n(t), \text{ a.e. } t \in [0,T]$$

so that

$$-\dot{u}_{f_n} - f_n \in \mathscr{A} u_{f_n},$$

where $\mathscr{A}$ is defined as in Proposition 6. As $(-\dot{u}_{f_n} - f_n)$ weakly converges to $-z - f$ in $L^2_H([0,T])$ and the graph of $\mathscr{A}$ is sequentially strongly-weakly closed in $L^2_H([0,T]) \times L^2_H([0,T])$ (since the extension $\mathscr{A} : L^2_H([0,T]) \to L^2_H([0,T])$ of the maximal monotone operator $A(t) := \partial\varphi(t,\cdot)$ is maximal monotone by Proposition 6), we deduce that

$$-z - f \in \mathscr{A} w$$

or, by using Proposition 7

$$-\dot{w}(t) \in \partial\varphi(t, w(t)) + f(t), \text{ a.e. } t \in [0,T].$$

Thus, w is the absolutely continuous solution to

$$-\dot{u}_f(t) \in \partial\varphi(t, u_f(t)) + f(t), \text{ a.e. } t \in [0,T],$$

i.e. $w = u_f \in \mathscr{X}$ and $\mathscr{X}$ is closed in $\mathscr{C}_H([0,T])$.

Theorem 2 *Assume that φ satisfies conditions (H_1) and (H_2), and that* dom $\varphi(t,\cdot)$ *is ball-compact, for every $t \in [0,T]$. Let $F : [0,T] \times H \to H$ be a convex weakly compact valued mapping satisfying:*

(1) *For each $e \in H$, the scalar function $\delta^*(e, F(\cdot,\cdot))$ is $\mathscr{L}([0,T]) \times \mathscr{B}(H)$-measurable,*
(2) *For each $t \in [0,T]$, $F(t,\cdot)$ is scalarly upper semicontinuous on H, i.e., for each $e \in H$, the scalar function $\delta^*(e, F(t,\cdot))$ is upper semicontinuous on H,*
(3) *$F(t,x) \subset \beta(t)(1+||x||)\overline{B}_H$, for all $(t,x) \in [0,T] \times H$ for some non-negative function $\beta \in L^2_{\mathscr{R}}([0,T])$.*

Then the set of absolutely continuous solutions to the inclusion

$$\begin{cases} -\dot{u}(t) \in \partial\varphi(t, u(t)) + F(t, u(t)), \text{ a.e. } t \in [0,T] \\ u(0) = u_0 \in \text{dom}\,\varphi(0,\cdot) \end{cases}$$

is nonempty and compact in $\mathscr{C}_H([0,T])$.

Proof

Step 1 We will use some arguments in the proof of Theorem 3.2. in [9].

Let $u_0 \in \text{dom}\, \varphi(0,\cdot)$. Let $u : [0,T] \to H$ be the unique absolutely continuous solution to

$$-\dot{u}(t) \in \partial\varphi(t, u(t))$$

$$u(0) = u_0 \in \text{dom}\, \varphi(0,\cdot).$$

Now, let $r : [0,T] \to \mathscr{R}^+$ be the unique absolutely continuous solution of the differential equation

$$\dot{r}(t) = \beta(t)(1 + r(t)) \quad \text{a.e. with } r(0) = \sup_{t\in[0,T]} ||u(t)||.$$

Since $\dot{r} \in L^2_{\mathscr{R}^+}([0,T])$, the set

$$K := \{h \in L^2_H([0,T]) : ||h(t)|| \le \dot{r}(t) \text{ a.e. } t \in [0,T]\},$$

is clearly convex $\sigma(L^2_H([0,T]), L^2_H([0,T]))$-compact. For any $h \in K$ denote by u_h the unique absolutely continuous solution to the perturbed evolution problem

$$\begin{cases} -\dot{u}_h(t) \in \partial\varphi(t, u_h(t)) + h(t) \text{ a.e. } t \in [0,T] \\ u_h(0) = u_0 \in \text{dom}\, \varphi(0,\cdot). \end{cases}$$

Using the monotonicity of $\partial\varphi(t,\cdot)$ for all $t \in [0,T]$, one obtains the estimate

$$\frac{1}{2}||u_h(t) - u(t)||^2 \le \int_0^t ||h(s)||\,||u_h(s) - u(s)||ds.$$

Thanks to Lemma A.5 in [6], it follows that

$$||u_h(t) - u(t)|| \le \int_0^t ||h(s)||ds$$

so that

$$||u_h(t)|| \le r(0) + \int_0^t \dot{r}(s)ds = r(t)).$$

Set $L := \{u_h : h \in K\}$. **Main fact** L is compact in $\mathscr{C}_H([0,T])$.

This follows from the above estimate and the compactness property in Lemma 1.

Step 2 By construction, for every $h \in K$, we note that

$$F(t, u_h(t)) \subset \beta(t)(1 + ||u_h(t)||)\overline{B}_H \subset \beta(t)(1 + r(t))\overline{B}_H = \dot{r}(t)\overline{B}_H.$$

Let us define

$$\Phi(h) = \left\{ f \in L^2_H([0, T]) : \ f(t) \in F(t, u_h(t)) \text{ a.e. } t \in [0, T] \right\}$$

where u_h is the unique absolutely continuous solution to the inclusion

$$\begin{cases} -\dot{u}_h(t) \in \partial\varphi(t, u_h(t)) + h(t) \text{ a.e. } t \in [0, T] \\ u_h(0) = u_0 \in \operatorname{dom} \varphi(0, \cdot). \end{cases}$$

It is clear that $\Phi(h)$ is nonempty with $\Phi(h) \subset K$. In fact $\Phi(h)$ is the set of $L^2_H([0, T])$-selections of the convex weakly compact valued scalarly measurable mapping $t \to F(t, u_h(t))$. Clearly, if h is a fixed point of Φ ($h \in \Phi(h)$), then u_h is an absolutely continuous solution to the inclusion under consideration, namely

$$\begin{cases} -\dot{u}_h(t) \in \partial\varphi(t, u_h(t)) + h(t), \ \text{ a.e. } t \in [0, T] \\ u_h(0) = u_0 \in \operatorname{dom} \varphi(0, \cdot), \end{cases}$$

with $h(t) \in F(t, u_h(t))$ a.e. Now $\Phi : K \to K$ is upper semicontinuous with convex $\sigma(L^2_H([0, T]), L^2_H([0, T]))$-compact values. By compactness, it is enough to show that the graph of Φ is sequentially weakly closed in $L^2_H([0, T])$. Let $(f_n) \subset \Phi(h_n)$ be such that (f_n) weakly converges in $L^2_H([0, T])$ to $f \in K$ and (h_n) weakly converges in $L^2_H([0, T])$ to $h \in K$. We show that $f \in \Phi(h)$. Recall that the set $L := \{u_h : h \in K\}$ of solutions to

$$\begin{cases} -\dot{u}_h(t) \in \partial\varphi(t, u_h(t)) + h(t), \ \text{ a.e. } t \in [0, T], \ h \in K \\ u_h(0) = u_0 \in \operatorname{dom} \varphi(0, \cdot) \end{cases}$$

is compact in $\mathscr{C}_H([0, T])$. Hence (u_{h_n}) uniformly converges to $u_h \in L$. Since $f_n(t) \in F(t, u_{h_n}(t))$, then for each $E \in \mathscr{L}([0, T])$ and for each $e \in H$

$$\int_E \langle e, f(t)\rangle dt = \lim_n \int_E \langle e, f_n(t)\rangle dt \le \limsup_n \int_E \delta^*(e, F(t, u_{h_n}(t)))dt$$

$$\le \int_E \limsup_n \delta^*(x, F(t, u_{h_n}(t)))dt \le \int_E \delta^*(e, F(t, u_h(t)))dt.$$

Consequently

$$\langle e, f(t)\rangle \le \delta^*(e, F(t, u_h(t))), \ \text{ a.e. } t \in [0, T].$$

By the separability of H and by (Castaing-Valadier [8], Proposition III.35), we get

$$f(t) \in F(t, u_h(t)), \text{ a.e. } t \in [0, T].$$

Applying Kakutani-Ky Fan fixed point theorem to the convex weakly compact valued upper semicontinuous mapping $\Phi : K \to K$ shows that Φ admits a fixed point, $h \in \Phi(h)$, thus proving the existence of at least one absolutely continuous solution to our inclusion.

Step 3 Compactness follows easily from the above arguments and the compactness in $\mathscr{C}_H([0, T])$ of L given in Lemma 1.

Corollary 1 *Assume that for every $t \in [0, T]$, φ satisfies conditions (H_1) and (H_2) and every* dom $\varphi(t, \cdot)$ *is ball-compact. Let $F : [0, T] \times H \to H_\sigma$*[2] *be a closed convex valued mapping satisfying:*

$(\mathscr{H}_F^1)$: The graph of F is closed in $[0, T] \times H \times H_\sigma$,
$(\mathscr{H}_F^2)$: $d(0, F(t, x)) < K$, for all $(t, x) \in [0, T] \times H$, for some $K > 0$.

Then the inclusion

$$\begin{cases} -\dot{u}(t) \in \partial\varphi(t, u(t)) + F(t, u(t)), \text{ a.e. } t \in [0, T] \\ u(0) = u_0 \in \text{dom } \varphi(0, \cdot) \end{cases}$$

admits at least one absolutely continuous solution.

Proof Let $F_K : [0, T] \times H \to H_\sigma$ be defined by $F_K(t, x) := F(t, x) \cap \overline{B}_H(0, K)$ for all $(t, x) \in [0, T] \times H$. Since F_K is upper semicontinuous and convex (weakly) compact valued, then by the preceding theorem, the inclusion

$$\begin{cases} -\dot{u}(t) \in \partial\varphi(t, u(t)) + F_K(t, u(t)), \text{ a.e. } t \in [0, T] \\ u(0) = u_0 \in \text{dom } \varphi(0, \cdot) \end{cases}$$

admits at least an absolutely continuous solution.

Now comes a direct application of Theorem 2 to a parabolic variational inequality.

Theorem 3 *Assume that for every $t \in [0, T]$, φ satisfies conditions (H_1) and (H_2) and every* dom $\varphi(t, \cdot)$ *is ball-compact. Let $C : [0, T] \times H \to H$ be a closed convex valued mapping satisfying*

$$|d(x, C(t, w)) - d(y, C(\tau, z))| \le ||x - y|| + k\big(|t - \tau| + ||w - z||\big)$$

[2] H_σ denotes the vector space H endowed with the weak topology.

for all $t, \tau \in [0, T]$ *and for all* $x, y, z, w \in H$, *where* $k > 0$ *is fixed. Then for all* $u_0 \in \operatorname{dom}\varphi(0, \cdot)$ *the problem*

$$\begin{cases} -\frac{du}{dt}(t) \in \partial\varphi(t, u(t)) + \partial[k.d_{C(t,u(t))}](u(t)), \text{ a.e. } t \in [0, T], \\ u(0) = u_0 \in \operatorname{dom}\varphi(0, \cdot) \end{cases}$$

admits an absolutely continuous solution u.

Assume further that $\operatorname{dom} \varphi(t, \cdot) \subset C(t, x)$ *for all* $(t, x) \in [0, T] \times \operatorname{dom}\varphi(t, \cdot)$; *then one has*

$$-\frac{du}{dt}(t) \in \partial\varphi(t, u(t)) + N_{C(t,u(t))}(u(t)), \text{ a.e. } t \in [0, T].$$

Proof Since the mapping $F(t, x) := \partial[k.d_{C(t,x)}](x)$ is bounded, convex weakly compact valued and scalarly upper semicontinuous, by Theorem 2 there is an absolutely continuous solution to

$$\begin{cases} -\frac{du}{dt}(t) \in \partial\varphi(t, u(t)) + \partial[k.d_{C(t,u(t))}](u(t)), \text{ a.e. } t \in [0, T]; \\ u(0) = u_0 \in \operatorname{dom} \varphi(0, \cdot). \end{cases}$$

If $\operatorname{dom} \varphi(t, \cdot) \subset C(t, x)$ for all $(t, x) \in [0, T] \times \operatorname{dom} \varphi(t, \cdot)$, then $u(t) \in \operatorname{dom} \varphi(t, \cdot) \subset C(t, u(t))$ so that by the characterization of the Clarke normal cone via the Clarke subdifferential of the distance function, the inclusion

$$\partial[k.d_{C(t,u(t))}](u(t)) \subset N_{C(t,u(t))}(u(t)), \text{ a.e. } t \in [0, T]$$

holds. As a consequence, u is a solution of the last differential inclusion.

Now, we proceed to our evolution inclusion with **mixed semicontinuous** perturbation F

$$\begin{cases} -\dot{u}(t) \in \partial\varphi(t, u(t)) + F(t, u(t)) \quad \text{a.e. } t \in I \\ u(0) = u_0 \in \operatorname{dom} \varphi(0, \cdot). \end{cases}$$

For this purpose, it seems convenient to recall the following selection theorem (see Theorem 6.6. [27]).

Theorem 4 *Let* Y *be a separable Banach space and* $J \subset \mathscr{R}$. *Let* $G : J \times Y \to Y$ *be a set-valued map with compact values that satisfies*

(j) G *is* $\mathscr{L}(J) \otimes \mathscr{B}(Y)$*-measurable;*
(jj) for every $t \in J$, *at each* $x \in Y$ *such that* $G(t, x)$ *is convex, the set-valued map* $G(t, \cdot)$ *is upper semicontinuous on* Y *and whenever* $G(t, x)$ *is not convex, the set-valued map* $G(t, \cdot)$ *is lower semi-continuous on some neighborhood of* x;

(jjj) *there exists a function* $g : J \times Y \longrightarrow \mathscr{R}^+$ *of Carathéodory type which is integrably bounded on bounded subsets of* Y *and which is such that*

$$G(t,x) \cap \overline{B}_Y\big(0, g(t,x)\big) \neq \emptyset \quad \textit{for any } x \in Y \textit{ and a.e. } t \in J.$$

Then, for any $\varepsilon > 0$ *and any compact set* $K \subset \mathscr{C}_Y(J)$*, there is a nonempty closed convex set valued map* $\Psi : K \to L^1_Y(J)$ *which has a sequentially closed graph with respect to the norm of uniform convergence in* K *and the weak topology* $\sigma(L^1_Y(J), L^\infty_Y(J))$ *in* $L^1_Y(J)$ *and which is such that for any* $x \in K$ *and* $h \in \Psi(x)$*, one has for a.e.* $t \in J$

$$h(t) \in G(t, x(t))$$

$$||h(t)|| \leq g(t, x(t)) + \varepsilon.$$

Theorem 5 *Let* $I := [0,1]$*, for simplicity. Assume that for every* $t \in I$*,* φ *satisfies conditions* (H_1) *and* (H_2) *and every* dom $\varphi(t,\cdot)$ *is ball-compact. Let* $F : I \times H \to H$ *be a compact set-valued map satisfying*

(i) F *is* $\mathscr{L}(I) \otimes \mathscr{B}(H)$*-measurable,*
(ii) for every $t \in I$*, at each* $x \in H$ *such that* $F(t,x)$ *is convex, the set-valued map* $F(t,\cdot)$ *is upper semicontinuous and whenever* $F(t,x)$ *is not convex,* $F(t,\cdot)$ *is lower semi-continuous on some neighborhood of* x*;*
(iii) $F(t,x) \cap \alpha(t)(1 + ||x||)\overline{B}_H \neq \emptyset$ *for all* $(t,x) \in I \times H$*, for some measurable function* α *with* $0 < \alpha(t) < 1$ *for all* $t \in I$*.*

Then, for any $u_0 \in$ dom $\varphi(0,\cdot)$*, there is an absolutely continuous solution* $u(\cdot)$ *of the differential inclusion*

$$\begin{cases} -\dot{u}(t) \in \partial\varphi(t, u(t)) + F(t, u(t)) & \text{a.e. } t \in I \\ u(0) = u_0 \in \text{dom } \varphi(0,\cdot). \end{cases}$$

Proof

Step 1 We will use some arguments in the proof of Theorem 3.2 in [9].

Let $u_0 \in$ dom $\varphi(0,\cdot)$, $\varepsilon > 0$. Let $u : [0,1] \to H$ be the unique absolutely continuous solution to

$$-\dot{u}(t) \in \partial\varphi(t, u(t))$$

$$u(0) = u_0 \in \text{dom } \varphi(0,\cdot).$$

Now, let $r : [0,1] \to \mathscr{R}^+$ be the unique absolutely continuous solution of the differential equation

$$\dot{r}(t) = \alpha(t)(1 + r(t)) + \varepsilon \quad \text{a.e. with } r(0) = \sup_{t\in[0,1]} ||u(t)||.$$

Since $\dot{r} + \varepsilon \in L^{\infty}_{\mathscr{R}^+}(I) \subset L^{1}_{\mathscr{R}^+}(I)$, the set

$$M := \{h \in L^1_H(I) : ||h(t)|| \le \dot{r}(t) + \varepsilon \text{ a.e. } t \in I\}$$

is clearly convex $\sigma(L^1_H(I), L^{\infty}_H(I))$-compact. For any $h \in M$ denote by u_h the unique absolutely continuous solution to the perturbed evolution problem

$$\begin{cases} -\dot{u}_h(t) \in \partial\varphi(t, u_h(t)) + h(t) & \text{a.e. } t \in I \\ u_h(0) = u_0 \in \operatorname{dom} \varphi(0, \cdot). \end{cases}$$

Using the monotonicity of $\partial\varphi(t, \cdot)$ for all $t \in I$, one obtains as in Theorem 2

$$\frac{1}{2}||u_h(t) - u(t)||^2 \le \int_0^t ||h(s)||||u_h(s) - u(s)||ds.$$

Thanks to Lemma A.5 in [6], it follows

$$||u_h(t) - u(t)|| \le \int_0^t ||h(s)||ds$$

so that

$$||u_h(t)|| \le r(0) + \int_0^t [\dot{r}(s) + \varepsilon]ds = r(t) + \varepsilon t \le r(t) + \varepsilon.$$

Set $K := \{u_h : h \in M\}$. **Main fact** K is compact in $\mathscr{C}_H(I)$. This follows from the above estimate and Lemma 1.

Step 2 Take any $\varepsilon > 0$. By Theorem 4 applied to the mixed semicontinuous mapping F, there is a nonempty closed convex set valued-map $\Psi : K \to L^1_H(I)$ whose graph is sequentially closed with respect to the topology of uniform convergence in K and the weak topology in $L^1_H(I)$ and such that, for any $x \in K$ and $y \in \Psi(x)$, for a.e. $t \in I$, one has

$$y(t) \in F(t, x(t)) \text{ and } ||y(t)|| \le g(t, x(t)) + \varepsilon = \alpha(t)(1 + ||x(t)||) + \varepsilon.$$

This, taken together with $u_h \in K$ and $y_h \in \Psi(u_h)$, implies that, for a.e $t \in I$

$$y_h(t) \in F(t, u_h(t)) \text{ and } ||y_h(t)|| \le \alpha(t)(1+||u_h(t)||)+\varepsilon \le \alpha(t)(1+r(t)+\varepsilon)+\varepsilon$$

$$= \alpha(t)(1 + r(t)) + \alpha(t)\varepsilon + \varepsilon \le \alpha(t)(1 + r(t)) + \varepsilon + \varepsilon = \dot{r}(t) + \varepsilon,$$

taking into account the fact $||u_h(t)|| \le r(t) + \varepsilon$. It is easy to see that $y_h \in M$, hence $\Psi(u_h) \subset M$ for all $u_h \in K$. Thus, there is a nonempty convex weakly-compact set-valued map $\Psi : K \to L^1_H(I)$ with the following properties (*j*) $\Psi(x) \subset \{h \in M : h(t) \in F(t, x(t)) \text{ a.e.}\} \subset M$, for each $x \in K$,

(*jj*) the set-valued map Ψ has closed graph, i.e., for any $y_n \in \Psi(x_n)$ with (x_n) uniformly converging to $x \in K$ and (y_n) $\sigma(L^1_H(I), L^\infty_H(I))$-converging in $L^1_H(I)$ to y, then $y \in \Psi(x)$; equivalently, $\Psi : K \to M$ is upper semi-continuous from $K \subset \mathscr{C}_H(I)$ to M endowed with the $\sigma(L^1_H(I), L^\infty_H(I))$ topology.

Step 3 We finish the proof by using Kakutani-Ky Fan fixed point theorem.

For each $h \in M$, let us set $\Phi(h) = \Psi(u_h)$ where

$$\begin{cases} -\dot{u}_h(t) \in \partial\varphi(t, u_h(t)) + h(t), \ h \in M, \ \text{a.e. } t \in I \\ u_h(0) = u_0 \in \text{dom } \varphi(0, \cdot). \end{cases}$$

Then, it is clear that Φ is a convex weakly compact set-valued map from M to M. We claim that Φ has a fixed point h, i.e., $h \in \Phi(h) = \Psi(u_h)$, then, by (j) $h(t) \in F(t, u_h(t))$ a.e., proving that

$$\begin{cases} -\dot{u}_h(t) \in \partial\varphi(t, u_h(t)) + F(t, u_h(t)), \ \text{a.e. } t \in I \\ u_h(0) = u_0 \in \text{dom } \varphi(0, \cdot) \end{cases}$$

has at least a solution. In order to use the Kakutani-Ky Fan fixed point theorem, we need to prove that $\Phi : M \to M$ is upper semi-continuous with nonempty convex weakly compact values from M into itself; equivalently, the graph of Φ is sequentially closed in $M \times M$, for M equipped with the $\sigma(L^1_H(I), L^\infty_H(I))$-topology.

Indeed, for each $n \in \mathcal{N}$, let $g_n \in \Phi(h_n) = \Psi(u_{h_n})$ such that

$$\begin{cases} -\dot{u}_{h_n}(t) \in \partial\varphi(t, u_{h_n}(t)) + h_n(t) \quad \text{a.e. } t \in I \\ u_{h_n}(0) = u_0 \in \text{dom } \varphi(0, \cdot) \end{cases}$$

with $u_{h_n} \in K$, by our definition of K and (g_n) which $\sigma(L^1_H(I), L^\infty_H(I))$-converges to $g \in M$. Furthermore, by compactness $(u_{h_n})_n$ uniformly converges to u_h with

$$\begin{cases} -\dot{u}_h(t) \in \partial\varphi(t, u_h(t)) + h(t) \quad \text{a.e. } t \in I \\ u_h(0) = u_0 \in \text{dom } \varphi(0, \cdot). \end{cases}$$

By the property (jj) of the set-valued map Ψ, we have $g \in \Psi(u_h) = \Phi(h)$. So, the graph of $\Phi : M \to M$ is closed, hence the convex weakly compact valued mapping Φ admits a fixed point by Kakutani-Ky Fan Theorem, $h \in \Phi(h) = \Psi(u_h)$. By the property (j), we get $h(t) \in F(t, u_h(t))$ a.e. which implies

$$\begin{cases} -\dot{u}_h(t) \in \partial\varphi(t, u_h(t)) + h(t), \ \text{a.e. } t \in I \\ u_h(0) = u_0 \in \text{dom } \varphi(0, \cdot) \end{cases}$$

with $h(t) \in F(t, u_h(t))$ a.e.

5.2 Some Variants with the Velocity in the Subdifferential

In this section,[3] we are interested in the existence of absolutely continuous solutions to the evolution inclusion

$$f(t) - Au(t) \in \partial\varphi(t, \frac{du}{dt}(t)), \text{ a.e. } t \in [0, T] \qquad (P_5)$$

where $\varphi : [0, T] \times H \to [0, +\infty]$ is a normal lower semicontinuous convex integrand, A is a linear bounded coercive operator in H and $f : [0, T] \to H$ is a bounded continuous mapping. Also, we present a second variant dealing with the evolution inclusion governed by a convex sweeping process

$$f(t) - Au(t) \in N_{C(t)}(\frac{du}{dt}(t)), \text{ a.e. } t \in [0, T] \qquad (P_6)$$

where $C : [0, T] \to H$ is a convex compact moving set. For the convenience of the reader, we recall and summarize two useful results, namely [3, Corollary 2.9 and Corollary 2.10], on which we build upon for the main proofs.

Corollary 2

(1) If $A : H \to H$ is linear continuous and coercive: $\langle Ax, x\rangle \geq \omega||x||^2$ for all $x \in H$ for some $\omega > 0$ and if $\varphi : [0, T]\times H \to]-\infty, +\infty]$ is lower semicontinuous convex proper, then, for $f \in H$, the problem $f \in Ay + \partial\varphi(y)$ admits a unique solution y.

(2) If $A : H \to H$ is linear continuous and coercive: $\langle Ax, x\rangle \geq \omega||x||^2$ for all $x \in H$ for some $\omega > 0$ and if K is a closed convex subset in H, then, for $f \in H$, the problem $f \in Ay + N_K(y)$ admits a unique solution y.

Now come the main results in this section.

Theorem 6 *Assume that $H = \mathscr{R}^d$.[4] Let K be a convex compact subset of H and denote $S^1_K := \{u \in L^1_H([0, T]) : u(t) \in K \text{ a.e.}\}$. Let $f : [0, T] \to H$ be a continuous mapping such that $||f(t)|| \leq \beta$ for all $t \in [0, T]$, let $v : [0, T] \to \mathscr{R}^+$ be a positive nondecreasing continuous function with $v(0) = 0$, let $\varphi : [0, T]\times K \to [0, +\infty]$ be a normal lower semicontinuous convex integrand satisfying*

(H_3) $\varphi(t, x) \leq \varphi(\tau, x) + |v(t) - v(\tau)|$ for all $t, \tau \in [0, T], x \in K$,
(H_4) $\{t \mapsto \varphi(t, u(t)) : u \in S^1_K\}$ is uniformly integrable in $L^1_{\mathscr{R}}([0, T])$.

---

[3]From now on, results and proofs are independent from those given above, so there is no risk of confusion.

[4]For infinite-dimensional Hilbert spaces, an analogous proof of a similar result would have to rely on some compactness, but we may not assume that a linear operator is coercive and compact, as these properties are not compatible in that setting.

Let $A : H \to H$ be a linear continuous coercive symmetric operator. Then, for any $u_0 \in H$, the evolution inclusion problem

$$f(t) - Au(t) \in \partial\varphi(t, \frac{du}{dt}(t)), \text{ a.e. } t \in [0, T], \quad u(0) = u_0 \qquad (P_5)$$

admits a unique absolutely continuous solution $u : [0, T] \to H$.

Proof **Uniqueness** Let u_1 and u_2 be two solutions of the above variational inequality with $u_1(0) = u_2(0) = u_0$. Then we have

$$\frac{du_1}{dt}(t) \in \partial\varphi^*(t, f(t) - Au_1(t)), \text{ a.e. } t \in [0, T]$$

$$\frac{du_2}{dt}(t) \in \partial\varphi^*(t, f(t) - Au_2(t)), \text{ a.e. } t \in [0, T].$$

By the monotonicity of subdifferential operator we have

$$\langle\frac{du_1}{dt}(t) - \frac{du_2}{dt}(t), Au_1(t) - Au_2(t)\rangle \leq 0, \text{ a.e. } t \in [0, T].$$

But since A is symmetric

$$\frac{d}{dt}\langle u_1(t) - u_2(t), Au_1(t) - Au_2(t)\rangle = 2[\langle\frac{du_1}{dt}(t) - \frac{du_2}{dt}(t), Au_1(t) - Au_2(t)\rangle],$$

for a.e. $t \in [0, T]$. By integrating

$$\omega||u_1(t) - u_2(t)||^2 \leq \langle u_1(t) - u_2(t), Au_1(t) - Au_2(t)\rangle \leq 0$$

and $u_1(t) = u_2(t)$, for all $t \in [0, T]$.

Existence Step 1 Like in [1, Theorem 5.1], we will use Moreau's caching-up algorithm. We consider for each $n \in \mathcal{N}$ the following partition of $[0, T]$:
$t_i^n = i\frac{T}{n} := i\eta_n$ for $0 \leq i \leq n$ and we set $I_i^n :=]t_i^n, t_{i+1}^n]$ for $0 \leq i \leq n - 1$.
We denote $u_0^n = u_0$ and $f_i^n = f(t_i^n)$ for all $i = 1, \cdots, n$.
By Corollary 2 (1), there is $z_1^n \in K$ such that

$$f_1^n - Au_0^n \in \eta_n Az_1^n + \partial\varphi(t_1^n, z_1^n).$$

Put $u_1^n = u_0^n + \eta_n z_1^n$. Suppose that $u_0^n, u_1^n, \cdots, u_i^n, z_1^n, z_2^n, \cdots, z_i^n$ are constructed. As above, by Corollary 2. (1) there exists $z_{i+1}^n \in K$ such that

$$f_{i+1}^n - Au_i^n \in \eta_n Az_{i+1}^n + \partial\varphi(t_{i+1}^n, z_{i+1}^n)$$

and we set $u_{i+1}^n = u_i^n + \eta_n z_{i+1}^n$. Then, by induction, there are finite sequences $(u_i^n)_{i=0}^n$ and $(z_i^n)_{i=1}^n$ such that

$$f_{i+1}^n - Au_i^n \in \eta_n A z_{i+1}^n + \partial\varphi(t_{i+1}^n, z_{i+1}^n)$$
$$u_{i+1}^n = u_i^n + \eta_n z_{i+1}^n.$$

From $(u_i^n)_{i=0}^n$, $(z_i^n)_{i=1}^n$ $(f_i^n)_{i=1}^n$, we construct two sequences (u_n) from $[0, T]$ to H, (f_n) from $[0, T]$ to H, by setting $f_n(0) = f_1^n$, $u_n(0) = u_0^n$ and for each $i = 0, \cdots, n-1$ we set $f_n(t) = f_{i+1}^n$ and for $t \in]t_i^n, t_{i+1}^n]$:

$$u_n(t) = u_i^n + \frac{t - t_i^n}{\eta_n}(u_{i+1}^n - u_i^n).$$

Clearly, the mapping $u_n(\cdot)$ is Lipschitz continuous on $[0, T]$, and any $\rho > 0$ such that $K \subset \rho\overline{B}_H$ is a Lipschitz constant of $u_n(\cdot)$ on $[0, T]$ since for $t \in]t_i^n, t_{i+1}^n[$

$$\dot{u}_n(t) = \frac{u_{i+1}^n - u_i^n}{\eta_n} = z_{i+1}^n \in K \subset \rho\overline{B}_H.$$

Furthermore, $u_n(t) = u_0 + \int_0^t \dot{u}_n(s)ds$ implies that $||u_n(t)|| \leq ||u_0|| + \rho T$, for every t. Using the linearity of A and the definition of u_n , we see that

$$f_{i+1}^n - Au_{i+1}^n \in \partial\varphi(t_{i+1}^n, z_{i+1}^n).$$

So, defining the function θ_n from $[0, T]$ to $[0, T]$ by $\theta_n(0) = t_1^n$ and $\theta_n(t) = t_{i+1}^n$ for any $t \in]t_i^n, t_{i+1}^n]$, the last inclusion becomes

$$f_n(t) - Au_n(\theta_n(t)) \in \partial\varphi(\theta_n(t), \dot{u}_n(t))$$

for a.e. $t \in [0, T]$ and we also note that $\sup_{t\in[0,T]} |\theta_n(t) - t| \to 0$ as $n \to \infty$. We note that $||u_n(t)|| \leq ||u_0|| + \rho T$, $||f_n(t)|| \leq \beta$ for all $t \in [0, T]$ and $u_n(t) = u_0 + \int_0^t \dot{u}_n(s)ds$ for all $t \in [0, T]$ with $\dot{u}_n(t) \in K$ a.e.

Step 2 Convergence of the Algorithm and Final Conclusion Let $S_K^1 := \{h \in L_H^1([0, T]) : h(t) \in K\ a.e.\}$ and let

$$\mathscr{X} := \{v : [0, T] \to H : v(t) = u_0 + \int_0^t \dot{v}(s)ds,\ t \in [0, T];\ \dot{v} \in S_K^1\}.$$

Then it is clear that S_K^1 is convex and weakly compact in $L_H^1([0, T])$ (see e.g. [2, 10] and the references therein) and that $\mathscr{X}$ is convex, equicontinuous and compact in $\mathscr{C}_H([0, T])$. As $(u_n) \subset \mathscr{X}$, one can extract from (u_n) a subsequence not relabelled which pointwise converges to $u : [0, T] \to H$ such that $u(t) = u_0 + \int_0^t \dot{u}(s)ds$, for all $t \in [0,T]$ and $(\dot u_n)$ $\sigma(L^1_H([0,T]), L^\infty_H([0,T]))$-converges to $\dot u \in S^1_K$. As consequence, $u_n(\theta_n(t)) \to u(t)$ pointwise in H. Then, $Au_n(\theta_n(t)) \to Au(t)$ pointwise in H. So we deduce that $g_n(t) := f_n(t) - Au_n(\theta_n(t)) \to f(t) - Au(t)$ pointwise in H. As φ is normal $t \mapsto \varphi(t, u(t))$ is measurable, (H_4) ensures the uniform integrability condition for the mappings $t \mapsto \varphi(t, u(t))$ for $u \in S^1_K$. As consequences the conjugate function $\varphi^* : [0,T] \times H \to \mathscr{R}$

$$\varphi^*(t, y) = \sup_{x \in K}[\langle x, y\rangle - \varphi(t, x)] \tag{2}$$

is normal, see e.g Castaing-Valadier [8] and satisfies

$$\varphi^*(t, y) \le \varphi^*(\tau, y) + |v(t) - v(\tau)|$$

for all $t, \tau \in [0,T]$, $y \in H$. By using (H_3) and (H_4) and the normality of φ, the mappings $t \mapsto \varphi(\theta_n(t), \dot u_n(t))$ and $t \mapsto \varphi(t, \dot u_n(t))$ are measurable and integrable. By construction we have

$$g_n(t) = f_n(t) - Au_n(\theta_n(t)) \in \partial\varphi(\theta_n(t), \dot u_n(t))$$

so that by the normality of φ^*, the mapping $t \mapsto \varphi^*(\theta_n(t), g_n(t))$ is measurable and integrable. Further by (2) and (H_3) we have

$$-\varphi(t, \dot u_n(t)) + \langle \dot u_n(t), g_n(t)\rangle \le \varphi^*(t, g_n(t)) \le \varphi^*(\theta_n(t), g_n(t)) + |v(t) - v(\theta_n(t))| \tag{3}$$

so that $t \mapsto -\varphi(t, \dot u_n(t)) + \langle \dot u_n(t), g_n(t)\rangle$ is uniformly integrable thank to (H_4). We note that $(g_n(t) = f_n(t) - Au_n(\theta_n(t)))$ is uniformly bounded and pointwise converges to $g(t) = f(t) - Au(t)$ in H. Hence $(g_n(\cdot) - g(\cdot))$ is uniformly bounded and pointwise converges to 0, so that it converges to 0 uniformly on any uniformly integrable subset of $L^1_H([0,T])$, in other terms it converges to 0 with respect to the Mackey topology $\tau(L^\infty_H([0,T]), L^1_H([0,T]))$ (see [7]),[5] so that, for every Lebesgue measurable set $B \subset [0,T]$,

$$\lim_{n\to\infty} \int_B \langle g_n(t) - g(t), \dot u_n(t)\rangle dt = 0$$

[5]If $H = \mathscr{R}^d$, one may invoke a classical fact that on bounded subsets of L^∞_H the topology of convergence in measure coincides with the topology of uniform convergence on uniformly integrable sets, i.e. on relatively weakly compact subsets, alias the Mackey topology. This is a lemma due to Grothendieck [12] [Ch.5 §4 no 1 Prop. 1 and exercise].

because $(\dot{u}_n)$ is uniformly integrable. Consequently

$$\lim_{n\to\infty}\int_B \langle g_n(t), \dot{u}_n(t)\rangle dt$$

$$= \lim_{n\to\infty}\int_B \langle g_n(t) - g(t), \dot{u}_n(t)\rangle dt + \lim_{n\to\infty}\int_B \langle g(t), \dot{u}_n(t)\rangle dt$$

$$= \lim_{n\to\infty}\int_B \langle g(t), \dot{u}_n(t)\rangle dt = \int_B \langle g(t), \dot{u}(t)\rangle dt. \tag{4}$$

Now, by (3) we have

$$\varphi^*(t, g_n(t)) \le \varphi^*(\theta_n(t), g_n(t)) + |v(t) - v(\theta_n(t))|$$

$$= \langle \dot{u}_n(t), g_n(t)\rangle - \varphi(\theta_n(t), \dot{u}_n(t)) + |v(t)) - v(\theta_n(t))|$$

$$\le \langle \dot{u}_n(t), g_n(t)\rangle + \varphi(t, \dot{u}_n(t)) + 2|v(t) - v(\theta_n(t))|.$$

Whence

$$\int_0^T \varphi^*(t, g_n(t))dt \le \sup_n \int_0^T [|\langle \dot{u}_n(t), g_n(t)\rangle| + \varphi(t, \dot{u}_n(t)) + 2|v(t) - v(\theta_n(t))|]dt$$

$$< Constant < \infty$$

because

$$t \mapsto [|\langle \dot{u}_n(t), g_n(t)\rangle| + \varphi(t, \dot{u}_n(t)) + 2|v(t) - v(\theta_n(t))|]$$

is bounded in $L^1_{\mathscr{R}}([0,T])$ so that by noting that (g_n) weakly converges to g in $L^1_H([0,T])$ and applying the lower semicontinuity of integral convex functional [10, Theorem 8.1.6] to φ^*, we deduce that

$$\int_B [-\varphi(t, \dot{u}(t)) + \langle \dot{u}(t), g(t)\rangle]dt \le \int_B \varphi^*(t, g(t))dt$$

$$\le \liminf_n \int_B \varphi^*(t, g_n(t))dt < Constant < \infty$$

as consequence

$$\int_B \varphi^*(t, g(t))dt \le \liminf_n \int_B \varphi^*(t, g_n(t))dt \le \liminf_n \int_B \varphi^*(\theta_n(t), g_n(t))dt \tag{5}$$

with $t \mapsto \varphi^*(t, g(t))$ integrable. From

$$0 \le \varphi(t, \dot{u}_n(t)) \le \varphi(\theta_n(t), \dot{u}_n(t)) + |v(t) - v(\theta_n(t))|$$

we deduce that

$$\liminf_n \int_B \varphi(t, \dot{u}_n(t))dt \le \liminf_n \int_B \varphi(\theta_n(t), \dot{u}_n(t))dt.$$

As $(\dot{u}_n)$ weakly converges to $\dot{u} \in L^1_H([0, T])$, by the lower semicontinuity theorem [10, Theorem 8.1.6] applied to the convex integral functional associated with φ, we deduce that

$$0 \le \int_B \varphi(t, \dot{u}(t))dt \le \liminf_n \int_B \varphi(t, \dot{u}_n(t))dt \le \liminf_n \int_B \varphi(\theta_n(t), \dot{u}_n(t))dt \tag{6}$$

with $\dot{u}(t) \in K$ a.e. and $t \mapsto \varphi(t, \dot{u}(t))$ is integrable. Now integrating on any Lebesgue measurable set B in $[0, T]$ the equality

$$\varphi(\theta_n(t), \dot{u}_n(t)) + \varphi^*(\theta_n(t), g_n(t)) = \langle \dot{u}_n(t), g_n(t)\rangle$$

gives

$$\int_B \varphi(\theta_n(t), \dot{u}_n(t))dt + \int_B \varphi^*(\theta_n(t), g_n(t))dt = \int_B \langle \dot{u}_n(t), g_n(t)\rangle dt.$$

By passing to the limit when n goes to ∞ in this equality using (4)–(6) gives

$$\int_B \varphi(t, \dot{u}(t))dt + \int_B \varphi^*(t, g(t))dt \le \int_B \langle \dot{u}(t), g(t)\rangle dt.$$

As $t \mapsto \varphi(t, \dot{u}(t)) + \varphi^*(t, g(t)) - \langle \dot{u}(t), g(t)\rangle$ is integrable, we deduce that

$$\varphi(t, \dot{u}(t)) + \varphi^*(t, g(t)) - \langle \dot{u}(t), g(t)\rangle \le 0$$

a.e. with $\dot{u}(t) \in K$ a.e. So we conclude that $\varphi(t, \dot{u}(t)) + \varphi^*(t, g(t)) = \langle \dot{u}(t), g(t)\rangle$ a.e., equivalently $g(t) = f(t) - Au(t) \in \partial\varphi(t, \dot{u}(t))$ a.e. and equivalently $\dot{u}(t) \in \partial\varphi^*(t, f(t) - Au(t))$ a.e.

Using the above tools, we present a variational limit result, which can be applied to further convex sweeping process, even if $\dim H = \infty$.

Proposition 8 *Let $(C_n, C)_{n\in\mathcal{N}}$ be a sequence of scalarly measurable convex weakly compact valued mappings such that*

(i) $C_n(t), C(t) \subset r(t)\overline{B}_H$ *for all* $n \in \mathcal{N}$, *for all* $t \in [0, T]$ *where* $r : [0, T] \to \mathcal{R}^+$ *is a positive integrable function.*

(ii) $d_H(C_n(t), C(t)) \le \rho_n(t)$ *for every* $t \in [0, T]$ *where* (ρ_n) *is a positive bounded sequence in* $L^\infty_{\mathcal{R}}([0, T])$ *with* $\rho_n(t) \le \alpha$, $(\alpha > 0)$ *for all* $n \in \mathcal{N}$ *such that* $\rho_n(t)$ *pointwise converges to* 0 *for each* $t \in [0, T]$.

Let $(f_n, f)_{n\in\mathcal{N}}$ *be bounded in* $L^\infty_H([0, T])$ *with* $||f_n(t)|| \le \beta$, $||f(t)|| \le \beta$ $(\beta > 0)$ *for all* $n \in \mathcal{N}$ *and let* $f_n(t)$ *converge to* $f(t)$ *for each* $t \in [0, T]$.

Let $(v_n, v)_{n\in\mathcal{N}}$ *be a bounded sequence in* $L^\infty_H([0, T])$ *with* $||v_n(t)|| \le \gamma$, $||v(t)|| \le \gamma$ $(\gamma > 0)$ *for all* $n \in \mathcal{N}$ *such that* $v_n(t)$ *pointwise converges to* $v(t)$ *for each* $t \in [0, T]$ *with respect to the weak topology in* H.

Let (u_n) *be an integrable sequence in* $L^1_H([0, T])$ *such that* $u_n(t) \in C_n(t)$ *for all* $n \in \mathcal{N}$ *and for all* $t \in [0, T]$.

Let A *be a linear continuous compact operator in* H.

Assume that $f_n(t) - Av_n(t) \in N_{C_n(t)}(u_n(t))$ *for all* $n \in \mathcal{N}$ *and a.e.* $t \in [0, T]$, *then there is a subsequence* (u_n) *that weakly converges in* $L^1_H([0, T])$ *to* u *with* $u(t) \in C(t)$ *a.e. and* $f(t) - Av(t) \in N_{C(t)}(u(t))$ *a.e.*

Proof From condition (ii), we deduce by the Hormander formula (see e.g. Castaing-Valadier [8]) that

$$\sup_{x\in\overline{B}_H} |\delta^*(x, C_n(t)) - \delta^*(x, C(t))| = d_H(C_n(t), C(t)) \le \rho_n(t).$$

As we have $f_n(t) - Av_n(t) \in N_{C_n(t)}(u_n(t))$ then

$$\delta^*(f_n(t) - Av_n(t), C_n(t)) \le \langle f_n(t) - Av_n(t), u_n(t)\rangle$$

with

$$u_n(t) \in C_n(t) \subset r(t)\overline{B}_H.$$

Hence (u_n) is uniformly integrable. So we may assume that (u_n) weakly converges in $L^1_H([0, T])$ to u. We first check that $u(t) \in C(t)$ a.e. Indeed, we have $\langle x, u_n(t)\rangle \le \delta^*(x, C_n(t))$, $\forall x \in H$. Then by integrating on any Lebesgue measurable set $B \subset [0, T]$

$$\int_B \langle x, u_n(t)\rangle dt \le \int_B \delta^*(x, C_n(t))dt.$$

By using (ii) and by passing to the limit when n goes to ∞ gives

$$\int_B \langle x, u(t)\rangle dt \le \limsup_n \int_B \delta^*(x, C_n(t))dt \le \int_B \delta^*(x, C(t))dt.$$

Then we deduce that $\langle x, u(t)\rangle \leq \delta^*(x, C(t))$ a.e. so that by Castaing-Valadier [8, Proposition III.35], we get $u(t) \in C(t)$ a.e. Recall that

$$g_n(t) = f_n(t) - Av_n(t) \in N_{C_n(t)}(u_n(t))$$

and that $(g_n(t) = f_n(t) - Av_n(t))$ is bounded and pointwise converges to $g(t) = f(t) - Av(t)$ (remember that A is a compact operator) and (u_n) weakly converges in $L^1_H([0, T])$ to u with $u(t) \in C(t)$ a.e. So as above by Castaing's trick, we have the **main fact**

$$\lim_n \int_B \langle g_n(t), u_n(t)\rangle dt = \int_B \langle g(t), u(t)\rangle dt \tag{*}$$

for any Lebesgue measurable set B in $[0, T]$. By integrating on B the inequality

$$\delta^*(f_n(t) - Av_n(t), C_n(t)) \leq \langle f_n(t) - Av_n(t), u_n(t)\rangle$$

we get

$$\int_B \delta^*(g_n(t), C_n(t))dt - \int_B \langle g_n(t), u_n(t)\rangle dt \leq 0.$$

But by (ii) we have the estimation

$$\int_B \delta^*(g_n(t), C(t))dt$$

$$\leq \int_B \delta^*(g_n(t), C_n(t))dt + Constant \int_B \rho_n(t)dt$$

with $\int_B \rho_n(t)dt \to 0$. So that by invoking the lower semicontinuity of the integral convex functional [10, Theorem 8.1.6] associated with the normal convex integrand $(t, x) \mapsto \delta^*(x, C(t))$ by noting that $\langle u(t), g_n(t)\rangle$ is uniformly integrable with $\langle u(t), g_n(t)\rangle \leq \delta^*(g_n(t), C(t))$ gives

$$\liminf_n \int_B \delta^*(g_n(t), C_n(t))dt \geq \liminf_n \int_B \delta^*(g_n(t), C(t))dt \geq \int_B \delta^*(g(t), C(t))dt \tag{**}$$

so that as a consequence using (*), (**) and passing to the limit when n goes to ∞ in the inequality

$$\int_B \delta^*(g_n(t), C_n(t))dt \leq \int_B \langle g_n(t), u_n(t)\rangle dt$$

gives

$$\int_B \delta^*(g(t), C(t))dt \le \int_B \langle g(t), u(t)\rangle dt,$$

so that

$$\delta^*(g(t), C(t)) \le \langle g(t), u(t)\rangle$$

a.e. with $u(t) \in C(t)$ a.e. So we conclude that

$$g(t) = f(t) - Av(t) \in N_{C(t)}(u(t)), \text{ a.e. } t \in [0, T].$$

Theorem 7 *Let L be a convex compact set in H. Let $f : [0, T] \to H$ be a continuous mapping such that $||f(t)|| \le \beta$ for all $t \in [0, T]$, let $v : [0, T] \to \mathscr{R}^+$ be a positive nondecreasing continuous function with $v(0) = 0$. Let $C : [0, T] \to L$ be a convex compact valued mapping such that $d_H(C(t), C(\tau)) \le |v(t) - v(\tau)|$ for all $t, \tau \in [0, T]$. Let $A : H \to H$ be a linear continuous coercive symmetric operator. Then, for any $u_0 \in H$, the evolution inclusion*

$$f(t) - Au(t) \in N_{C(t)}(\frac{du}{dt}(t)), \text{ a.e. } t \in [0, T] \qquad (P_7)$$

$$u(0) = u_0$$

admits a unique absolutely continuous solution $u : [0, T] \to H$.

Proof **Uniqueness**, as in the proof of Theorem 6, follows from the fact that the normal cone is monotone and A is coercive symmetric.

Existence follows from the Moreau's catching-up algorithm and tools developed in Step 1 of the proof of Theorem 6. involving Corollary 2. (2). We note that by Hormander formula (see e.g. Castaing-Valadier [8])

$$|\delta^*(x, C(t)) - \delta^*(x, C(\tau))| \le ||x||d_H(C(t), C(\tau)) \le ||x|||v(t) - v(\tau)|.$$

As a consequence C is scalarly continuous, a fortiori scalarly upper semicontinuous. We want to show that the above techniques allow to prove the existence of absolutely continuous solution to the inclusion

$$f(t) - Au(t) \in N_{C(t)}(\dot{u}(t)), u(0) = u_0.$$

Applying the results and notations of the algorithm via Corollary 2. (2) given in Step 1 of Theorem 6., provides u_n absolutely continuous satisfying

$$f_n(t) - Au_n(\theta_n(t)) \in N_{C(\theta_n(t))}(\dot{u}_n(t)).$$

Hence

$$\delta^*(f_n(t) - Au_n(\theta_n(t)), C(\theta_n(t))) \leq \langle f_n(t) - Au_n(\theta_n(t)), \dot{u}_n(t)\rangle$$

with

$$\dot{u}_n(t) \in C(\theta_n(t)) \subset L$$

so that $\dot{u}_n \in S^1_L$ where $S^1_L := \{v \in L^1_H([0, T]) : v(t) \in L\ a.e.\}$ and $u_n(t) = u_0 + \int_0^t \dot{u}_n(s)ds$ for all $t \in [0, T]$ with $\dot{u}_n(t) \in L$ a.e. We note that S^1_L is a convex weakly compact set of $L^1_H([0, T])$ (see e.g. [2, 10] and the references therein). Let

$$\mathscr{X} := \{v : [0, T] \to H : v(t) = u_0 + \int_0^t \dot{v}(s)ds,\ t \in [0, T];\ \dot{v} \in S^1_L\}.$$

Then it is clear that $\mathscr{X}$ is convex, equicontinuous and compact in $\mathscr{C}_H([0, T])$. As $(u_n) \subset \mathscr{X}$, one can extract from (u_n) a subsequence not relabelled which pointwise converges to $u : [0, T] \to H$ such that $u(t) = u_0 + \int_0^t \dot{u}(s)ds$, for all $t \in [0, T]$ and $(\dot{u}_n)$ $\sigma(L^1_H([0, T]), L^\infty_H([0, T]))$-converges to $\dot{u} \in S^1_L$. As consequence, $u_n(\theta_n(t)) \to u(t)$ pointwise in H. Our first task is to prove the inclusion $\dot{u}(t) \in C(t)$ a.e. Indeed, for every Lebesgue measurable set B in $[0, T]$ and for any $x \in H$, the function $1_B x \in L^\infty_H([0, T])$. We have

$$\langle x, \dot{u}_n(t)\rangle \leq \delta^*(x, C(\theta_n(t))).$$

Integrating on B gives

$$\int_B \langle 1_B(t)x, \dot{u}_n(t)\rangle dt = \int_B \langle x, \dot{u}_n(t)\rangle dt \leq \int_B \delta^*(x, C(\theta_n(t)))dt.$$

Passing to the limit in this inequality

$$\int_B \langle 1_B(t)x, \dot{u}(t)\rangle dt \leq \limsup_n \int_B \delta^*(x, C(\theta_n(t)))dt$$

$$\leq \int_B \limsup_n \delta^*(x, C(\theta_n(t)))dt \leq \int_B \delta^*(x, C(t))dt$$

using the fact that C is scalarly upper semicontinuous. So we deduce that

$$\langle x, \dot{u}(t)\rangle \leq \delta^*(x, C(t))$$

a.e. By the separability of H and by Castaing-Valadier [8, Proposition III.35], we get $\dot{u}(t) \in C(t)$ a.e. Recall that

$$g_n(t) = f_n(t) - Au_n(\theta_n(t)) \in N_{C(\theta_n(t))}(\dot{u}_n(t))$$

and that $(g_n(t) = f_n(t) - Au_n(\theta_n(t)))$ is bounded and pointwise converges to $g(t) = f(t) - Au(t)$ with respect to the strong topology and $(\dot{u}_n)$ is uniformly integrable and weakly converges to $\dot{u} \in S^1_L$. So as above by Castaing's trick, we get the **main fact**

$$\lim_n \int_B \langle g_n(t), \dot{u}_n(t)\rangle dt = \int_B \langle g(t), \dot{u}(t)\rangle dt \tag{*}$$

for every Lebesgue measurable set B in $[0, T]$. From

$$g_n(t) = f_n(t) - Au_n(\theta_n(t)) \in N_{C(\theta_n(t))}(\dot{u}_n(t))$$

we have

$$\delta^*(g_n(t), C(\theta_n(t))) - \langle g_n(t), \dot{u}_n(t)\rangle \leq 0.$$

By integrating on B we get

$$\int_B \delta^*(g_n(t), C(\theta_n(t)))dt - \int_B \langle g_n(t), \dot{u}_n(t)\rangle dt \leq 0.$$

But we have the estimation

$$\int_B \langle \dot{u}(t), g_n(t)\rangle dt \leq \int_B \delta^*(g_n(t), C(t))dt$$

$$\leq \int_B \delta^*(g_n(t), C(\theta_n(t)))dt + Constant \int_B |v(\theta_n(t)) - v(t)| dt.$$

So that by invoking the lower semicontinuity of the integral convex functional [10, Theorem 8.1.6] associated with the normal lower semicontinuous convex integrand $(t, x) \mapsto \delta^*(x, C(t))$ by noting that $(\langle \dot{u}(t), g_n(t)\rangle)$ is uniformly integrable gives

$$\liminf_n \int_B \delta^*(g_n(t), C(\theta_n(t)))dt \geq \liminf_n \int_B \delta^*(g_n(t), C(t))dt \geq \int_B \delta^*(g(t), C(t))dt \tag{**}$$

so that as a consequence using (*), (**) and passing to the limit when n goes to ∞ in the inequality

$$\int_B \delta^*(g_n(t), C(\theta_n(t)))dt \leq \int_B \langle g_n(t), \dot{u}_n(t)\rangle dt$$

gives

$$\int_B \delta^*(g(t), C(t))dt \le \int_B \langle g(t), \dot{u}(t)\rangle dt$$

so that

$$\delta^*(g(t), C(t)) \le \langle g(t), \dot{u}(t)\rangle$$

a.e. with $\dot{u}(t) \in C(t)$ a.e. So we conclude that

$$g(t) = f(t) - Au(t) \in N_{C(t)}(\dot{u}(t)), \text{ a.e. } t \in [0, T].$$

6 Applications

6.1 A Skorokhod Problem

We present a new version of the Skorokhod problem in Castaing et al. [11] dealing with the sweeping process associated with an absolutely continuous (or continuous) closed convex moving set $C(t)$ in H.

Theorem 8 *Let $I := [0, 1]$, for simplicity, and $H = \mathscr{R}^d$. Let $v : I \to \mathscr{R}^+$ be a positive nondecreasing continuous function with $v(0) = 0$. Let $C : I \to H$ be a convex compact valued mapping such that*

(i) $C(t) \subset M\overline{B}_H$ for all $t \in I$ where M is a positive constant.
(ii) $d_H(C(t), C(\tau)) \le |v(t) - v(\tau)|$ for all $t, \tau \in I$.

Let $A : H \to H$ be a linear continuous coercive symmetric operator and $b : I \times H \to H$ be a Carathéodory mapping satisfying:

(j) $||b(t, x)|| \le M, \forall (t, x) \in I \times H$,
(jj) $||b(t, x) - b(t, y)|| \le M||x - y||, \forall (t, x, y) \in I \times H \times H$.

Let $a \in H$. Then there exist an absolutely continuous function $X : I \to H$ and an absolutely continuous function $Y : I \to H$ satisfying

$$\begin{cases} X(0) = Y(0) = a \\ X(t) = \int_0^t b(s, X(s))ds + Y(t), \ \forall t \in I \\ \int_0^t b(s, X(s))ds - AY(t) \in N_{C(t)}(\dot{Y}(t)), \text{ a.e. } t \in I. \end{cases}$$

Proof We use some arguments developed in [11, Theorem 4.2]. Let us set for all $t \in I := [0, 1]$

$$X^0(t) = a, \ h^1(t) = \int_0^t b(s, a)ds;$$

then h^1 is bounded continuous with $||h^1(t)|| \leq M$ for all $t \in I$. By Theorem 7, there is a unique absolutely continuous solution Y^1 to

$$h^1(t) - AY^1(t) \in N_{C(t)}(\dot{Y}^1(t)) \text{ a.e. } t \in I, \ Y^1(0) = a$$

with

$$\dot{Y}^1(t) \in C(t)$$

a.e. Set for all $t \in I$

$$X^1(t) = h^1(t) + Y^1(t) = \int_0^t b\big(s, X^0(s)\big)ds + Y^1(t),$$

so that X^1 is absolutely continuous with

$$\int_0^t b\big(s, X^0(s)\big)ds - AY^1(t) \in N_{C(t)}(\dot{Y}^1(t)), \ \dot{Y}^1(t) \in C(t) \text{ a.e.}$$

Now, we construct X^n by induction as follows. Let for all $t \in I$

$$h^n(t) = \int_0^t b\big(s, X^{n-1}(s)\big)ds.$$

Then h^n is bounded continuous with $||h^n(t)|| \leq M$, for all $t \in I$. By Theorem 7, there is a unique absolutely continuous solution Y^n to

$$h^n(t) - AY^n(t) \in N_{C(t)}(\dot{Y}^n(t)) \text{ a.e.} t \in I, \ Y^n(0) = a$$

with $\dot{Y}^n(t) \in C(t) \subset M\overline{B}_H$ a.e. Let us set

$$X^n(t) = h^n(t) + Y^n(t) = \int_0^t b\big(s, X^{n-1}(s)\big)ds + Y^n(t), \ \forall t \in I,$$

so that X^n is absolutely continuous and

$$\int_0^t b\big(s, X^{n-1}(s)\big)ds - AY^n(t) \in N_{C(t)}(\dot{Y}^n(t)), \ \dot{Y}^n(t) \in C(t) \text{ a.e. } t \in I. \quad (*)$$

As (Y^n) is equicontinuous, we may assume that (Y^n) uniformly converges to an absolutely continuous mapping $Y : I \to H$. Using $\dot{Y}^n(t) \in C(t) \subset M\overline{B}_H$ a.e., we may also assume that $(\dot{Y}^n)$ weakly converges in $L^1_H(I)$ to $\dot{Y}$, and by (j) $(s \mapsto b(s, X^{n-1}(s)))$ weakly converges to $Z \in L^1_H(I)$. Hence $\int_0^t b(s, X^{n-1}(s))ds \to$

$\int_0^t Z(s)ds$ for each $t \in I$. So we have

$$\lim_{n\to\infty} X^n(t) := X(t) = \lim_{n\to\infty} \left(\int_0^t b(s, X^{n-1}(s))ds + Y^n(t) \right) = \int_0^t Z(s)ds + Y(t).$$

As $(X^n(t))$ pointwise converges to $X(t)$ on I, $s \mapsto b\big(s, X^{n-1}(s)\big)$ is uniformly bounded and pointwise converges to $s \mapsto b\big(s, X(s)\big)$, then $\int_0^t b(s, X^{n-1}(s))ds \to \int_0^t b(s, X(s))ds$ **uniformly** on I. Indeed we have

$$\sup_{t\in I} \left|\left| \int_0^t b\big(s, X^{n-1}(s)\big)ds - \int_0^t b(s, X(s))ds \right|\right| \le \int_0^1 M||X^{n-1}(s) - X(s)||ds \to 0$$

as $n \to \infty$, by Lebesgue's theorem. By identifying the limits we have

$$X(t) = \int_0^t b\big(s, X(s)\big)ds + Y(t), \forall t \in I.$$

As $\int_0^t b\big(s, X^{n-1}(s)\big)ds \to \int_0^t b\big(s, X(s)\big)ds$ uniformly in I, $(\dot{Y}^n)$ weakly converges in $L^1_H(I)$ to $\dot{Y}$ with $\dot{Y}^n(t) \in C(t)$ a.e. and by (*)

$$\int_0^t b\big(s, X^{n-1}(s)\big)ds - AY^n(t) \in N_{C(t)}(\dot{Y}^n(t)), \text{ a.e. } t \in I$$

we apply the arguments given in Proposition 8 to get the inclusions

$$\dot{Y}(t) \in C(t), \text{ a.e. } t \in I$$

$$\int_0^t b\big(s, X(s)\big)ds - AY(t) \in N_{C(t)}(\dot{Y}(t)), \text{ a.e. } t \in I.$$

The proof is therefore complete.

By combining the tools developed in Theorem 6 and Theorem 8 we obtain the following variant. We omit the proof for shortness.

Theorem 9 *Let $I := [0, 1]$ and $H = \mathscr{R}^d$. Let K be a convex compact in H with $K \subset M\overline{B}_H$ and $S^1_K := \{u \in L^1_H(I) : u(t) \in K \text{ a.e.}\}$. Let $v : I \to \mathscr{R}^+$ be a positive nondecreasing continuous function with $v(0) = 0$ and let $\varphi : [0, 1] \times K \to [0, +\infty]$ be a normal convex lower semicontinuous integrand satisfying*

(H_3) $\varphi(t, x) \le \varphi(\tau, x) + |v(t) - v(\tau)|$ for all $t, \tau \in I, x \in K$,
(H_4) $\{t \mapsto \varphi(t, u(t)) : u \in S^1_K\}$ is uniformly integrable in $L^1_{\mathscr{R}}(I)$.

Let $A : H \to H$ be linear continuous coercive and symmetric. Suppose that $b : I \times H \to H$ is a Carathéodory mapping satisfying:

(j) $||b(t,x)|| \leq M, \forall (t,x) \in I \times H,$
(jj) $||b(t,x) - b(t,y)|| \leq M||x-y||, \forall (t,x,y) \in I \times H \times H.$

Let $a \in H$. Then there exist an absolutely continuous function $X : I \to H$ and an absolutely continuous function $Y : I \to H$ satisfying

$$\begin{cases} X(0) = Y(0) = a \\ X(t) = \int_0^t b(s, X(s))ds + Y(t), \ \forall t \in I \\ \int_0^t b(s, X(s))ds - AY(t) \in \partial\varphi(t, \dot{Y}(t)), \ \text{a.e. } t \in I. \end{cases}$$

6.2 *Towards a Couple of Evolution Inclusion Involving Fractional Differential Equation*

We assume that $H = \mathscr{R}^d$. For the sake of completeness, we present some needed properties in fractional calculus [16, 17]. Throughout we assume $\alpha \in]1,2]$. Let $f \in L^1_H([0,T])$. The fractional Bochner integral of order $\gamma > 0$ of the function f is defined by

$$I^\gamma f(t) := \frac{1}{\Gamma(\gamma)} \int_0^t (t-s)^{\gamma-1} f(s)ds, \ \ t \in [0,T].$$

The Caputo fractional derivative of order $\gamma > 0$ of a function $h : [0,T] \to H$, ${}^cD^\gamma h : [0,T] \to H$, is defined by

$${}^cD^\gamma h(t) = \frac{1}{\Gamma(n-\gamma)} \int_0^t \frac{h^{(n)}(s)}{(t-s)^{1-n+\gamma}} ds.$$

Here $n = [\gamma] + 1$ and $[\gamma]$ denotes the integer part of γ. Denote by

$$\mathscr{C}^1_H([0,T]) = \{u \in \mathscr{C}_H([0,T]) : \frac{du}{dt} \in \mathscr{C}_H([0,T])\}$$

where $\frac{du}{dt}$ is the derivative of u,

$$W^{\alpha,\infty}_H([0,T]) = \{u \in \mathscr{C}^1_H([0,T]) : {}^cD^{\alpha-1}u \in \mathscr{C}_H([0,T]); \ {}^cD^\alpha u \in L^\infty_H([0,T])\}$$

where ${}^cD^{\alpha-1}u$ and ${}^cD^\alpha u$ are the fractional Caputo derivatives of order $\alpha - 1$ and α of u, respectively.

The following lemma is basic:

Lemma 2 *Let $f \in L_H^\infty([0,T])$ and $x \in H$. Then the function $u : [0,T] \to H$ is a $W_H^{\alpha,\infty}([0,T])$-solution to the fractional differential equation (FDE)*

$$\begin{cases} {}^cD^\alpha u(t) = f(t), \text{ a.e. } t \in [0,T] \\ u(0) = x \end{cases}$$

if and only if $u(t) = x + \int_0^t \frac{(t-s)^{\alpha-1}}{\Gamma(\alpha)} f(s)ds, \ t \in [0,T]$.

Theorem 10 *Let K be a convex compact subset of $H = \mathscr{R}^d$ and denote its set of selections $S_K^1 := \{u \in L_H^1([0,T]) : u(t) \in K \text{ a.e.}\}$. Let $v : [0,T] \to \mathscr{R}^+$ be positive nondecreasing and continuous, with $v(0) = 0$, and let $\varphi : [0,T] \times K \to [0,+\infty]$ be a normal convex lower semicontinuous integrand satisfying*

(H_3) $\varphi(t,x) \le \varphi(\tau,x) + |v(t) - v(\tau)|$ for all $t, \tau \in [0,T], x \in K$,
(H_4) $\{t \mapsto \varphi(t,u(t)) : u \in S_K^1\}$ is uniformly integrable in $L_{\mathscr{R}}^1([0,T])$.

Let $A : H \to H$ be linear continuous symmetric and coercive. Let $f : [0,T] \times H \to H$ be a bounded continuous mapping such that $||f(t,x)|| \le M$ for all $(t,x) \in [0,T] \times H$ for some positive constant M. Then for $(u_0, x_0) \in H \times H$, there is an absolutely continuous mapping $u : [0,T] \to H$, and a $W_H^{\alpha,\infty}([0,T])$ mapping $x : [0,T] \to H$ satisfying

$$x(0) = x_0 \in H$$

$${}^cD^\alpha x(t) = u(t), \text{ a.e. } t \in [0,T]$$

$$u(0) = u_0 \in H$$

$$f(t, x(t)) - Au(t) \in \partial\varphi(t, \frac{du}{dt}(t)), \text{ a.e. } t \in [0,T].$$

Proof By Theorem 6 and the assumptions on f, for any bounded continuous mapping $h : [0,T] \to H$ there is a unique absolutely continuous solution v_h to the inclusion

$$\begin{cases} v_h(0) = u_0 \in H \\ f(t, h(t)) - Av_h(t) \in \partial\varphi(t, \frac{dv_h}{dt}(t)), \text{ a.e. } t \in [0,T] \end{cases}$$

with $\frac{dv_h}{dt}(t) \in K$ a.e. so that $||v_h(t)|| \le L, \ t \in [0,T]$ for some positive constant L. Let us consider the closed convex subset $\mathscr{X}$ in the Banach space $\mathscr{C}_H([0,T])$ defined by

$$\mathscr{X} := \{u_f : [0,T] \to H : u_f(t) = x_0 + \int_0^t \frac{(t-s)^{\alpha-1}}{\Gamma(\alpha)} f(s)ds, \ f \in S^1_{L\overline{B}_H}, \ t \in [0,T]\}$$

where $S^1_{L\overline{B}_H}$ denotes the set of all integrable selections of the convex compact valued constant multifunction $L\overline{B}_H$. Now for each $h \in \mathscr{X}$ let us consider the mapping defined by

$$\Phi(h)(t) := x_0 + \int_0^t \frac{(t-s)^{\alpha-1}}{\Gamma(\alpha)} v_h(s)ds \in x_0 + \int_0^t \frac{(t-s)^{\alpha-1}}{\Gamma(\alpha)} L\overline{B}_H ds,$$

for $t \in [0, T]$. Then it is clear that $\Phi(h) \in \mathscr{X}$.

Since $L\overline{B}_H$ is a convex compact valued and integrably bounded (constant) multifunction, the right-hand side is convex compact valued, so that $\Phi(\mathscr{X})$ is equicontinuous and relatively compact in the Banach space $\mathscr{C}_H([0, T])$. Now we check that Φ is continuous. It is sufficient to show that, if (h_n) uniformly converges to h in $\mathscr{X}$, then the absolutely continuous solution v_{h_n} associated with h_n

$$\begin{cases} v_{h_n}(0) = u_0 \in H \\ f(t, h_n(t)) - Av_{h_n}(t) + \partial\varphi(t, \frac{dv_{h_n}}{dt}(t)), \text{ a.e. } t \in [0, T] \end{cases}$$

uniformly converges to the absolutely continuous solution v_h associated with h

$$\begin{cases} v_h(0) = u_0 \in H \\ f(t, h(t)) - Av_h(t) + \partial\varphi(t, \frac{dv_h}{dt}(t)), \text{ a.e. } t \in [0, T]. \end{cases}$$

As (v_{h_n}) is equi-absolutely continuous we may assume that (v_{h_n}) uniformly converges to an absolutely continuous mapping z. Since $v_{h_n}(t) = u_0 + \int_{]0,t]} \frac{dv_{h_n}}{ds}(s)ds$, $t \in [0, T]$ and $\frac{dv_{h_n}}{ds}(s) \in K$, $a.e.\ s \in [0, T]$, we may assume that $(\frac{dv_{h_n}}{dt})$ weakly converges in $L^1_H([0, T])$ to $w \in L^1_H([0, T])$ with $w(t) \in K,\ t \in [0, T]$ so that

$$\lim_n v_{h_n}(t) = u_0 + \int_0^t w(s)ds := u(t),\ t \in [0, T].$$

By identifying the limits, we get

$$u(t) = z(t) = u_0 + \int_0^t w(s)ds.$$

Therefore $f(t, h(t)) - Au(t) \in \partial\varphi(t, \frac{du}{dt}(t))$ a.e. by repeating the arguments developed in Theorem 6, with $u(0) = u_0 \in H$, so that by uniqueness $u = v_h$. Since $h_n \to h$, we have

$$\begin{aligned}\Phi(h_n)(t) - \Phi(h)(t) &= \int_0^t \frac{(t-s)^{\alpha-1}}{\Gamma(\alpha)} v_{h_n}(s)ds - \int_0^t \frac{(t-s)^{\alpha-1}}{\Gamma(\alpha)} v_h(s)ds \\ &= \int_0^t \frac{(t-s)^{\alpha-1}}{\Gamma(\alpha)} [v_{h_n}(s) - v_h(s)]ds.\end{aligned}$$

As $||v_{h_n}(\cdot) - v_h(\cdot)|| \to 0$ pointwise and $||v_{h_n}(\cdot) - v_h(\cdot)|| \le 2L$, we conclude that

$$\sup_{t\in[0,T]} ||\Phi(h_n)(t) - \Phi(h)(t)|| \le \int_0^T \frac{(t-s)^{\alpha-1}}{\Gamma(\alpha)} ||v_{h_n}(\cdot) - v_h(\cdot)||ds \to 0$$

by the dominated convergence theorem, so that $\Phi(h_n) \to \Phi(h)$ in $\mathscr{C}_H([0,T])$. Since $\Phi : \mathscr{X} \to \mathscr{X}$ is continuous and $\Phi(\mathscr{X})$ is relatively compact in $\mathscr{C}_H([0,T])$, by Idzik [13], O'Regan [21], and Park [22] Φ has a fixed point, say $h = \Phi(h) \in \mathscr{X}$. This means that

$$h(t) = \Phi(h)(t) = x_0 + \int_0^t \frac{(t-s)^{\alpha-1}}{\Gamma(\alpha)} v_h(s)ds,$$

with

$$\begin{cases} v_h(0) = u_0 \in H \\ f(t, h(t)) - Av_h(t) \in \partial\varphi(t, \frac{dv_h}{dt}(t)), \text{ a.e. } t \in [0,T]. \end{cases}$$

Coming back to Lemma 2 and applying the above notation, this means that we have just shown that there exists a mapping $h \in W_H^{\alpha,\infty}([0,T])$ satisfying

$$\begin{cases} h(0) = x_0 \\ {}^cD^\alpha h \text{ is absolutely continuous} \\ \frac{d}{dt}[{}^cD^\alpha h(\cdot)] \in S_K^1 \subset L_H^1([0,T]) \\ f(t, h(t)) - A[{}^cD^\alpha h(t)] \in \partial\varphi(t, \frac{d}{dt}[{}^cD^\alpha h(t)]), \text{ a.e. } t \in [0,T]. \end{cases}$$

6.3 A Second Order Evolution Inclusion

Further applications of the above stated results are available. For shortness we mention some direct applications involving second order evolution inclusions as follows. Let $H = \mathscr{R}^d$.

Theorem 11 *Let $v : [0,T] \to \mathscr{R}^+$ be a positive nondecreasing continuous function with $v(0) = 0$ and $C : [0,T] \to H$ be a convex compact valued mapping such that*

(i) $C(t) \subset M\overline{B}_H$ for all $t \in [0,T]$ where M is a positive constant.
(ii) $d_H(C(t), C(\tau)) \le |v(t) - v(\tau)|$ for all $t, \tau \in [0,T]$.

Let $A : H \to H$ be a linear continuous symmetric and coercive operator. Let $f : [0,T] \times H \to H$ be a bounded continuous mapping such that $||f(t,x)|| \le M$ for all $(t,x) \in [0,T] \times H$. Then for $(u_0, x_0) \in H \times H$, there exist absolutely *continuous mappings* $u : [0, T] \to H$ *and* $x : [0, T] \to H$ *satisfying*

$$x(0) = x_0 \in H$$

$$x(t) = x_0 + \int_0^t u(s)ds, \ t \in [0, T]$$

$$u(0) = u_0 \in H$$

$$f(t, x(t)) - Au(t) \in N_{C(t)}(\frac{du}{dt}(t)), \text{ a.e. } t \in [0, T].$$

Theorem 12 *Let* K *be a nonempty convex compact subset of* $H = \mathscr{R}^d$ *with* $S^1_K := \{u \in L^1_H([0, T]) : u(t) \in K \ a.e.\}$. *Let* $v : [0, T] \to \mathscr{R}^+$ *be a positive nondecreasing continuous function with* $v(0) = 0$ *and let* $\varphi : [0, T] \times K \to [0, +\infty]$ *be a normal convex lower semicontinuous positive integrand satisfying*

(H_3) $\varphi(t, x) \leq \varphi(\tau, x) + |v(t) - v(\tau)|$ *for all* $t, \tau \in [0, T], x \in K$,
(H_4) $\{t \mapsto \varphi(t, u(t)) : u \in S^1_K\}$ *is uniformly integrable in* $L^1_{\mathscr{R}}([0, T])$.

Let $A : H \to H$ *be linear continuous symmetric and coercive and* $f : [0, T] \times H \to H$ *be a bounded continuous mapping such that* $||f(t, x)|| \leq M$ *for all* $(t, x) \in [0, T] \times H$, *for some positive constant* M. *Then for any* $(u_0, x_0) \in H \times H$, *there exist absolutely continuous mappings* $u : [0, T] \to H$ *and* $x : [0, T] \to H$ *satisfying*

$$x(0) = x_0 \in H$$

$$x(t) = x_0 + \int_0^t u(s)ds, \ t \in [0, T]$$

$$u(0) = u_0 \in H$$

$$f(t, x(t)) - Au(t) \in \partial\varphi(t, \frac{du}{dt}(t)), \text{ a.e. } t \in [0, T].$$

Conclusions We have established existence results for sweeping processes and for evolution variational inequalities involving time-dependent subdifferential operators. Our results contain novelties. However, there remain several issues that need full developments, for instance, Skorokhod problems with a non convex set C, or with a general normal convex integrand $\varphi(\cdot, \cdot)$ and different types of perturbation.

Acknowledgement M. D. P. Monteiro Marques was partially supported by the Fundação para a Ciência e a Tecnologia, grant UID/MAT/04561/2019.

References

1. Adly S, Haddad T, Thibault L (2014) Convex sweeping process in the framework of measure differential inclusions and evolution variational inequalities. Math Program 148(1–2):5–47
2. Amrani A, Castaing C, Valadier M (1992) Méthodes de troncature appliquées à des problèmes de convergence faible ou forte dans L^1. Arch Ration Mech Anal 117:167–191
3. Barbu V (2010) Nonlinear differential equations of monotone types in Banach spaces. Springer Monographs in Mathematics. Springer, Berlin
4. Benabdellah H, Castaing C, Salvadori A (1997) Compactness and discretization methods for differential inclusions and evolution problems. Atti Semin Mat Fis Univ Modena XLV:9–51
5. Bourbaki N (1965) Intégration. Chapitre I-II-III-IV. Herman, Paris
6. Brezis H (1973) Opérateurs maximaux monotones et semi-groupes de contractions dans les espaces de Hilbert. Lecture Notes in Mathmatics. North-Holland, Amsterdam
7. Castaing C (1980) Topologie de la convergence uniforme sur les parties uniformément intégrables de L^1_E et théorèmes de compacité faible dans certains espaces du type Köthe-Orlicz. Travaux Sém Anal Convexe 10(1):5
8. Castaing C, Valadier M (1977) Convex analysis and measurable multifunctions. Lecture Notes in Mathematics. Springer, Berlin, p 580
9. Castaing C, Faik A, Salvadori A (2000) Evolution equations governed by m-accretive and subdifferential operators with delay. Int J Appl Math 2(9):1005–1026
10. Castaing C, Raynaud de Fitte P, Valadier M (2004) Young measures on topological spaces with applications in control theory and probability theory. Kluwer Academic Publishers, Dordrecht
11. Castaing C, Monteiro Marques MDP, Raynaud de Fitte P (2016) A Skorokhod problem governed by a closed convex moving set. J Convex Anal 23(2):387–423
12. Grothendieck A (1964) Espaces Vectoriels Topologiques. Publicação da Sociedade de Matemática de São Paulo
13. Idzik A (1988) Almost fixed points theorems. Proc Am Math Soc 104:779–784
14. Kenmochi N (1975) Some nonlinear parabolic variational inequalities. Israel J Math 22:304–331
15. Kenmochi N (1981) Solvability of nonlinear evolution equations with time-dependent constraints and applications. Bull Fac Edu 30:1–87
16. Kilbas AA, Srivastava HM, Trujillo JJ (2006) Theory and applications of fractional differential equations. North-Holland mathematics studies, vol 204. North-Holland, Amsterdam
17. Miller KS, Ross B (1993) An introduction to the fractional calculus and fractional differential equations. Willey, NewYork
18. Moreau JJ (1965) Proximité et dualité dans un espace hilbertien. Bull Soc Math France 93:273–299
19. Moreau JJ (1969) Un cas d'addition des sous-différentiels. Fac. Sci. Montpellier, Séminaire d'analyse unilatérale 2:3
20. Moreau JJ (1977) Evolution problem associated with a moving convex set in a Hilbert space. J Differ Equ 26:347–374
21. O'Regan D (2000) Fixed point theorem for weakly sequentially closed maps. Arch Math (Brno) 36:61–70
22. Park S (2006) Fixed points of approximable or Kakutani maps. J Nonlinear Convex Anal 7(1):1–17
23. Peralba JC (1973) Équations d'évolution dans un espace de Hilbert, associées à des opérateurs sous-différentiels. Thèse de doctorat de spécialité. Montpellier
24. Saïdi S, Thibault L, Yarou MF (2013) Relaxation of optimal control problems involving time dependent subdifferential operators. Numer Funct Anal Optim 34(10):1156–1186
25. Saïdi S, Yarou MF (2015) On a time-dependent subdifferential evolution inclusion with Carathéodory perturbation. Ann Pol Math 114(2):133–146
26. Saïdi S, Yarou MF (2015) Set-valued perturbation for time dependent subdifferential operator. Topol Methods Nonlinear Anal 46(1):447–470

27. Tolstonogov AA (2000) Differential inclusions in a Banach space. Springer, Dordrecht
28. Tolstonogov AA (2017) Polyhedral sweeping processes with unbounded nonconvex valued perturbation. J Differ Equ 263(11):7965–7983
29. Tolstonogov AA (2017) Existence and relaxation of solutions for a subdifferential inclusion with unbounded perturbation. J Math Anal Appl 447(1):269–288

Infinite Horizon Optimal Control of Non-Convex Problems Under State Constraints

Hélène Frankowska

Abstract We consider the undiscounted infinite horizon optimal control problem under state constraints in the absence of convexity/concavity assumptions. Then the value function is, in general, nonsmooth. Using the tools of set-valued and nonsmooth analysis, the necessary optimality conditions and sensitivity relations are derived in such a framework. We also investigate relaxation theorems and uniqueness of solutions of the Hamilton–Jacobi–Bellman equation arising in this setting.

Keywords Infinite horizon · Value function · Relaxation theorem · Maximum principle · Sensitivity relations · Hamilton–Jacobi–Bellman equation

Article type: Research Article
Received: August 19, 2018
Revised: March 28, 2019

JEL Classification: C02
Mathematics Subject Classification (2010): 49J53, 49K15, 49L20, 49L25

H. Frankowska (✉)
CNRS, Institut de Mathématiques de Jussieu – Paris Rive Gauche, Sorbonne Université, Paris, France
e-mail: helene.frankowska@imj-prg.fr

T. Maruyama (ed.), *Advances in Mathematical Economics*, Advances in Mathematical Economics 23, https://doi.org/10.1007/978-981-15-0713-7_2

1 Introduction

The following infinite horizon optimal control problem is often present in models of mathematical economics and also in some engineering problems, (like, for instance, the general model of capital accumulation or design of asymptotically stabilizing controls),

$$\text{maximize} \int_0^\infty e^{-\lambda t}\ell(x(t), u(t))\, dt$$

over all trajectory-control pairs (x, u) of the autonomous control system

$$\begin{cases} x'(t) = f(x(t), u(t)), & u(t) \in U \quad \text{for a.e. } t \geq 0 \\ x(0) = x_0, \end{cases}$$

subject to the state constraint $x(t) = (x_1(t), \ldots, x_n(t)) \geq 0$, where $e^{-\lambda t}$ is the discount factor for a given $\lambda > 0$. Its history goes back to Ramsey [39]. Note that in the engineering problems maximization is often replaced by minimization, where results are similar after obvious adaptation of the involved data.

The literature addressing this problem deals with traditional questions of existence of optimal solutions, necessary and sufficient optimality conditions, sensitivity analysis, regularity of the value function, uniqueness of solutions of the associated Hamilton–Jacobi equation, ergodic theory, etc. The above problem is well investigated under the convexity/concavity assumptions implying concavity of the value function (also called the utility function). Then the powerful duality theory of convex analysis can be applied to get both necessary and sufficient optimality conditions and to show differentiability of the value function, see for instance [12, 41]. Indeed, when data are convex/concave, necessary optimality conditions are also sufficient and the value function is differentiable whenever optimal trajectories are unique for every initial condition. This is, for instance, the case of strictly concave problems. In this way a clear picture of optimal solutions can be obtained.

In the general nonlinear case, however, typically the optimal solutions are not unique and necessary conditions are no longer sufficient for optimality. Even when state constraints are not involved, one can expect, at most, local Lipschitz continuity of the value function. Furthermore, if the discount factor is absent, the situation worsens, the value function being, in general, at most upper semicontinuous possibly taking infinite values. Thus, the classical tools can not be used any longer and have to be replaced by notions coming from the set-valued and non-smooth analysis. For instance, solutions of the Hamilton–Jacobi equation have to be understood in a generalized sense, e.g. viscosity solutions. Sensitivity relations become also more complex and involve generalized differentials, instead of derivatives.

Recently, while investigating necessary optimality conditions (in the absence of state constraints), a number of authors addressed more general setting of the infinite horizon problems, not involving the discount factor.

In the present paper, we consider the nonautonomous infinite horizon optimal control problem

$$V(t_0, x_0) = \sup \int_{t_0}^{\infty} L(t, x(t), u(t))\, dt \tag{1}$$

over all trajectory-control pairs (x, u) of the control system

$$\begin{cases} x'(t) = f(t, x(t), u(t)), & u(t) \in U(t) \text{ for a.e. } t \geq t_0 \\ x(t_0) = x_0, \end{cases} \tag{2}$$

satisfying the state constraint

$$x(t) \in K \text{ for all } t \geq t_0, \tag{3}$$

where $L : \mathbb{R}_+ \times \mathbb{R}^n \times \mathbb{R}^m \to \mathbb{R}$, $f : \mathbb{R}_+ \times \mathbb{R}^n \times \mathbb{R}^m \to \mathbb{R}^n$ are given mappings, $\mathbb{R}_+ = [0, +\infty)$, $t_0 \in \mathbb{R}_+$, $x_0 \in \mathbb{R}^n$, $U : \mathbb{R}_+ \rightsquigarrow \mathbb{R}^m$ is a measurable set-valued map with closed nonempty images and $K \subset \mathbb{R}^n$ is nonempty and closed. Selections $u(t) \in U(t)$ are supposed to be Lebesgue measurable and are called controls. The above setting subsumes the classical infinite horizon optimal control problem when f and U are time independent, $L(t, x, u) = e^{-\lambda t}\ell(x, u)$ for some mapping $\ell : \mathbb{R}^n \times \mathbb{R}^m \to \mathbb{R}_+$ and $\lambda > 0$, $t_0 = 0$.

Infinite horizon problems exhibit many phenomena not arising in the context of finite horizon problems and their study is still going on, even in the absence of state constraints, see [1–6, 33, 34, 36–38, 40, 42, 44] and their bibliographies. Among such phenomena let us recall that already in 1970s Halkin, see [32] and also [36], observed that in the necessary optimality conditions for an infinite horizon problem it may happen that the co-state of the maximum principle is different from zero at infinity and that only abnormal maximum principles hold true (even for problems without state constraints). Such phenomena do not occur for the finite horizon problems in the absence of final point constraints.

The presence of state-constraints drastically changes the maximum principle: as in the case of finite horizon problems, a trajectory-control pair satisfying simultaneously the unconstrained Pontryagin maximum principle and the state constraint may be unique and not necessarily optimal. As a consequence, one has to work with discontinuous co-states and more complex adjoint systems.

While the existence theories for problems with or without state constraints are essentially the same (on the domain $dom(V)$ of the value function), this is no longer the case in what concerns optimality conditions and sensitivity relations. Since 1970s many paths were exploited in the literature to derive necessary optimality conditions for the infinite horizon problem when $K = \mathbb{R}^n$. The most immediate one consists in replacing the infinite horizon problem by a family of (finite horizon) Bolza problems on intervals $[t_0, T]$ for $T > t_0$ (that is substituting ∞ by T in the definition of the cost (1)) and using the known results for the Bolza problem. In particular, the first order necessary condition for each Bolza problem takes the form of the maximum principle: if $(\bar{x}, \bar{u})$ is an optimal trajectory-control pair for the Bolza problem at the initial condition (t_0, x_0), then the solution $p_T := p$ of the adjoint system

$$-p'(t) = f_x(t, \bar{x}(t), \bar{u}(t))^* p(t) + L_x(t, \bar{x}(t), \bar{u}(t)), \quad p(T) = 0$$

satisfies the maximality condition

$$\langle p(t), f(t, \bar{x}(t), \bar{u}(t))\rangle + L(t, \bar{x}(t), \bar{u}(t)) = H(t, \bar{x}(t), p(t)) \text{ a.e. in } [t_0, T],$$

where the Hamiltonian $H : \mathbb{R}_+ \times \mathbb{R}^n \times \mathbb{R}^n \to \mathbb{R}$ is defined by

$$H(t, x, p) := \sup_{u \in U(t)} (\langle p, f(t, x, u)\rangle + L(t, x, u)).$$

We underline that the transversality condition $p(T) = 0$ is due to the fact that there is no additional cost term depending on the final state $x(T)$ in the considered Bolza problems.

Recall that if $H(t, \cdot, \cdot)$ is differentiable, then the adjoint system and the maximality condition can be equivalently written in the form of the Hamiltonian system: for a.e. $t \in [t_0, T]$

$$\begin{cases} -p'(t) = H_x(t, \bar{x}(t), p(t)), & p(T) = 0 \\ \bar{x}'(t) = H_p(t, \bar{x}(t), p(t)), & \bar{x}(t_0) = x_0. \end{cases}$$

In general, however, $H(t, \cdot, \cdot)$ is not differentiable and one writes instead a Hamiltonian differential inclusion involving generalized gradients of $H(t, \cdot, \cdot)$, see [19, 45].

Then, taking limits of co-states $p_T(\cdot)$ when $T \to \infty$ is expected to lead to the maximum principle of the infinite horizon problem. This approach requests however some important modifications, due to the fact that the restrictions of an optimal solution $(\bar{x}, \bar{u})$ of the infinite horizon problem to the finite intervals $[t_0, T]$ may be not optimal for the Bolza problems. To overcome this difficulty, some authors add the end point constraint $x(T) = \bar{x}(T)$. With such an additional constraint the restriction of $(\bar{x}, \bar{u})$ to the time interval $[t_0, T]$ becomes optimal for the above Bolza problem. This leads, however, to possibly abnormal maximum principles for finite horizon problems, and, *in fine*, admits necessary optimality conditions not involving the cost function L. Also the transversality condition at time T does disappear, becoming $-p(T) \in N_{\{\bar{x}(T)\}}(\bar{x}(T)) = \mathbb{R}^n$ (normal cone to the singleton $\{\bar{x}(T)\}$ at $\bar{x}(T)$).

Another way to deal with this issue is to modify the very definition of optimal solution, cf. [3, 4, 17, 32, 46]. However, the notions like overtaking (or weakly overtaking) optimal controls do not have appropriate existence theory. More precisely, no specific sufficient conditions were proposed to guarantee the existence of overtaking optimal controls. This is the reason why we prefer to stick to the classical notion of optimality, where sufficient condition for the existence of optimal controls are well understood.

An alternative approach to the discounted infinite horizon problems consists in modification of the cost function in such a way that restrictions of $(\bar{x}, \bar{u})$ to intervals $[t_0, T]$ are locally optimal for the Bolza problems, cf. [35, 40, 47]. The presence of the discount factor allows then to pass to the limit of the (finite horizon) maximum principles and to conclude that a co-state satisfies the adjoint system, the maximality condition and vanishes at infinity. We would like to underline here that such a "terminal" transversality condition at infinity is a consequence of the assumptions on data. This differs substantially from the finite horizon settings, where the transversality condition at the terminal time is an independent requirement. This approach exploits the value function V. Actually, in [35] V is supposed to be C^1 (which is a too strong request) to get these conclusions, while in [47] it is Lipschitz continuous. Furthermore, for the discounted problems considered in [7, 35, 40, 47] the sensitivity relations helped to write a transversality condition also at the initial time.

Another way to derive the maximum principle (still when there are no state constraints) relies on the duality theory on weighted Sobolev spaces with respect to the measure $e^{-\lambda t}dt$ (or more general measures), cf. [7, 34, 36, 37, 44].

In the absence of the discount factor, the question of necessary conditions is quite challenging, because, unlike for classical finite horizon problems, transversality conditions are not immediate. We refer to [2] for an extended overview of the literature devoted to transversality conditions and for bibliographical comments and also to [3] for a further discussion.

The major difficulties in dealing with state constrained problems are due to the fact that for a given optimal solution $(\bar{x}, \bar{u})$ of (1)–(3), small perturbations of the initial state (t_0, x_0) or of the control $\bar{u}$ may result in trajectories violating state constraints. This creates obstacles for the direct application of classical variational methods (as for instance needle perturbations) to derive necessary optimality conditions. In addition, it may happen that the value function V takes infinite values and is discontinuous. For this reason the classical tools of optimal control theory like Hamilton–Jacobi equation and its viscosity solutions are no longer adapted.

Recall that (continuous) *viscosity* solutions to first-order partial differential equations were introduced in [20, 21] by Crandall, Evans, and Lions to investigate Hamilton–Jacobi equations not admitting classical solutions. In particular, given $T > 0$ and a continuous "terminal" function $g_T(\cdot)$, they proved existence and uniqueness of continuous solutions to

$$-\partial_t V + H(t, x, -\nabla_x V) = 0 \quad \text{on } (0, T) \times \mathbb{R}^n, \quad V(T, \cdot) = g_T(\cdot), \tag{4}$$

when the Hamiltonian H is continuous. In the absence of state constraints, under mild assumptions, the value function of the Bolza problem is the unique viscosity solution of (4) provided it is bounded and uniformly continuous. Some sufficient conditions (in the form of an *inward pointing assumption*) for continuity of the value function for a discounted state constrained infinite horizon problem can be found in [43], when f, ℓ are time independent and K is a compact set having smooth boundary. It is shown in [43] that the value function is the unique viscosity solution to a corresponding *stationary* Hamilton–Jacobi equation. However such a framework leaves aside the conical state constraints and the time dependent case, because, as it was shown later on, arguments of [43] no longer apply in the non-autonomous case whenever the time dependence is merely continuous. Some extensions, when K is a locally compact set with possibly nonsmooth boundary, can be found in [28].

In the class of finite horizon state-constrained problems, the Mayer one has been successfully investigated by many authors, see for instance [13, 15, 19, 26, 30, 45] and their bibliographies. Also, it is well known that the Bolza problem can be stated, in an equivalent way, as the Mayer one (without loosing optimality of solutions). This created a favorable background to approach the infinite horizon problems under state constraints.

In [14], in the absence of state constraints, we proposed to use systematically the dynamic programming and to add to the integral functional of the Bolza problem defined on $[0, T]$ the discontinuous (in general) cost function $V(T, \cdot)$. That is for $g_T(\cdot) := V(T, \cdot)$ we considered the Bolza problem

$$\text{maximize } \left(g_T(x(T)) + \int_{t_0}^{T} L(t, x(t), u(t))\, dt \right)$$

over all trajectory-control pairs (x, u) of (2) defined on $[t_0, T]$.

The above finite horizon Bolza problem enjoys the following crucial property: restrictions of an optimal solution $(\bar{x}, \bar{u})$ of the infinite horizon problem to the finite intervals $[t_0, T]$ with $T > t_0$ are optimal for the Bolza problems. Also, for every $T > t_0$ the value function of the Bolza problem coincides with V on $[t_0, T] \times \mathbb{R}^n$.

Let us underline however that this new problem involves the, possibly discontinuous, cost function $V(T, \cdot)$ and for this reason one needs nonsmooth maximum principles derived for finite horizon problems to express necessary optimality conditions.

Such an approach allowed us, by passing to the limit when $T \to \infty$, to get the maximum principle and sensitivity relations and has lead to the transversality condition at the initial time. In particular, this result contains the maximum principle (the sensitivity relation and the initial time transversality condition providing an additional information). Furthermore, we have shown the validity of the relaxation theorems whenever the value function of the *relaxed problem* is continuous with respect to the state variable. It could be interesting to extend also the second order sensitivity relations from [16] to the case of infinite horizon problems.

The present paper discusses results of the same nature but in the presence of state constraints. We start by proving the upper semicontinuity of the value function under the classical assumptions guaranteeing existence of optimal solutions. These assumptions involve convexity of sets

$$F(t, x) := \{ (f(t, x, u), L(t, x, u) - r) : u \in U(t) \text{ and } r \geq 0 \}.$$

We would like to emphasize that this condition (of Cesari-Olech type) is classical in the existence theory of optimal control and does not yield concavity of the value function. To investigate uniqueness of solutions to the Hamilton–Jacobi equation, we also need a sufficient condition for V to vanish at infinity. For this aim we shall impose the following assumption

(H_0): There exists $S > 0$ such that $|L(t, x, u)| \leq \alpha(t)$ for a.e. $t \geq S$ and all $x \in K$, $u \in U(t)$, where $\alpha : [S, +\infty) \to \mathbb{R}_+$ is integrable on $[S, +\infty)$ (see Sect. 4 for more details.)

Under this assumption

$$\lim_{t\to\infty} \sup_{x\in\mathrm{dom}(V(t,\cdot))} |V(t, x)| = 0.$$

This "terminal" condition at infinity replaces the final condition of (4).

When sets $F(t, x)$ are not convex, then it is usual to consider the so-called relaxed problems and speak about generalized solutions. They can be stated either by using the *probability measures* and relaxed controls or, equivalently, by considering the *convexified* infinite horizon problem

$$V^{rel}(t_0, x_0) = \sup \int_{t_0}^{\infty} \left(\sum_{i=0}^{n} \lambda_i(t) L(t, x(t), u_i(t)) \right) dt$$

over all trajectory-control pairs of the relaxed constrained control system

$$\begin{cases} x'(t) = \sum_{i=0}^{n} \lambda_i(t) f(t, x(t), u_i(t)), \ u_i(t) \in U(t), \ \lambda_i(t) \geq 0, \ \sum_{i=0}^{n} \lambda_i(t) = 1 \\ x(t_0) = x_0, \ \ x(t) \in K \ \ \forall t \geq t_0, \end{cases}$$

where $u_i(\cdot)$, $\lambda_i(\cdot)$ are Lebesgue measurable on $\mathbb{R}_+$ for $i = 0, \ldots, n$. Clearly $V^{rel} \geq V$. Furthermore, for the relaxed problem the corresponding sets $F(t, x)$ are convex. When $K = \mathbb{R}^n$, assumption (H_0) allows to prove a relaxation theorem whenever the sets $\{(f(t, x, u), L(t, x, u)) : u \in U(t)\}$ are compact and to identify a more complex relaxed problem when they are neither closed nor bounded, see Sect. 5 below for more details. In this way we extend the relaxation theorem from [14] to the case of *unbounded* sets $\{(f(t, x, u), L(t, x, u)) : u \in U(t)\}$.

For the finite horizon problems, one can find in the literature some relaxation theorems concerned with a single relaxed solution $x(\cdot)$ of a differential inclusion whose right-hand side has compact integrably bounded values on a tubular neighborhood of $x(\cdot)$, cf. [19, 45]. In our finite horizon relaxation Theorem 4 below we show that for control systems such assumptions can be skipped. Even though Theorem 4 concerns all the relaxed trajectories, its proof can be localized to a tubular neighborhood of a fixed relaxed trajectory. Then assumptions can be localized as well without requiring compactness and integral boundedness imposed in [19, 45]. In this respect, Theorem 4 would imply a new result also in the setting of [19, 45].

The situation changes drastically, even in the finite horizon framework, when state constraints are present. For instance the original control system may not have any feasible trajectories, while the relaxed system does. Then, in order to get relaxation theorems, one needs the so called *relaxed inward pointing condition* (IPC^{rel}) linking f with tangents to K at the boundary points of K, that we recall in Sect. 6. This condition expresses the compatibility of dynamics with state constraints. It was introduced in [26] to derive the normal maximum principle and the sensitivity relations for a state-constrained Mayer problem with locally Lipschitz cost function.

In Sect. 8 we show that it can be exploited as well to get similar results for the state constrained infinite horizon problem, provided $V(t, \cdot)$ is locally Lipschitz on K for all large t. (IPC^{rel}) is an alternative to the *inward pointing condition* from [43] to prove the so called *Neighboring Feasible Trajectory theorem*, when dynamics depend on time. The inward pointing condition is much simpler than (IPC^{rel}), but, unfortunately, is not convenient to work with data depending measurably (or even continuously) on time and state constraints having nonsmooth boundaries. It is not difficult to realize that under assumptions imposed in [43], the inward pointing condition is equivalent to (IPC^{rel}).

Recall that for problems without state constraints the sensitivity relation $p(t) = V_x(t, \bar{x}(t))$ has a significant economic interpretation (see for instance [1], [41]): the co-state p (of the maximum principle) is the *shadow price* describing the contribution to the value function of a unit increase of capital x.

When the value function is merely locally Lipschitz on $\mathbb{R}_+ \times K$, the sensitivity relation is more complex and takes the form

$$(-H(t, \bar{x}(t), q(t)), q(t)) \in \partial V(t, \bar{x}(t)),$$

where q is the adjoint state (of bounded variation) and ∂V denotes the generalized gradient of V (defined in Sect. 8 taking into consideration the state constraints). In Sect. 8 we derive the maximum principle augmented by the above sensitivity relation for a state constrained infinite horizon problem. An important future of the obtained here necessary optimality condition is its *normality*.

Local Lipschitz continuity of V for infinite horizon problems under state constraints was recently investigated in [10, 11]. On the other hand, uniqueness of upper semicontinuous solutions of the associated Hamilton–Jacobi equation was studied in [9]. In Sect. 9 we show uniqueness of locally Lipschitz solutions of the Hamilton–Jacobi equation by arguments simpler than those in [9].

The outline of the paper is as follows. In Sect. 2, we recall some definitions from set-valued and nonsmooth analysis. In Sect. 3, we introduce the value function V and basic assumptions that imply in Sect. 4 the upper semicontinuity of V. In Sects. 5 and 6 we discuss the relaxation results for problems without and with state constraints, respectively, and in Sect. 7 we describe the link between the finite and infinite horizon problems. Sect. 8 is devoted to the maximum principle and sensitivity relations, while Sect. 9 deals with the uniqueness of solutions to the Hamilton–Jacobi equation.

2 Preliminaries and Notations

Denote by $L^1_{\text{loc}}(\mathbb{R}_+; \mathbb{R}_+)$ the set of all locally integrable functions $\psi : \mathbb{R}_+ \to \mathbb{R}_+$. For any $\psi \in L^1_{\text{loc}}(\mathbb{R}_+; \mathbb{R}_+)$ and $\sigma > 0$ define

$$\theta_\psi(\sigma) = \sup\left\{\int_J \psi(\tau)\,d\tau \;:\; J \subset \mathbb{R}_+,\; \mathcal{M}(J) \leqslant \sigma\right\},$$

where $\mathcal{M}(J)$ stands for the Lebesgue measure of J. Denote by $\mathcal{L}_{\text{loc}}$ the subset of all $\psi \in L^1_{\text{loc}}(\mathbb{R}_+; \mathbb{R}_+)$ such that $\lim_{\sigma\to 0}\theta_\psi(\sigma) = 0$. Notation $W^{1,1}_{\text{loc}}(\mathbb{R}_+; \mathbb{R}^n)$ stands for the set of locally absolutely continuous functions on $\mathbb{R}_+$. For $\psi : \mathbb{R}_+ \to \mathbb{R}$ and $t_0 \geq 0$ define

$$\int_{t_0}^{\infty} \psi(t)dt = \lim_{T\to+\infty} \int_{t_0}^{T} \psi(t)dt,$$

provided the above limit does exist.

Let X be a normed space, $B(x, R)$ be the closed ball in X centered at $x \in X$ with radius $R > 0$ and set $B := B(0, 1)$. For a nonempty subset $C \subset X$ we denote its interior by $\operatorname{int}(C)$, its boundary by $\operatorname{bd}(C)$, its convex hull by $\operatorname{co} C$, its closed convex hull by $\overline{\operatorname{co}}\, C$, and the distance from $x \in X$ to C by $d_C(x) := \inf\{|x - y|_X : y \in C\}$. If $X = \mathbb{R}^n$, the *negative polar cone* to C is given by $C^- = \{p \in \mathbb{R}^n : \langle p, c\rangle \leqslant 0 \quad \forall c \in C\}$, where $\langle\cdot,\cdot\rangle$ is the scalar product in $\mathbb{R}^n$. The unit sphere in $\mathbb{R}^n$ is denoted by S^{n-1}.

Let $\mathcal{T} \subset \mathbb{R}^k$ be nonempty and $\{A_\tau\}_{\tau\in\mathcal{T}}$ be a family of subsets of $\mathbb{R}^n$. The *upper and lower limits*, in the Péano-Kuratowski sense, of A_τ at $\tau_0 \in \mathcal{T}$ are the closed sets defined respectively by

$$\begin{aligned}\operatorname{Limsup}_{\tau\to_{\mathcal{T}}\tau_0} A_\tau &= \left\{v \in \mathbb{R}^n : \liminf_{\tau\to_{\mathcal{T}}\tau_0} d_{A_\tau}(v) = 0\right\},\\ \operatorname{Liminf}_{\tau\to_{\mathcal{T}}\tau_0} A_\tau &= \left\{v \in \mathbb{R}^n : \limsup_{\tau\to_{\mathcal{T}}\tau_0} d_{A_\tau}(v) = 0\right\},\end{aligned}$$

where $\to_{\mathcal{T}}$ stands for the convergence in $\mathcal{T}$ and $d_{A_\tau}(v) = +\infty$ whenever $A_\tau = \emptyset$. See for instance [8] for properties of these set limits.

Let $K \subset \mathbb{R}^n$ and $x \in K$. The contingent cone to K at x consists of all $v \in \mathbb{R}^n$ such that for some sequences $h_i \to 0+$, $v_i \to v$ we have $x + h_i v_i \in K$. The limiting normal cone to a closed subset $K \subset \mathbb{R}^n$ at $x \in K$ is given by

$$N^L_K(x) := \operatorname{Limsup}_{y\to_K x} T_K(y)^-.$$

It is well known that if x lies on the boundary of K, then $N^L_K(x)$ is not reduced to zero. The Clarke tangent and normal cones to K at x are defined by $C_K(x) = \left(N^L_K(x)\right)^-$ and $N_K(x) = C_K(x)^-$, respectively. Note that $N_K(x) = \overline{\operatorname{co}} N^L_K(x)$ and set $N^1_K(x) := N_K(x) \cap S^{n-1}$.

For $\varphi : \mathbb{R}^n \to \mathbb{R} \cup \{\pm\infty\}$ denote by $dom(\varphi)$ the domain of φ, that is the set of all $x \in \mathbb{R}^n$ such that $\varphi(x)$ is finite and by $epi(\varphi)$ and $hyp(\varphi)$, respectively, its epigraph and hypograph. For any $x \in dom(\varphi)$ the upper and lower (contingent) directional derivatives of φ at x in the direction $y \in \mathbb{R}^n$ are defined respectively by

$$D_{\downarrow}\varphi(x)y = \limsup_{z \to y, h \to 0+} \frac{\varphi(x+hz)-\varphi(x)}{h},$$
$$D_{\uparrow}\varphi(x)y = \liminf_{z \to y, h \to 0+} \frac{\varphi(x+hz)-\varphi(x)}{h}$$

and the Fréchet superdifferential $\partial^+\varphi(x)$ (resp. subdifferential $\partial^-\varphi(x)$) of φ at x by

$$p \in \partial^+\varphi(x) \iff \limsup_{y \to x} \frac{\varphi(y) - \varphi(x) - \langle p, y - x \rangle}{|y - x|} \leq 0$$

and

$$p \in \partial^-\varphi(x) \iff \liminf_{y \to x} \frac{\varphi(y) - \varphi(x) - \langle p, y - x \rangle}{|y - x|} \geq 0.$$

By [22], $p \in \partial^+\varphi(x)$ if and only if $(-p, +1) \in T_{hyp(\varphi)}(x, \varphi(x))^-$ and $p \in \partial^-\varphi(x)$ if and only if $(p, -1) \in T_{epi(\varphi)}(x, \varphi(x))^-$. Furthermore,

$$\partial^+\varphi(x) = \{p \ : \ \langle p, y \rangle \geq D_{\downarrow}\varphi(x)y \ \ \forall \ y \in \mathbb{R}^n\},$$
$$\partial^-\varphi(x) = \{p \ : \ \langle p, y \rangle \leq D_{\uparrow}\varphi(x)y \ \ \forall \ y \in \mathbb{R}^n\}.$$

To compare with sub and superdifferentials used in the theory of viscosity solutions, let us emphasize that the very same arguments as those of [21, Proof of Proposition 1.1] imply the following result.

Proposition 1 *Let $\varphi : \mathbb{R}^n \to \mathbb{R} \cup \{-\infty\}$ be Lebesgue measurable. For any $x \in dom(\varphi)$, a vector $p \in \partial^+\varphi(x)$ if and only if there exists a continuous mapping $\psi : \mathbb{R}^n \to \mathbb{R}$ such that $\psi(x) = \varphi(x)$, $\psi(y) > \varphi(y)$ for all $y \neq x$ and the Fréchet derivative of ψ at x exists and is equal to p.*

Actually, in [21] φ is continuous and $\psi \in C^1$. However for discontinuous mappings, in general, ψ constructed in [21] is not continuously differentiable. A similar result can be also stated for the subdifferential $\partial^-\varphi(x)$.

Clearly, for all $p \in \mathbb{R}^n$ and $q \in \mathbb{R}$ satisfying $(p, q) \in N^L_{epi(\varphi)}(x, \varphi(x))$ we have $q \leq 0$. Furthermore, if $q < 0$, then $(p, q) \in N^L_{epi(\varphi)}(x, \varphi(x))$ if and only if $(p/|q|, -1) \in N^L_{epi(\varphi)}(x, \varphi(x))$. Any $p \in \mathbb{R}^n$ satisfying $(p, -1) \in N^L_{epi(\varphi)}(x, \varphi(x))$ is called a *limiting subgradient* of φ at x. The set of all limiting subgradients of φ at x is denoted by $\partial^{L,-}\varphi(x)$.

Consider $\varphi : \mathbb{R}^n \to \mathbb{R}$, Lipschitz around a given $x \in \mathbb{R}^n$, and denote by $\nabla\varphi(\cdot)$ its gradient, which, by the Rademacher theorem, exists a.e. in a neighborhood of x.

The *Clarke generalized gradient* of $\varphi(\cdot)$ at x is defined by

$$\partial\varphi(x) := co\,\mathrm{Limsup}_{y\to x}\{\nabla\varphi(y)\}.$$

It is well known that $\partial\varphi(x) = co\,\partial^{L,-}\varphi(x)$, see [19].

3 Value Function of the Infinite Horizon Problem

Consider the non-autonomous infinite horizon optimal control problem (1)–(3) with data as described in the introduction. In particular, $U(\cdot)$ is Lebesgue measurable and has closed nonempty images. Every Lebesgue measurable $u : \mathbb{R}_+ \to \mathbb{R}^m$ satisfying $u(t) \in U(t)$ a.e. is called a control and the set of all controls is denoted by $\mathcal{U}$. Note that to state (2) we need controls to be defined only on $[t_0, +\infty)$. However, since throughout the paper the time interval varies, to avoid additional notations and without any loss of generality, we consider controls defined on $[0, +\infty)$. We underline that, by the measurable selection theorem, for any measurable selection $u(t) \in U(t)$ for $t \in [t_0, \infty)$ we can find $\tilde{u} \in \mathcal{U}$ such that $\tilde{u} = u$ on $[t_0, \infty)$.

Assumptions **(H1)**:

(i) There exists $c \in L^1_{\mathrm{loc}}(\mathbb{R}_+; \mathbb{R}_+)$ such that for a.e. $t \geq 0$,

$$2\langle f(t,x,u), x\rangle \leq c(t)(1+|x|^2), \quad |f(t,0,u)| \leq c(t) \;\; \forall\, x \in \mathbb{R}^n,\; u \in U(t);$$

(ii) For every $R > 0$, there exist $c_R \in L^1_{\mathrm{loc}}(\mathbb{R}_+; \mathbb{R}_+)$ and a modulus of continuity $\omega_R : \mathbb{R}_+ \times \mathbb{R}_+ \to \mathbb{R}_+$ such that for a.e. $t \in \mathbb{R}_+$, $\omega_R(t, \cdot)$ is increasing, $\lim_{r\to 0+} \omega_R(t, r) = 0$ and for every $u \in U(t)$ and $x, y \in B(0, R)$,

$$|f(t,x,u) - f(t,y,u)| \leq c_R(t)|x-y|, \; |L(t,x,u) - L(t,y,u)| \leq \omega_R(t, |x-y|);$$

(iii) The mappings f, L are Carathéodory, that is measurable in t and continuous in x, u;

(iv) There exists $\beta \in L^1_{\mathrm{loc}}(\mathbb{R}_+; \mathbb{R}_+)$ and an increasing function $\phi : \mathbb{R}_+ \to \mathbb{R}_+$ such that for a.e. $t \in \mathbb{R}_+$,

$$|L(t,x,u)| \leq \beta(t)\phi(|x|), \quad \forall\, x \in K,\; u \in U(t);$$

(v) There exists $S > 0$ such that $L(t,x,u) \leq \alpha(t)$ for a.e. $t \geq S$ and all $x \in K$, $u \in U(t)$, where $\alpha : [S, +\infty) \to \mathbb{R}_+$ is integrable on $[S, +\infty)$.

(vi) For a.e. $t \in \mathbb{R}_+$ and for all $x \in \mathbb{R}^n$ the set

$$F(t,x) := \Big\{\big(f(t,x,u), L(t,x,u) - r\big) : u \in U(t) \text{ and } r \geq 0\Big\}$$

is closed and convex.

Remark 1

(a) In some results below (H1) (i) or (vi) will be skipped.
(b) Assumptions (H1) (i)–(iii) imply that for every $(t_0, x_0) \in \mathbb{R}_+ \times \mathbb{R}^n$ and $u \in \mathcal{U}$ there exists a unique solution defined on $[t_0, \infty)$ of the system

$$x' = f(t, x, u(t)) \tag{5}$$

satisfying $x(t_0) = x_0$.
(c) Note that assumption (H1) (i) holds true whenever

$$\exists\, \theta \in L^1_{\text{loc}}(\mathbb{R}_+; \mathbb{R}_+),\ |f(t, x, u)| \leq \theta(t)(|x| + 1),\ \forall\, t \in \mathbb{R}_+,\ x \in \mathbb{R}^n,\ u \in U(t). \tag{6}$$

(d) Note that (iv) and (v) imply that for every feasible trajectory control pair of (5) defined on $[t_0, \infty)$ the limit $\lim_{T\to\infty} \int_{t_0}^{T} L(t, x(t), u(t))\, dt$ does exist and belongs to $[-\infty, \infty)$.
(e) Even though it may seem, at first glance, that conditions like (i), (ii), (iv), (vi) yield compactness of sets $U(t)$, since $f(t, x, \cdot)$ is genuinely nonlinear and merely continuous, (H1) does not imply boundedness of sets $U(t)$.

Consider $0 \leq a \leq b$ and $u \in \mathcal{U}$. An absolutely continuous function $x : [a, b] \to \mathbb{R}^n$ satisfying $x'(t) = f(t, x(t), u(t))$ a.e. in $[a, b]$ is called a solution of (5) corresponding to the control $u(\cdot)$ and (x, u) is called a trajectory-control pair of (5) on $[a, b]$. If moreover $x(t) \in K$ for all $t \in [a, b]$, then such a solution x is called feasible and (x, u) is called a feasible trajectory-control pair on $[a, b]$. It may happen that for some control $u(\cdot)$ and $(t_0, x_0) \in \mathbb{R}_+ \times K$ the system (5) does not have any feasible solution satisfying $x(t_0) = x_0$. If $(x, u) : [a, \infty) \to \mathbb{R}^n \times \mathbb{R}^m$ is so that for every $b > a$, the restriction of (x, u) to $[a, b]$ is a trajectory-control pair of (5) on $[a, b]$, then (x, u) is called a trajectory-control pair on $[a, \infty)$. It is feasible, if $x(t) \in K$ for every $t \geq a$.

The function $V : \mathbb{R}_+ \times K \to \mathbb{R}_+$ defined by (1)–(3) is called the value function of the infinite horizon problem. If for a given $(t_0, x_0) \in \mathbb{R}_+ \times K$ no trajectory of (2) and (3) does exist, we set $V(t_0, x_0) = -\infty$. This choice is dictated by the fact that, in general, V is discontinuous and, under the (classical) assumptions (H1), it is upper semicontinuous on $dom(V) \subset \mathbb{R}_+ \times K$, see the next section. To preserve its upper semicontinuity on $\mathbb{R}_+ \times \mathbb{R}^n$ we set $V = -\infty$ on $(\mathbb{R}_+ \times \mathbb{R}^n) \backslash dom(V)$. Clearly, under assumption (H1) (iv), if for some $T \geq 0$ the set $dom(V(T, \cdot)) \neq \emptyset$, then $dom(V(t, \cdot)) \neq \emptyset$ for every $t \geq T$.

Denote by $x(\cdot; t_0, x_0, u)$ the trajectory of (5) corresponding to the control u and satisfying $x(t_0) = x_0$. By Gronwall's lemma and (H1) (i), for all $(t_0, x_0) \in \mathbb{R}_+ \times \mathbb{R}^n$,

$$|x(t; t_0, x_0, u)|^2 \leq \Big(|x_0|^2 + \int_{t_0}^{t} c(s)\, ds\Big) e^{\int_{t_0}^{t} c(s)\, ds} \quad \forall\, t \geq t_0.$$

Moreover, setting

$$M_t(T,r)^2 = \left(r^2 + \int_t^T c(s)\,ds\right)e^{\int_t^T c(s)\,ds} \quad \forall\, T \geq t \geq 0,\; r \geq 0,$$

the following holds true: for all $r \geq 0$ and $u \in \mathcal{U}$

$$|x_0| \leq r \implies |x(t; t_0, x_0, u)| \leq M_{t_0}(t,r) \quad \forall\, t \geq t_0. \tag{7}$$

The above bound, together with the assumption (H1) (ii) and the Gronwall lemma, yield the local Lipschitz dependence of trajectories on the initial conditions: for all $r,\ T > 0$, for every $t_0 \in [0,T]$ and $x_0,\ x_1 \in B(0,r)$,

$$|x(t; t_0, x_1, u) - x(t; t_0, x_0, u)| \leq |x_1 - x_0|\, e^{\int_{t_0}^t c_{M_{t_0}(T,r)}(s)\,ds} \quad \forall\, t \in [t_0, T].$$

Given a feasible trajectory-control pair (x,u), define

$$\int_{t_0}^{\infty} L(s, x(s), u(s))\,ds = \lim_{t\to\infty} \int_{t_0}^{t} L(s, x(s), u(s))\,ds.$$

We claim that (H1) (iv), (v) imply that the above limit does exist and belongs to $[-\infty, \infty)$. Indeed, since $L(s,x,u) \leq \alpha(s)$ for a.e. $s \geq S$ and all $x \in K$, $u \in U(s)$, the mapping $t \mapsto \int_S^t (L(s,x(s),u(s)) - \alpha(s))\,ds$ is nonincreasing on $[S,\infty)$ (with respect to t) and so it has a limit when $t \to \infty$ (possibly equal to $-\infty$). Since

$$\int_S^t (L(s,x(s),u(s)) - \alpha(s))\,ds = \int_S^t L(s,x(s),u(s))\,ds - \int_S^t \alpha(s)\,ds \leq 0$$

and $\lim_{t\to\infty} \int_S^t \alpha(s)\,ds$ does exist and is finite, it follows that also the limit $\lim_{t\to\infty} \int_S^t L(s,x(s),u(s))\,ds$ does exist and is different from $+\infty$. On the other hand, by (H1) (iv) the integral $\int_{t_0}^S L(s,x(s),u(s))\,ds$ is finite, proving our claim.

For any $(t_0, x_0) \in \mathbb{R}_+ \times K$, a feasible trajectory-control pair $(\bar{x}, \bar{u})$ on $[t_0,\infty)$ is called optimal for the infinite horizon problem at (t_0, x_0) if $\bar{x}(t_0) = x_0$ and for every feasible trajectory-control pair (x,u) on $[t_0,\infty)$ satisfying $x(t_0) = x_0$ we have

$$\int_{t_0}^{\infty} L(t, \bar{x}(t), \bar{u}(t))\,dt \geq \int_{t_0}^{\infty} L(t, x(t), u(t))\,dt.$$

It is not difficult to realize that if (H1) (i), (iv), (v) are satisfied, then V is locally bounded from the above on $dom(V)$ and that V takes values in $[-\infty, \infty)$.

4 Upper Semicontinuity of the Value Function

The question of existence of optimal controls is pretty well understood. The standard proofs rely on taking limits of maximizing subsequences of trajectories and weak limits of their derivatives. Cesari-Olech type convexity and upper semicontinuity assumptions are needed to justify that the limiting trajectory is optimal. The very same arguments can be applied as well to study the upper semicontinuity of the value function.

Theorem 1 *Assume (H1). Then V is upper semicontinuous, takes values in $[-\infty, \infty)$ and for every $(t_0, x_0) \in dom(V)$, there exists a feasible trajectory-control pair $(\bar{x}, \bar{u})$ satisfying $V(t_0, x_0) = \int_{t_0}^{\infty} L(t, \bar{x}(t), \bar{u}(t))\, dt$.*

Moreover, if

$$\begin{cases} \exists\, \bar{S} > 0 \text{ and an integrable } \delta : [\bar{S}, \infty) \to \mathbb{R}_+ \text{ such that} \\ L(t, x, u) \geq -\delta(t) \text{ for a.e. } t > \bar{S},\ \forall\, x \in K,\ u \in U(t), \end{cases} \tag{8}$$

then

$$\lim_{t \to \infty} \sup_{x \in \mathrm{dom}(V(t, \cdot))} |V(t, x)| = 0. \tag{9}$$

Proof Assumptions (H1) (iv), (v) and (7) yield that V never takes value $+\infty$. The arguments for proving the existence of optimal solutions are well known. We recall them because similar ones will be also exploited in the other results of this paper.

Let $(t_0, x_0) \in dom(V)$. Consider a maximizing sequence of feasible trajectory-control pairs (x_i, u_i) satisfying $x_i(t_0) = x_0$. That is

$$\lim_{i \to \infty} \int_{t_0}^{\infty} L(t, x_i(t), u_i(t))dt = V(t_0, x_0).$$

In particular, $x_i(t) \in K$ for all $t \geq t_0$ and $i \geq 1$. By (7), for every $T > t_0$, the restrictions of x_i to $[t_0, T]$ are equibounded.

We construct the optimal trajectory control pair $(\bar{x}, \bar{u})$ of the infinite horizon problem using the induction argument. Let $R > 0$ be such that for every i,

$$\sup_{t \in [t_0, t_0+1]} |x_i(t)| \leq R.$$

Then

$$\begin{aligned} |f(t, x_i(t), u_i(t))| &\leq |f(t, 0, u_i(t))| + |f(t, x_i(t), u_i(t)) - f(t, 0, u_i(t))| \\ &\leq c(t) + c_R(t)R \end{aligned} \tag{10}$$

Thus, (H1) (iv) and the Dunford–Pettis theorem, imply that, taking a subsequence and keeping the same notation, we may assume that for some integrable functions $y : [t_0, t_0 + 1] \to \mathbb{R}^n$, $\gamma : [t_0, t_0 + 1] \to \mathbb{R}_+$, the restrictions of $(x_i'(\cdot), L(\cdot, x_i(\cdot), u_i(\cdot)))$ to $[t_0, t_0 + 1]$ converge weakly in $L^1([t_0, t_0 + 1]; \mathbb{R}^n \times \mathbb{R})$ to (y, γ). Define

$$z_i(t) = \int_{t_0}^{t} L(t, x_i(t), u_i(t))\,dt, \quad \forall t \in [t_0, t_0 + 1].$$

Since

$$x_i(t) = x_0 + \int_{t_0}^{t} f(t, x_i(t), u_i(t))\,dt, \quad \forall t \in [t_0, t_0 + 1],$$

we deduce that functions (x_i, z_i) converge pointwise on $[t_0, t_0 + 1]$ when $i \to \infty$ to the function $(\bar{x}, z) : [t_0, t_0 + 1] \to \mathbb{R}^n \times \mathbb{R}$ defined by

$$\bar{x}(t) = x_0 + \int_{t_0}^{t} y(t)\,dt, \quad z(t) = \int_{t_0}^{t} \gamma(t)\,dt, \quad \forall t \in [t_0, t_0 + 1].$$

Hence $(\bar{x}, z)$ is absolutely continuous on $[t_0, t_0 + 1]$. Furthermore, (7), (10), (H1) (iv) and the Ascoli–Arzelà theorem imply that (x_i, z_i) converge uniformly to $(\bar{x}, z)$. Moreover, at every Lebesgue point t of $y(\cdot)$ we have $\bar{x}'(t) = y(t)$.

Observe next that, by (H1) (ii), for a.e. $t \in [t_0, t_0 + 1]$,

$$(x_i'(t), L(t, x_i(t), u_i(t))) \in F(t, x_i(t)) \subset F(t, \bar{x}(t)) + (c_R(t)|x_i(t) - \bar{x}(t)| + \omega_R(t, |x_i(t) - \bar{x}(t)|))B.$$

Let $\varepsilon > 0$ and $i_0 \geq 1$ be such that $\sup_{t \in [t_0, t_0+1]} |x_i(t) - \bar{x}(t)| \leq \varepsilon$ for all $i \geq i_0$. The set

$$F_\varepsilon(t, \bar{x}(t)) := F(t, \bar{x}(t)) + (c_R(t)\varepsilon + \omega_R(t, \varepsilon))B$$

being convex and closed, also the set

$$\mathcal{F}_\varepsilon := \{(v, w) \in L^1([t_0, t_0 + 1]; \mathbb{R}^n \times \mathbb{R}) \mid (v(t), w(t)) \in F_\varepsilon(t, \bar{x}(t)) \text{ for a.e. } t\}$$

is convex and closed. Thus, by the Mazur theorem, it is weakly closed and therefore $(\bar{x}', \gamma) \in \mathcal{F}_\varepsilon$. So

$$(\bar{x}'(t), \gamma(t)) \in F(t, \bar{x}(t)) + (c_R(t)\varepsilon + \omega_R(t, \varepsilon))B \quad \text{for a.e. } t \in [t_0, t_0 + 1].$$

By the arbitrariness of $\varepsilon > 0$, $(\bar{x}'(t), \gamma(t)) \in F(t, \bar{x}(t))$ for a.e. $t \in [t_0, t_0 + 1]$.

From the measurable selection theorem [8, Theorem 8.2.10] we deduce that there exist a control $\bar{u}(\cdot)$ and a measurable function $r : [t_0, t_0 + 1] \to \mathbb{R}_+$ such that for a.e. $t \in [t_0, t_0 + 1]$,

$$\bar{x}'(t) = f(t, \bar{x}(t), \bar{u}(t)), \quad \gamma(t) = L(t, \bar{x}(t), \bar{u}(t)) - r(t).$$

Since $x_i(t) \in K$ for all $t \geq t_0$ and K is closed we know that $\bar{x}([t_0, t_0 + 1]) \subset K$. Furthermore,

$$\begin{aligned}\lim_{i\to\infty} \int_{t_0}^{t_0+1} L(t, x_i(t), u_i(t))dt &= \int_{t_0}^{t_0+1} (L(t, \bar{x}(t), \bar{u}(t)) - r(t))\, dt \\ &\leq \int_{t_0}^{t_0+1} L(t, \bar{x}(t), \bar{u}(t)))\, dt.\end{aligned}$$

We extend next $(\bar{x}, \bar{u})$ on $[t_0, \infty)$. Set $(\bar{x}^1(t), \bar{u}^1(t)) := (\bar{x}(t), \bar{u}(t))$ for $t \in [t_0, t_0 + 1]$. Let us assume that for some $k \geq 1$ we have constructed a subsequence $\{(x^k_{i_j}, u^k_{i_j})\}_j$ of $\{(x_i, u_i)\}_i$, a trajectory control pair $(\bar{x}^k, \bar{u}^k)$ on $[t_0, t_0 + k]$ and an absolutely continuous function $z^k \in W^{1,1}([t_0, t_0+k]; \mathbb{R})$ such that $\bar{x}^k([t_0, t_0+k]) \subset K$ and for

$$z^k_{i_j}(t) := \int_{t_0}^{t} L(s, x^k_{i_j}(s), u^k_{i_j}(s))\, ds, \quad \forall\, t \in [t_0, t_0 + k]$$

the following holds true:

$(x^k_{i_j}, z^k_{i_j})$ converge uniformly on $[t_0, t_0 + k]$ to $(\bar{x}^k, z^k)$, $((x^k_{i_j})', (z^k_{i_j})')$ converge weakly in $L^1([t_0, t_0 + k]; \mathbb{R}^n \times \mathbb{R})$ to $((\bar{x}^k)', (z^k)')$ and if $k \geq 2$

$$(\bar{x}^k(t), \bar{u}^k(t), z^k(t)) = (\bar{x}^{k-1}(t), \bar{u}^{k-1}(t), z^{k-1}(t)) \;\; \forall t \in [t_0, t_0 + k - 1].$$

Consider the interval $[t_0, t_0+k+1]$. By the same arguments, we find a subsequence $\{(x^k_{i_{j_\ell}}, u^k_{i_{j_\ell}}, z^k_{i_{j_\ell}})\}_\ell$ of $\{(x^k_{i_j}, u^k_{i_j}, z^k_{i_j})\}_j$, a trajectory-control pair $(\bar{x}^{k+1}, \bar{u}^{k+1})$ on $[t_0, t_0 + k + 1]$ and an absolutely continuous $z^{k+1} : [t_0, t_0 + k + 1] \to \mathbb{R}$, such that $(x^k_{i_{j_\ell}}, z^k_{i_{j_\ell}})$ converge uniformly on $[t_0, t_0 + k + 1]$ to $(\bar{x}^{k+1}, z^{k+1})$, $((x^k_{i_{j_\ell}})', (z^k_{i_{j_\ell}})')$ converge weakly in $L^1([t_0, t_0 + k + 1]; \mathbb{R}^n \times \mathbb{R})$ to $((\bar{x}^{k+1})', (z^{k+1})')$,

$$\lim_{\ell\to\infty} \int_{t_0}^{t_0+k+1} L(t, x^k_{i_{j_\ell}}(t), u^k_{i_{j_\ell}}(t))dt \leq \int_{t_0}^{t_0+k+1} L(t, \bar{x}^{k+1}(t), \bar{u}^{k+1}(t))\, dt$$

and $\bar{x}^{k+1}([t_0, t_0 + k + 1]) \subset K$,

$$(\bar{x}^{k+1}(t), \bar{u}^{k+1}(t), z^{k+1}(t)) = (\bar{x}^k(t), \bar{u}^k(t), z^k(t)) \;\; \forall t \in [t_0, t_0 + k].$$

Rename $(x^k_{i_{j_\ell}}, u^k_{i_{j_\ell}})$ by $(x^{k+1}_{i_j}, u^{k+1}_{i_j})$ and set $(\bar{x}(t), \bar{u}(t)) = (\bar{x}^{k+1}(t), \bar{u}^{k+1}(t))$ for $t \in [t_0, t_0 + k + 1]$. Applying the induction argument with respect to k we obtain a trajectory-control pair $(\bar{x}, \bar{u})$ defined on $[t_0, \infty)$.

To show that it is optimal, fix $\varepsilon > 0$. By (H1) (v) for all large T and for every feasible trajectory-control pair (x, u) on $[t_0, \infty)$,

$$\int_T^{\infty} L(s, x(s), u(s))ds \leq \varepsilon.$$

Consequently for any fixed sufficiently large k, using the same notation as before, we get

$$\lim_{j\to\infty} \int_{t_0}^{\infty} L(t, x_{i_j}^k(t), u_{i_j}^k(t))\, dt \leq \lim_{j\to\infty} \int_{t_0}^{t_0+k} L(t, x_{i_j}^k(t), u_{i_j}^k(t))\, dt + \varepsilon$$
$$\leq \int_{t_0}^{t_0+k} L(t, \bar{x}(t), \bar{u}(t))dt + \varepsilon.$$

We proved that for every $\varepsilon > 0$ and all large k,

$$V(t_0, x_0) \leq \int_{t_0}^{t_0+k} L(t, \bar{x}(t), \bar{u}(t))dt + \varepsilon$$

Taking the limit when $k \to \infty$ we obtain

$$V(t_0, x_0) \leq \int_{t_0}^{\infty} L(t, \bar{x}(t), \bar{u}(t))dt + \varepsilon.$$

Since $\varepsilon > 0$ is arbitrary, this inequality implies that $(\bar{x}, \bar{u})$ is optimal.

To prove the upper semicontinuity of V, consider a sequence $(t_0^i, x_0^i) \in \mathbb{R}_+ \times K$ converging to some (t_0, x_0) when $i \to \infty$. We have to show that $\limsup_{i\to\infty} V(t_0^i, x_0^i) \leq V(t_0, x_0)$. If for all large i, $V(t_0^i, x_0^i) = -\infty$, then we are done. So it is enough to consider the case when $(t_0^i, x_0^i) \in dom(V)$ for all i.

Let (x_i, u_i) be an optimal trajectory-control pair corresponding to the initial condition (t_0^i, x_0^i). If $t_0^i > t_0$, then we extend x_i on $[t_0, t_0^i]$ by setting $x_i(s) = x_0^i$ for $s \in [t_0, t_0^i]$. Using exactly the same arguments as before we construct a trajectory-control pair $(\bar{x}, \bar{u})$ such that $\bar{x}(t_0) = x_0$ and for every $\varepsilon > 0$ and all large k,

$$\limsup_{i\to\infty} \int_{t_i}^{\infty} L(t, x_i(t), u_i(t))\, dt \leq \int_{t_0}^{t_0+k} L(t, \bar{x}(t), \bar{u}(t))dt + \varepsilon.$$

Taking the limit when $k \to \infty$ and using that ε is arbitrary, the upper semicontinuity of V at (t_0, x_0) follows.

Suppose next that (8) is satisfied. Thus, if t is sufficiently large and $x_t \in \mathrm{dom}(V(t, \cdot))$, then $|V(t, x_t))| \leq \int_t^{\infty}(\alpha(s) + \delta(s))ds$. Then (9) follows from the equality $\lim_{t\to\infty} \int_t^{\infty}(\alpha(s) + \delta(s))ds = 0$.

In the above proof we needed assumption (H1) (i) to deduce (7). A different assumption involving bounds on the growth of $|f|$ with respect to L is convenient as well. Its advantage lies in the fact that we do not request anymore the sets $f(t, x, U(t))$ to be bounded.

Theorem 2 *Assume (H1) (ii)–(vi) and (8). If there exist $c > 0$, $r > 0$ and $\theta \in L^1_{\mathrm{loc}}(\mathbb{R}_+; \mathbb{R}_+)$ such that for a.e. $t \geq 0$,*

$$|f(t,x,u)|^{1+r} \leq \theta(t) + cL(t,x,u) \;\; \forall\, x \in \mathbb{R}^n,\; u \in U(t), \tag{11}$$

then (9) holds true and for every $(t_0, x_0) \in dom(V)$, there exists a feasible trajectory-control pair $(\bar{x}, \bar{u})$ satisfying $V(t_0, x_0) = \int_{t_0}^{\infty} L(t, \bar{x}(t), \bar{u}(t))\, dt$.

Furthermore, if V is locally bounded from the above, then it is upper semicontinuous.

Proof Conclusion (9) follows as in the proof of Theorem 1. Let $(t_0, x_0) \in dom(V)$. Consider a maximizing sequence of feasible trajectory-control pairs (x_i, u_i) satisfying $x_i(t_0) = x_0$. By our assumptions, for a.e. $t \geq t_0$ and all $i \geq 1$,

$$|f(t, x_i(t), u_i(t))|^{1+r} \leq \theta(t) + cL(t, x_i(t), u_i(t)).$$

Hence for every $t \geq t_0$,

$$\sup_{i\geq 1} \int_{t_0}^{t} |f(s, x_i(s), u_i(s))|^{1+r} ds \leq \sup_{i \geq 1} \int_{t_0}^{t} (\theta(s) + cL(s, x_i(s), u_i(s)))\, ds.$$

Observe next that there exists $M > 0$ such that for all large $t > t_0$ and every i,

$$\int_{t_0}^{t} L(s, x_i(s), u_i(s)) ds \leq \int_{t_0}^{\infty} L(s, x_i(s), u_i(s)) ds + \int_{t}^{\infty} \delta(s) ds < M.$$

Consequently $\{f(\cdot, x_i(\cdot), u_i(\cdot))\}_i$ is bounded in $L^{1+r}([t_0, t]; \mathbb{R}^n)$ and therefore it is also bounded in $L^{1+r}([t_0, t_0 + 1]; \mathbb{R}^n)$. Since $L^{1+r}([t_0, t_0 + 1]; \mathbb{R}^n)$ is reflexive, taking a subsequence and keeping the same notation, we may assume that for some integrable function $y : [t_0, t_0 + 1] \to \mathbb{R}^n$, the restrictions of $x_i'(\cdot)$ to $[t_0, t_0 + 1]$ converge weakly in $L^{1+r}([t_0, t_0 + 1]; \mathbb{R}^n)$ to y.

Set $p = (1 + r)/r$. By the Hölder inequality, for any i and for all $t_0 \leq a < b \leq t_0 + 1$,

$$|x_i(b) - x_i(a)| \leq (b-a)^{\frac{1}{p}} \left(\int_a^b |f(s, x_i(s), u_i(s))|^{1+r} ds \right)^{\frac{1}{1+r}}.$$

Therefore $\{x_i\}_i$ are equicontinuous on $[t_0, t_0 + 1]$. Starting at this point the same arguments as those in the proof of Theorem 1 can be applied to get the existence of an optimal solution at the initial condition (t_0, x_0). The upper semicontinuity of V can be proved in a similar way using that V is locally bounded from the above.

5 Relaxation in the Absence of State Constraints

In the previous section we have shown that the value function is upper semicontinuous and that the optimal trajectories do exist assuming that the sets $F(t, x)$ are closed and convex. If the convexity assumption (H1) (vi) is not imposed, then, in the literature, one usually considers the so-called relaxed problems.

In this section we restrict our attention to the case when $K = \mathbb{R}^n$, that is without state constraints. Consider the relaxed infinite horizon problem

$$V^{rel}(t_0, x_0) = \sup \int_{t_0}^{\infty} \left(\sum_{i=0}^{n} \lambda_i(t) L(t, x(t), u_i(t)) \right) dt \tag{12}$$

over all trajectory-control pairs of

$$\begin{cases} x'(t) = \sum_{i=0}^{n} \lambda_i(t) f(t, x(t), u_i(t)), \; u_i(t) \in U(t), \; \lambda_i(t) \geq 0, \; \sum_{i=0}^{n} \lambda_i(t) = 1 \\ x(t_0) = x_0, \end{cases} \tag{13}$$

where $u_i(\cdot)$, $\lambda_i(\cdot)$ are Lebesgue measurable on $\mathbb{R}_+$ for $i = 0, \ldots, n$. Then $V^{rel} \geq V$. For $v = (u_0, \ldots, u_n)$, $\Lambda = (\lambda_0, \ldots, \lambda_n)$ define

$$\hat{f}(t, x, v, \Lambda) = \sum_{i=0}^{n} \lambda_i f(t, x, u_i), \quad \hat{L}(t, x, v, \Lambda) = \sum_{i=0}^{n} \lambda_i L(t, x, u_i)$$

and

$$\widehat{U}(t) := \underbrace{U(t) \times \ldots \times U(t)}_{n+1} \times \{(\lambda_0, \ldots, \lambda_n) \mid \lambda_i \geq 0 \;\forall i, \; \Sigma_{i=0}^{n} \lambda_i = 1\}.$$

Thus the relaxed problem is of type (1) and (2) with f, L replaced by $\hat{f}$, $\hat{L}$ and $U(t)$ replaced by $\widehat{U}(t)$.

Our first relaxation result addresses a case where the celebrated Filippov-Ważewski theorem can be applied.

Theorem 3 *Assume (H1) (i)–(v) with $\omega_R(t, r) = \bar{c}_R(t) r$, where for all $R > 0$, $\bar{c}_R : \mathbb{R}_+ \to \mathbb{R}_+$ is locally integrable, and that for a.e. $t \in \mathbb{R}_+$ and all $x \in \mathbb{R}^n$, the set*

$$G(t, x) := \{ (f(t, x, u), L(t, x, u)) : u \in U(t) \}$$

is compact. Further assume that (8) is satisfied.

Then $V^{rel} = V$ *on* $\mathbb{R}_+ \times \mathbb{R}^n$ *and for every* $(t_0, x_0) \in \mathbb{R}_+ \times \mathbb{R}^n$*, there exists* $(\bar{x}(\cdot), \bar{v}(\cdot) = (\bar{u}_0(\cdot), \dots, \bar{u}_n(\cdot)), \bar{\Lambda}(\cdot) = (\bar{\lambda}_0(\cdot), \dots, \bar{\lambda}_n(\cdot)))$ *satisfying (13) such that*

$$V^{rel}(t_0, x_0) = \int_{t_0}^{\infty} \hat{L}(t, \bar{x}(t), \bar{v}(t), \bar{\Lambda}(t))\, dt.$$

Furthermore, (9) holds true.

In particular, if a trajectory-control pair $(\bar{x}, \bar{u})$ *is optimal for (1), (2), then it is also optimal for the relaxed problem (12), (13).*

Remark 2

(a) If the set $U(t)$ is compact, then so is $G(t, x)$.
(b) In [14] a similar relaxation result was proved under slightly different assumptions. It can be shown that (8) and (H1) (i)–(v) imply that $V^{rel}(t, \cdot)$ is continuous and so the proof could be done using the same scheme as in [14]. However, as we show below, under our assumptions it can be simplified avoiding the use of V^{rel}.

Proof It is not difficult to realize that $\hat{f}$, $\hat{L}$, $\widehat{U}$ satisfy (H1). Let $(t_0, x_0) \in \mathbb{R}_+ \times \mathbb{R}^n$. Since $K = \mathbb{R}^n$, Theorem 1 and (8) imply that $V^{rel} \neq \pm\infty$. By Theorem 1 applied to $\hat{f}$, $\hat{L}$, $\widehat{U}$, there exists $(\bar{x}(\cdot), \bar{v}(\cdot), \bar{\Lambda}(\cdot))$ satisfying (13) such that

$$V^{rel}(t_0, x_0) = \int_{t_0}^{\infty} \hat{L}(t, \bar{x}(t), \bar{v}(t), \bar{\Lambda}(t))\, dt.$$

Fix $\varepsilon > 0$ and let $k > \max\{t_0, S, \bar{S}\}$ be so that $\int_k^{\infty} (\alpha(t) + \delta(t))dt \leq \varepsilon/3$.
Then

$$\begin{aligned} V^{rel}(t_0, x_0) &= \textstyle\int_k^{\infty} \hat{L}(t, \bar{x}(t), \bar{v}(t), \bar{\Lambda}(t))dt + \int_{t_0}^k \hat{L}(t, \bar{x}(t), \bar{v}(t), \bar{\Lambda}(t))dt \\ &\leq \tfrac{\varepsilon}{3} + \textstyle\int_{t_0}^k \hat{L}(t, \bar{x}(t), \bar{v}(t), \bar{\Lambda}(t))dt. \end{aligned}$$

By the Filippov-Ważewski relaxation theorem and the measurable selection theorem, see for instance [23], there exists $u_\varepsilon \in \mathcal{U}$ such that the solution x_ε of the system

$$x' = f(t, x, u_\varepsilon(t)), \;\; x(t_0) = x_0$$

satisfies

$$\left| \int_{t_0}^{k} \hat{L}(t, \bar{x}(t), \bar{v}(t), \bar{\Lambda}(t))dt - \int_{t_0}^{k} L(t, x_\varepsilon(t), u_\varepsilon(t))dt \right| < \frac{\varepsilon}{3}.$$

Consider any trajectory-control pair (x, u) of (5) on $[k, \infty)$ with $x(k) = x_\varepsilon(k)$. Thus

$$\int_k^\infty L(t, x(t), u(t))dt \geq \int_k^\infty (-\delta(t))dt$$

We extend the trajectory-control pair $(x_\varepsilon, u_\varepsilon)$ on the time interval $[k, \infty)$ by setting $(x_\varepsilon(s), u_\varepsilon(s)) = (x(s), u(s))$ for $s > k$. Hence

$$\begin{aligned} V^{rel}(t_0, x_0) &\leq \tfrac{2\varepsilon}{3} + \textstyle\int_{t_0}^k L(t, x_\varepsilon(t), u_\varepsilon(t))dt \\ &\leq \tfrac{2\varepsilon}{3} + \textstyle\int_{t_0}^\infty L(t, x_\varepsilon(t), u_\varepsilon(t))dt + \int_k^\infty \delta(t)dt \leq \varepsilon + V(t_0, x_0). \end{aligned}$$

This yields $V^{rel}(t_0, x_0) \leq V(t_0, x_0) + \varepsilon$. Since $\varepsilon > 0$ and (t_0, x_0) are arbitrary, we get $V^{rel} = V$.

The above result has a restrictive assumption of compactness of sets $G(t, x)$ because in the proof we used the Filippov-Ważewski relaxation theorem dealing with compact valued maps. In the case of control systems this theorem can be stated without such compactness assumption.

Theorem 4 (Finite Horizon Relaxation Theorem) *Let $T > 0$, $U : [0, T] \rightsquigarrow \mathbb{R}^m$ be measurable, with closed nonempty images, $g : [0, T] \times \mathbb{R}^n \times \mathbb{R}^m \to \mathbb{R}^n$ be a Carathéodory function such that for every $R > 0$, there exists $c_R \in L^1([0, T]; \mathbb{R}_+)$ satisfying for a.e. $t \in [0, T]$,*

$$|g(t, x, u) - g(t, y, u)| \leq c_R(t)|x - y| \;\; \forall x, y \in B(0, R), \;\; u \in U(t).$$

Further assume that $\psi(t) := \inf_{u \in U(t)} |g(t, 0, u)|$ is integrable on $[0, T]$.

Then for any $\varepsilon > 0$ and any absolutely continuous $x : [0, T] \to \mathbb{R}^n$ satisfying

$$x'(t) \in \overline{co}\, g(t, x(t), U(t)) \quad a.e. \text{ in } [0, T], \tag{14}$$

there exists a measurable selection $u(t) \in U(t)$ for $t \in [0, T]$ and an absolutely continuous function $x_\varepsilon : [0, T] \to \mathbb{R}^n$ such that

$$x_\varepsilon'(t) = g(t, x_\varepsilon(t), u(t)) \;\; a.e. \text{ in } [0, T], \quad x_\varepsilon(0) = x(0)$$

and $\max_{t \in [0,T]} |x_\varepsilon(t) - x(t)| < \varepsilon$.

Remark 3 We would like to underline that in the above theorem the sets $g(t, x, U(t))$ are neither closed nor bounded.

Proof We first observe that $g(t, 0, U(t)) \cap B(0, \psi(t) + \varepsilon) \neq \emptyset$ for every $\varepsilon > 0$ and every $t \in [0, T]$. By the inverse image theorem [8, Theorem 8.2.9], there exists a measurable selection $u_0(t) \in U(t)$ for $t \in [0, T]$ such that $|g(t, 0, u_0(t))| \leq \psi(t) + \varepsilon$ a.e. Thus the function $k : [0, T] \to \mathbb{R}_+$ defined by $k(t) := |g(t, 0, u_0(t))|$ is integrable. By the Castaing representation theorem, see for instance [8], there exist measurable selections $u_i(t) \in U(t),\ i \geq 1$ for $t \in [0, T]$ such that $U(t) = \overline{\bigcup_{i\geq 1}\{u_i(t)\}}$ for every $t \in [0, T]$. Fix an integer $j \geq 1$ and define

$$u_{ij}(t) = \begin{cases} u_i(t) & \text{if} \quad |g(t, 0, u_i(t))| \leq j \cdot k(t) \\ u_0(t) & \text{otherwise,} \end{cases}$$

$$U_j(t) = \{u_0(t)\} \cup \bigcup_{1\leq i\leq j} \{u_{ij}(t)\}.$$

Observe that for every t, the family of finite sets $U_j(t)$ is increasing (with respect to j) and $U(t) = \overline{\bigcup_{j\geq 1} U_j(t)}$. By the continuity of $g(t, x, \cdot)$, for a.e. $t \in [0, T]$ and all $x \in \mathbb{R}^n$,

$$\overline{\bigcup_{j\geq 1} co\, g(t, x, U_j(t))} = \overline{co}\, g(t, x, U(t)).$$

Notice that for every t, the set $g(t, x, U_j(t))$ is compact. Let $x(\cdot) \in W^{1,1}([0, T]; \mathbb{R}^n)$ satisfy (14). Then

$$\begin{aligned} \gamma_j(t) &:= d_{co\, g(t,x(t),U_j(t))}(x'(t)) \leq |x'(t) - g(t, 0, u_0(t))| + c_r(t)|x(t)| \\ &\leq |x'(t)| + |g(t, 0, u_0(t))| + c_r(t)|x(t)|, \end{aligned}$$

where $r > 0$ is so that $\max_{t\in[0,T]} |x(t)| < r$. Hence $\{\gamma_j\}_{j\geq 1}$ are bounded by an integrable function. Moreover $\lim_{j\to\infty} \gamma_j(t) = 0$ for a.e. $t \in [0, T]$. Let

$$t_j = \max \left\{ t \in [0, T] \ :\ e^{\int_0^T c_{2r}(s)ds} \int_0^t \gamma_j(s)ds \leq r \right\}.$$

Since $\int_0^T \gamma_j(s)ds$ converge to zero when $j \to \infty$, we deduce that $t_j = T$ for all j larger than some j_0. By the Filippov theorem, see for instance [23, Theorem 1.2] and the remark following it, for every $j \geq j_0$, there exists an absolutely continuous function $x_j : [0, T] \to \mathbb{R}^n$ such that $x_j'(t) \in co\, g(t, x_j(t), U_j(t))$ a.e. in $[0, T]$, $x_j(0) = x(0)$ and

$$\sup_{t\in[0,T]} |x_j(t) - x(t)| \leq e^{\int_0^T c_{2r}(s)ds} \int_0^T \gamma_j(s)ds.$$

Let $\varepsilon > 0$ and consider $j \geq j_0$ such that $e^{\int_0^T c_{2r}(s)ds} \int_0^T \gamma_j(s)ds < \frac{\varepsilon}{2}$. By the Filippov-Ważewski relaxation theorem there exists an absolutely continuous function $x_\varepsilon : [0, T] \to \mathbb{R}^n$ such that

$$x_\varepsilon'(t) \in g(t, x_\varepsilon(t), U_j(t)) \ \text{ a.e. in } [0, T], \ x_\varepsilon(0) = x(0)$$

and $\sup_{t\in[0,T]} |x_\varepsilon(t) - x_j(t)| < \varepsilon/2$. By the measurable selection theorem we can find a measurable selection $u_\varepsilon(t) \in U_j(t)$ such that for a.e. $t \in [0, T]$ we have $x'_\varepsilon(t) = g(t, x_\varepsilon(t), u_\varepsilon(t))$. Since

$$\sup_{t\in[0,T]} |x_\varepsilon(t) - x(t)| \leq \sup_{t\in[0,T]} (|x_j(t) - x(t)| + |x_\varepsilon(t) - x_j(t)|) < \varepsilon,$$

the proof is complete.

We next apply the above result to the infinite horizon relaxation problem with possibly unbounded and not necessarily closed sets $f(t, x, U(t))$. Theorems 5 and 6 below are new.

Theorem 5 *Assume (H1) (ii)–(v) with* $\omega_R(t, r) = \bar{c}_R(t)r$, *where for all* $R > 0$, $\bar{c}_R : \mathbb{R}_+ \to \mathbb{R}_+$ *is locally integrable, and that for some* $T_0 > 0$ *and all* $t_0 \geq T_0$, $x_0 \in \mathbb{R}^n$ *there exists a trajectory of (2) defined on* $[t_0, \infty)$. *Further assume that (8) holds true and that the function*

$$\psi(t) := \inf \{ |f(t, 0, u)| + |L(t, 0, u)| \ : \ u \in U(t) \}$$

is locally integrable on $\mathbb{R}_+$. *Then (9) is satisfied and* $V^{rel} = V$.

Remark 4 The assumption that for some $T_0 > 0$ and all $t_0 \geq T_0$, $x_0 \in \mathbb{R}^n$ there exists a trajectory of (2) defined on $[t_0, \infty)$ holds true, for instance, whenever the function $\psi(\cdot)$ is locally integrable on $\mathbb{R}_+$ and there exists $T_0 > 0$ such that $c_R(t)$ do not depend on R for all $t \geq T_0$. Indeed, by the proof of Theorem 4, we know that there exists a measurable selection $u_0(t) \in U(t)$ such that $|f(t, 0, u_0(t))| \leq \psi(t) + 1$ for every $t \in \mathbb{R}_+$. Furthermore, setting $c(t) = c_R(t)$ for $t \geq T_0$ we obtain $|f(t, x, u_0(t))| \leq \psi(t) + 1 + c(t)|x|$. Given $t_0 \geq T_0$, $x_0 \in \mathbb{R}^n$, it is enough to consider the solution of

$$x' = f(t, x, u_0(t)), \ x(t_0) = x_0.$$

Proof of Theorem 5 Clearly $V \leq V^{rel}$. Therefore, if $V^{rel}(t_0, x_0) = -\infty$, then $V(t_0, x_0) = -\infty$. We first show that for every $(t_0, x_0) \in dom(V^{rel})$ we have $V^{rel}(t_0, x_0) \leq V(t_0, x_0)$. Fix $(t_0, x_0) \in dom(V^{rel})$, $\varepsilon > 0$ and consider a trajectory-control pair (x, v, Λ) of the relaxed system satisfying $V^{rel}(t_0, x_0) \leq \int_{t_0}^{\infty} \hat{L}(t, x(t), v(t), \Lambda(t))dt + \frac{\varepsilon}{4}$.

By our assumptions, for every sufficiently large $k > 0$ and any trajectory-control pair $(\tilde{x}, \tilde{u})$ defined on $[k, \infty)$, we have

$$\int_k^\infty |L(t, \tilde{x}(t), \tilde{u}(t))|dt \leq \frac{\varepsilon}{4}, \quad \int_k^\infty |\hat{L}(t, x(t), v(t), \Lambda(t))|dt \leq \frac{\varepsilon}{4}.$$

By Theorem 4 applied to the function $g(t, x, u) = (f(t, x, u), L(t, x, u))$ and the time interval $[t_0, k]$ with $k > t_0$, there exists a trajectory-control pair $(x_\varepsilon, u_\varepsilon)$ of the control system (2) defined on $[t_0, k]$ such that

$$\left| \int_{t_0}^{k} \hat{L}(t, x(t), v(t), \Lambda(t))dt - \int_{t_0}^{k} L(t, x_\varepsilon(t), u_\varepsilon(t))dt \right| < \frac{\varepsilon}{4}.$$

Consider $k > 0$ sufficiently large and a trajectory-control pair (x, u) of (5) defined on $[k, \infty)$ and satisfying $x(k) = x_\varepsilon(k)$. Set $(x_\varepsilon(t), u_\varepsilon(t)) = (x(t), u(t))$ for $t > k$. Then

$$V^{rel}(t_0, x_0) \le \int_{t_0}^{\infty} \hat{L}(t, x(t), v(t), \Lambda(t))dt + \tfrac{\varepsilon}{4} \le \int_{t_0}^{k} \hat{L}(t, x(t), v(t), \Lambda(t))dt + \tfrac{\varepsilon}{2}$$
$$\le \int_{t_0}^{k} L(t, x_\varepsilon(t), u_\varepsilon(t))dt + \tfrac{3\varepsilon}{4} \le \int_{t_0}^{\infty} L(t, x_\varepsilon(t), u_\varepsilon(t))dt + \varepsilon \le V(t_0, x_0) + \varepsilon.$$

Hence $V^{rel}(t_0, x_0) \le V(t_0, x_0)$, by the arbitrariness of $\varepsilon > 0$.

It remains to consider the case $V^{rel}(t_0, x_0) = +\infty$. Then there exist trajectory-control pairs (x^i, v^i, Λ^i) of the relaxed problem such that

$$\lim_{i \to \infty} \int_{t_0}^{\infty} \hat{L}(t, x^i(t), v^i(t), \Lambda^i(t))dt = +\infty.$$

Applying the same arguments as before, for every $k > 0$ and $\varepsilon > 0$, we can find trajectory-control pairs $(x_\varepsilon^i, u_\varepsilon^i)$ of (5) defined on $[t_0, \infty)$, satisfying $x_\varepsilon^i(t_0) = x_0$ and

$$\left| \int_{t_0}^{k} \hat{L}(t, x^i(t), v^i(t), \Lambda^i(t))dt - \int_{t_0}^{k} L(t, x_\varepsilon^i(t), u_\varepsilon^i(t))dt \right| < \frac{\varepsilon}{3}.$$

Moreover, for $k > 0$ large enough and every i we have

$$\int_{k}^{\infty} |\hat{L}(t, x^i(t), v^i(t), \Lambda^i(t))|dt < \frac{\varepsilon}{3}, \quad \int_{k}^{\infty} |L(t, x_\varepsilon^i(t), u_\varepsilon^i(t))|dt < \frac{\varepsilon}{3}.$$

Combining the above inequalities we get

$$\int_{t_0}^{\infty} \hat{L}(t, x^i(t), v^i(t), \Lambda^i(t))dt < \int_{t_0}^{\infty} L(t, x_\epsilon^i(t), u_\varepsilon^i(t))dt + \varepsilon.$$

Hence

$$\lim_{i \to \infty} \int_{t_0}^{\infty} L(t, x_\epsilon^i(t), u_\varepsilon^i(t))dt = +\infty$$

and therefore $V(t_0, x_0) = +\infty$.

Finally we would like to observe that, in general, when the sets $G(t, x)$ are not compact, the sets $co\, G(t, x)$ may be not closed. For this reason optimal solutions of the relaxed problem (12) and (13) may not exist. To get the existence, without changing the value function, the correct relaxed problem in the case of unbounded, not necessarily closed sets $G(t, x)$, takes the following less familiar form. Define

$$\overline{V}^{rel}(t_0, x_0) = \sup \int_{t_0}^{\infty} \ell(t)\, dt$$

over all $x \in W^{1,1}_{loc}([t_0, \infty); \mathbb{R}^n)$ and $\ell \in L^1_{loc}([t_0, \infty); \mathbb{R})$ satisfying

$$\begin{cases} (x'(t), \ell(t)) \in \overline{co}\, G(t, x(t)) \ \text{ a.e. in } [t_0, \infty) \\ x(t_0) = x_0. \end{cases} \tag{15}$$

Such a pair (x, ℓ) will be called below a solution of (15). In the above we set $\overline{V}^{rel}(t_0, x_0) = -\infty$ if (15) does not have solutions defined on $[t_0, \infty)$.

Theorem 6 *Under all the assumptions of Theorem 5 suppose that there exist* $c > 0$, $r > 0$ *and* $\theta \in L^1_{loc}([0, \infty); \mathbb{R}_+)$ *such that (11) holds true. Then for every* $(t_0, x_0) \in dom(\overline{V}^{rel})$*, there exists a solution* $(\bar{x}(\cdot), \bar{\ell}(\cdot))$ *of (15) satisfying*

$$\overline{V}^{rel}(t_0, x_0) = \int_{t_0}^{\infty} \bar{\ell}(t)\, dt. \tag{16}$$

Furthermore, $\overline{V}^{rel} = V$ *and (9) holds true. Moreover, if* V *is locally bounded from the above, then it is upper semicontinuous.*

Remark 5 Theorem 6 allows to avoid assumption (H1) (vi) to claim that V is upper semicontinuous and satisfies (9).

Proof By our assumptions, $(\hat{f}, \hat{L}, \widehat{U})$ satisfy (11). Thus for every $(y, \ell) \in co\, G(t, x)$ we have $|y|^{1+r} \le \theta(t) + c\ell$. Then the same inequality holds true also for any $(y, \ell) \in \overline{co}\, G(t, x)$.

The same arguments as those in the proof of Theorem 2 imply (9) and that for every $(t_0, x_0) \in dom(\overline{V}^{rel})$, there exists a solution $(\bar{x}(\cdot), \bar{\ell}(\cdot))$ of (15) satisfying (16). Also, as before, if $\overline{V}^{rel}$ is locally bounded from the above, then it is upper semicontinuous. It remains to show that $\overline{V}^{rel} = V$. By Theorem 5 it is sufficient to verify that $\overline{V}^{rel} = V^{rel}$. Fix $(t_0, x_0) \in \mathbb{R}_+ \times \mathbb{R}^n$. If $\overline{V}^{rel}(t_0, x_0) = -\infty$, then also $V^{rel}(t_0, x_0) = -\infty$. Assume next that $(t_0, x_0) \in dom(\overline{V}^{rel})$ and consider $(\bar{x}(\cdot), \bar{\ell}(\cdot))$ solving the inclusion (15) and satisfying (16). By the Castaing representation theorem there exist measurable selections $u_i(t) \in U(t)$, $i \ge 1$ defined on $\mathbb{R}_+$ such that $U(t) = \overline{\bigcup_{i \ge 1}\{u_i(t)\}}$ for every $t \ge 0$. Consider $u_0 \in \mathcal{U}$ such that the function defined by $k(t) := |f(t, 0, u_0(t))| + |L(t, 0, u_0(t))|$ is locally integrable on $\mathbb{R}_+$. Let the sets $U_j(t)$ be defined in the same way as in the proof of Theorem 4 for $g = (f, L)$.

Then for a.e. $t \ge t_0$

$$(\bar{x}'(t), \bar{\ell}(t)) \in \overline{\bigcup_{j \ge 1} co\,\{(f(t, \bar{x}(t), u), L(t, \bar{x}(t), u)) \ : \ u \in U_j(t)\}}.$$

Define

$$\gamma_j(t) := d_{co\{(f(t,\bar{x}(t),u),L(t,\bar{x}(t),u)) : u \in U_j(t)\}}(\bar{x}'(t), \bar{\ell}(t))$$

and observe that for a.e. $t \geq t_0$ we have $\lim_{j\to\infty} \gamma_j(t) = 0$.

Fix $\varepsilon > 0$. By our assumptions there exists $\tau > T_0$ such that for every j and any solution (x, ℓ) of the inclusion

$$(x'(t), \ell(t)) \in co\{(f(t, x(t), u), L(t, x(t), u)) \ : \ u \in U_j(t)\} \text{ for a.e. } t \geq t_0 \tag{17}$$

we have $\int_\tau^\infty |\ell(t)|dt < \varepsilon/3$ and $\int_\tau^\infty |\bar{\ell}(t)|dt < \varepsilon/3$.

Then, as in the proof of Theorem 4 applied with $T = \tau$ and the initial time t_0 instead of zero, it follows that for every sufficiently large j there exist an absolutely continuous $x_j : [t_0, \tau] \to \mathbb{R}^n$ and $\ell_j \in L^1([t_0, \tau]; \mathbb{R})$ solving (17) on $[t_0, \tau]$ and satisfying

$$\left| \int_{t_0}^{\tau} \ell_j(t)dt - \int_{t_0}^{\tau} \bar{\ell}(t)dt \right| < \frac{\varepsilon}{3}.$$

Consider any trajectory-control pair (x, u) of (2) with t_0 replaced by τ and x_0 by $x_j(\tau)$. We extend the trajectory-control pair (x_j, u_j) on the time interval (τ, ∞) by the pair (x, u) and set $\ell_j(t) = L(t, x(t), u(t))$ for all $t > \tau$. Hence

$$\overline{V}^{rel}(t_0, x_0) < \int_{t_0}^{\tau} \bar{\ell}(t)dt + \frac{\varepsilon}{3} < \int_{t_0}^{\tau} \ell_j(t)dt + \frac{2\varepsilon}{3} < \int_{t_0}^{\infty} \ell_j(t)dt + \varepsilon.$$

By the measurable selection theorem, any solution of (17) satisfies the relaxed system (13) for some measurable $\{(u_i, \lambda_i)\}_{i=0}^n$. Therefore

$$\overline{V}^{rel}(t_0, x_0) \leq V^{rel}(t_0, x_0) + \varepsilon.$$

Hence, by the arbitrariness of $\varepsilon > 0$, we get $\overline{V}^{rel}(t_0, x_0) \leq V^{rel}(t_0, x_0)$.

It remains to consider the case $\overline{V}^{rel}(t_0, x_0) = +\infty$. Then there exist $(\bar{x}^s, \bar{\ell}^s)$ solving (15) such that $\lim_{s\to\infty} \int_{t_0}^\infty \bar{\ell}^s(t)dt = +\infty$. By the same arguments as above we can find (x^s, v^s, Λ^s) satisfying (13) such that

$$\lim_{s\to\infty} \int_{t_0}^{\infty} \sum_{i=0}^{n} \lambda_i^s(t) L(t, x^s(t), u_i^s(t))\, dt = +\infty.$$

Hence $V^{rel}(t_0, x_0) = +\infty$. This completes the proof.

6 Relaxation in the Presence of State Constraints

In this section we consider the problem (12), (13), (3).

When $K \neq \mathbb{R}^n$, in general, it may happen that dom(V) is strictly contained in dom(V^{rel}) even under all the assumptions of the previous section. In fact one needs to impose some geometric restrictions on f on the boundary of K, the so called *Relaxed Inward Pointing Condition*, to obtain the relaxation theorem.

We denote by (IPC^{rel}) the following assumption:

$$\begin{cases} \forall\, t \in [0, \infty), \forall\, x \in \mathrm{bd}\,(K), \\ \forall\, v \in \mathrm{Limsup}_{(s,y)\to(t,x)}\, f(s, y, U(s)) \text{ with } \max_{n\in N_K^1(x)} \langle n, v\rangle \geq 0, \\ \exists\, w \in \mathrm{Liminf}_{(s,y)\to(t,x)}\, co\, f(s, y, U(s)) \text{ with } \max_{n\in N_K^1(x)} \langle n, w - v\rangle < 0. \end{cases}$$

Remark 6 If f is continuous and the set-valued map $U(\cdot)$ is continuous and has compact nonempty images, then the above condition takes a simpler form:

$$\begin{cases} \forall\, t \in [0, \infty), \forall\, x \in \mathrm{bd}\,(K),\ \forall u \in U(t) \text{ with } \max_{n\in N_K^1(x)} \langle n, f(t, x, u)\rangle \geq 0, \\ \exists\, w \in co\, f(t, x, U(t)) \text{ with } \max_{n\in N_K^1(x)} \langle n, w - f(t, x, u)\rangle < 0. \end{cases}$$

Theorem 7 *Under all the assumptions of Theorem 3 with (H1) (i) replaced by (6), suppose (IPC^{rel}) and that f, L are locally bounded on $\mathbb{R}_+ \times bd(K)$.*

Then $dom(V^{rel}) = \mathbb{R}_+ \times K$, $V^{rel} = V$ and for every $(t_0, x_0) \in \mathbb{R}_+ \times K$, there exists $(\bar{x}(\cdot), \bar{v}(\cdot) = (\bar{u}_0(\cdot), \ldots, \bar{u}_n(\cdot)), \bar{\Lambda}(\cdot) = (\bar{\lambda}_0(\cdot), \ldots, \bar{\lambda}_n(\cdot)))$ satisfying (13) such that $\bar{x}(t) \in K$ for every $t \geq t_0$ and

$$V^{rel}(t_0, x_0) = \int_{t_0}^{\infty} \hat{L}(t, \bar{x}(t), \bar{v}(t), \bar{\Lambda}(t))\, dt.$$

Remark 7 We replaced (H1) (i) by (6) just to fit the assumptions of [26]. However in [26] assumption (6) is needed only to get uniform bounds on trajectories of a differential inclusion on a finite time interval. By what precedes we know that such bounds follow also from (H1) (i). Hence the above result is valid as well with assumption (H1) (i) instead of (6).

Proof From [26, Theorem 3.3] applied with $\alpha = 0$, $\beta = 0$ and finite time intervals $[t_0 + k, t_0 + k + 1]$ instead of $[0, 1]$, we deduce, using the induction argument, that for every $(t_0, x_0) \in \mathbb{R}_+ \times K$, there exists a feasible solution of (13) defined on $[t_0, \infty)$. Hence, in the same way as before, $dom(V^{rel}) = \mathbb{R}_+ \times K$. Since $\hat{f}$, $\hat{L}$, $\widehat{U}$ satisfy (H1) the third statement follows from Theorem 1. To prove the second one, observe that $V^{rel} \geq V$. Let $(t_0, x_0) \in \mathbb{R}_+ \times K$ and $(\bar{x}(\cdot), \bar{v}(\cdot), \bar{\Lambda}(\cdot))$ be optimal for the relaxed problem. Fix any $\varepsilon > 0$ and $T > t_0$ such that $\int_T^\infty (\alpha(s)+\delta(s))ds \le \varepsilon/2$. Then

$$V^{rel}(t_0, x_0) \le \frac{\varepsilon}{2} + \int_{t_0}^{T} \hat{L}(t, \bar{x}(t), \bar{v}(t), \bar{\Lambda}(t))dt.$$

From [26, Corollary 3.4] we deduce that there exists a trajectory-control pair $(x_\varepsilon, u_\varepsilon)$ of (2) and (3) defined on $[t_0, T]$ such that

$$\left| \int_{t_0}^{T} \hat{L}(t, \bar{x}(t), \bar{v}(t), \bar{\Lambda}(t))dt - \int_{t_0}^{T} L(t, x_\varepsilon(t), u_\varepsilon(t))dt \right| < \frac{\varepsilon}{2}.$$

Applying [26, Theorem 3.3 and Corollary 3.4] and an induction argument, we extend the feasible trajectory-control pair $(x_\varepsilon, u_\varepsilon)$ on the time interval $[T, \infty)$.

The proof ends in the same way as the one of Theorem 3.

To investigate locally Lipschitz solutions of the Hamilton–Jacobi equation we need to recall the result below proved in [10].

Consider the infinite horizon optimal control problem $\mathcal{B}_\infty$:

$$\text{maximize} \int_{t_0}^{\infty} e^{-\lambda t} l(t, x(t), u(t))\, dt$$

over all feasible trajectory-control pairs $(x(\cdot), u(\cdot))$ of (2), (3) defined on $[t_0, \infty)$, where $\lambda > 0$, f, U, K are as in the introduction and $l : \mathbb{R}_+ \times \mathbb{R}^n \times \mathbb{R}^m \to \mathbb{R}$ is a given function. Let V_λ be the corresponding value function and V_λ^{rel} be the value function of the relaxed problem (12) and (13) under state constraint (3) for $L(t, x, u) = e^{-\lambda t} l(t, x, u)$.

We denote by **(H2)** the following assumptions on f and l:

(i) there exists $q \in \mathcal{L}_{\text{loc}}$ such that for a.e. $t \in \mathbb{R}_+$,

$$\sup_{u \in U(t)} (|f(t, x, u)| + |l(t, x, u)|) \leqslant q(t), \quad \forall x \in \text{bd}\,(K);$$

(ii) for all $(t, x) \in \mathbb{R}_+ \times \mathbb{R}^n$ the set $\{(f(t, x, u), l(t, x, u)) : u \in U(t)\}$ is compact;

(iii) l is bounded and there exist $c \in L^1_{\text{loc}}(\mathbb{R}_+; \mathbb{R}_+)$ and $k \in \mathcal{L}_{\text{loc}}$ such that for a.e. $t \in \mathbb{R}_+$ and for all $x,\ y \in \mathbb{R}^n$ and $u \in U(t)$,

$$|f(t, x, u) - f(t, y, u)| + |l(t, x, u) - l(t, y, u)| \leqslant k(t)|x - y|, \ |f(t, 0, u)| \leqslant c(t);$$

(iv) $\limsup_{t\to\infty} \frac{1}{t} \int_0^t (c(s) + k(s))\, ds < \infty$;

(v) the mappings f, l are Carathéodory;

We also need the following (stronger than before) Inward Pointing Condition linking the dynamics to the state constraints: (IPC^{∞})

$$\begin{cases} \exists\, \eta > 0,\ \rho > 0,\ M \geqslant 0 \text{ such that for a.e. } t \in \mathbb{R}_+,\ \forall\, y \in \mathrm{bd}(K) + \eta B, \\ \forall\, u \in U(t) \text{ with } \sup_{n \in N^1_{y,\eta}} \langle n, f(t,y,u)\rangle \geqslant 0, \\ \exists\, w \in \{w' \in U(t) \,:\, |f(t,y,w') - f(t,y,u)| \leqslant M\} \text{ such that} \\ \sup_{n \in N^1_{y,\eta}} \{\langle n, f(t,y,w)\rangle,\ \langle n, f(t,y,w) - f(t,y,u)\rangle\} \leqslant -\rho, \end{cases}$$

where $N^1_{y,\eta} := \{n \in N^1_K(x) \,:\, x \in \mathrm{bd}\,(K) \cap B(y,\eta)\}$.

Theorem 8 *Assume (H2) and* (IPC$^{\infty}$). *Then there exist $C > 1$ and $\kappa > 0$ such that for every $\lambda > \kappa$ and every $t \geqslant 0$ the function $V_\lambda(t,\cdot)$ is $Ce^{-(\lambda-\kappa)t}$-Lipschitz continuous on K.*

Furthermore, if f is also bounded, then for every $\lambda > \kappa$, V_λ is locally Lipschitz on $\mathbb{R}_+ \times K$.

Remark 8 In [10] the constants C and κ are obtained explicitly. They depend solely on ρ, M, q, c, k.

Theorem 9 *Assume (H2) and* (IPC$^{\infty}$). *Then, for all large $\lambda > 0$, $V_\lambda = V^{rel}_\lambda$ on $\mathbb{R}_+ \times K$.*

Proof of Theorem 8 is lengthly and highly technical. The presence of the discount factor plays an important role there. The main idea is to use the so called Neighboring Feasible Trajectory (NFT) theorem from [26]. Theorem 9 is proved thanks to the continuity of $V_\lambda(t,\cdot)$ and the technique developed in [14].

7 Relation to a Finite Horizon Bolza Problem

Let $t_0 \geq 0$ and $T > t_0$. Define $g_T(x) = V(T,x)$ for all $x \in K$, $g_T(x) = -\infty$ for all $x \notin K$ and consider the Bolza problem $\mathcal{B}_T$

$$\text{maximize}\left(g_T(x(T)) + \int_{t_0}^{T} L(t,x(t),u(t))\,dt\right)$$

over all trajectory-control pairs (x,u) of the system

$$\begin{cases} x'(t) = f(t,x(t),u(t)),\ u(t) \in U(t) & \text{for a.e. } t \in [t_0,T] \\ x(t_0) = x_0, & x(t) \in K \quad \forall t \in [t_0,T]. \end{cases}$$

The value function V^B of the Bolza problem $\mathcal{B}_T$ is given by: for every $(s_0, y_0) \in [t_0, T] \times K$

$$V^B(s_0, y_0) = \sup\left(g_T(x(T)) + \int_{s_0}^{T} L(t, x(t), u(t))\, dt\right)$$

over all feasible trajectory-control pairs (x, u) of (5) defined on $[s_0, T]$ with $x(s_0) = y_0$. Again, if there is no feasible trajectory-control pair of (5) defined on $[s_0, T]$ and satisfying $x(s_0) = y_0$, then we set $V^B(s_0, y_0) = -\infty$.

If (H1) holds true, then, by Theorem 1, $V(T, \cdot)$ is upper semicontinuous and, by the well known existence theorems for finite horizon problems, for every $y_0 \in dom(V^B(s_0, \cdot))$, the above Bolza problem has an optimal solution.

Proposition 2 *Assume (H1) (i)–(v). Then* $V^B(s_0, y_0) = V(s_0, y_0)$ *for every* $(s_0, y_0) \in [t_0, T] \times K$. *Moreover, if* $(\bar{x}, \bar{u})$ *is optimal for the infinite horizon problem at* (t_0, x_0), *then the restriction of* $(\bar{x}, \bar{u})$ *to* $[t_0, T]$ *is optimal for* $\mathcal{B}_T$.

Proof By (7) and (H1) (iv), for any $R > 0$ there exists $M > 0$ such that for any trajectory-control pair (x, u) of (5) defined on $[s_0, T]$ with $T \geq s_0 \geq 0$, satisfying $x(s_0) = y_0$ and $T + |y_0| \leq R$ we have $|x(T)| \leq M$ and $\int_{s_0}^{T} |L(t, x(t), u(t))|\, dt \leq M$. As in the proof of Theorem 1 it can be shown that V^B and V take values in $[-\infty, \infty)$.

Let $(s_0, y_0) \in [t_0, T] \times K$. If $V^B(s_0, y_0) = -\infty$, then $V^B(s_0, y_0) \leq V(s_0, y_0)$. If it is finite, then fix $\varepsilon > 0$ and consider a feasible trajectory-control pair $(x_\varepsilon, u_\varepsilon)$ with $x_\varepsilon(s_0) = y_0$ defined on $[s_0, T]$ such that

$$V^B(s_0, y_0) \leq g_T(x_\varepsilon(T)) + \int_{s_0}^{T} L(s, x_\varepsilon(s), u_\varepsilon(s))ds + \frac{\varepsilon}{2}.$$

Since $g_T(x_\varepsilon(T))$ is finite, we deduce that $(T, x_\varepsilon(T)) \in dom(V)$. Consider a feasible trajectory-control pair (x, u) of (5) on $[T, \infty)$ such that $x(T) = x_\varepsilon(T)$ and $V(T, x_\varepsilon(T)) \leq \int_T^\infty L(s, x(s), u(s))ds + \varepsilon/2$. Set $(x_\varepsilon(s), u_\varepsilon(s)) = (x(s), u(s))$ for $s > T$. Then,

$$\begin{aligned} V^B(s_0, y_0) &\leq \textstyle\int_{s_0}^{T} L(s, x_\varepsilon(s), u_\varepsilon(s))ds + \int_T^\infty L(s, x_\varepsilon(s), u_\varepsilon(s))ds + \varepsilon \\ &\leq V(s_0, y_0) + \varepsilon. \end{aligned}$$

Since $\varepsilon > 0$ is arbitrary, $V^B(s_0, y_0) \leq V(s_0, y_0)$. On the other hand, if $V(s_0, y_0) = -\infty$, then $V(s_0, y_0) \leq V^B(s_0, y_0)$. If it is finite, then consider a feasible trajectory-control pair $(x_\varepsilon, u_\varepsilon)$ of (5) on $[s_0, \infty)$ satisfying $x_\varepsilon(s_0) = y_0$ and such that $V(s_0, y_0) \leq \int_{s_0}^\infty L(s, x_\varepsilon(s), u_\varepsilon(s))ds + \varepsilon$. Then

$$\begin{aligned} V(s_0, y_0) &\leq \textstyle\int_{s_0}^{T} L(s, x_\varepsilon(s), u_\varepsilon(s))ds + \int_T^\infty L(s, x_\varepsilon(s), u_\varepsilon(s))ds + \varepsilon \\ &\leq \textstyle\int_{s_0}^{T} L(s, x_\varepsilon(s), u_\varepsilon(s))ds + V(T, x_\varepsilon(T)) + \varepsilon \leq V^B(s_0, y_0) + \varepsilon. \end{aligned}$$

By the arbitrariness of $\varepsilon > 0$, this yields $V(s_0, y_0) \leq V^B(s_0, y_0)$. Hence $V(s_0, y_0) = V^B(s_0, y_0)$. The point $(s_0, y_0) \in [t_0, T] \times K$ being arbitrary, we deduce the first statement. The second one is a simple consequence of the dynamic programming principle.

8 Maximum Principle and Sensitivity Relation

When state constraints are present, then, in general, the familiar maximum principle does not hold unless the optimal trajectory takes values in the interior of K only. Furthermore, for problems under state constraints the adjoint state may be discontinuous, even for finite horizon problems.

In this section we prove the maximum principle using the Hamiltonian H defined in the introduction. Recall that $H(t, x, \cdot)$ is convex and if $f(t, \cdot, u)$, $L(t, \cdot, u)$ are locally Lipschitz uniformly in $u \in U(t)$, then $H(t, \cdot, \cdot)$ is locally Lipschitz. Then we denote by $\partial H(t; y, q)$ the generalized gradient of $H(t, \cdot, \cdot)$ at (y, q). It has convex compact images and the set-valued map $(y, q) \rightsquigarrow \partial H(t; y, q)$ is upper semicontinuous.

If V is locally Lipschitz on $\mathbb{R}_+ \times K$, then define for $(t, x) \in \mathbb{R}_+ \times K$,

$$\partial^{int} V(t, x) := co \operatorname{Limsup}_{\substack{(s, y) \to (t, x) \\ y \in \operatorname{int}(K)}} \{\nabla V(s, y)\}.$$

Observe that the set $\partial^{int} V(t, x)$ is compact. It coincides with the generalized gradient of V at (t, x) whenever $x \in \operatorname{int}(K)$ and $t > 0$.

Below we denote by $\partial_x^{L,-} V(t_0, x_0)$ the limiting subdifferential of $V(t_0, \cdot)$ at x_0. Recalling the notation introduced in Sect. 4,

$$G(t, x) := \{(f(t, x, u), L(t, x, u)) : u \in U(t)\},$$

we state the maximum principle under state constraints.

Theorem 10 *Assume (IPC^{rel}), (H1) (ii)–(v) with $\omega_R(t, r) = \bar{c}_R(t)r$, where for all $R > 0$, $\bar{c}_R \in L^1_{\text{loc}}(\mathbb{R}_+; \mathbb{R}_+)$, that $G(t, x)$ is compact for every $(t, x) \in \mathbb{R}_+ \times \mathbb{R}^n$ and that there exists $c > 0$ such that for a.e. $t \geq 0$,*

$$\sup_{u \in U(t)} (|f(t, x, u)| + |L(t, x, u)|) \leq c(|x| + 1) \;\; \forall x \in \mathbb{R}^n.$$

Then

(I) *if for all large $j \in \mathbb{N}$, the function $V(j, \cdot)$ is locally Lipschitz on K, then V is locally Lipschitz on $\mathbb{R}_+ \times K$.*

(II) *if* $(\bar{x}, \bar{u})$ *is optimal for the infinite horizon problem at* $(t_0, x_0) \in \mathbb{R}_+ \times int(K)$, *then there exist* $p \in W^{1,1}_{loc}(t_0, \infty; \mathbb{R}^n)$, *a positive Borel measure* μ *on* $[t_0, \infty)$, *and a Borel measurable selection* $\nu(t) \in N_K(\bar{x}(t)) \cap B$ *defined on* $[t_0, \infty)$ *such that for* $q(t) := p(t) + \eta(t)$ *with*

$$\eta(t) := \begin{cases} \int_{[t_0,t]} \nu(s)\, d\mu(s) & t \in (t_0, \infty) \\ 0 & t = t_0, \end{cases}$$

the following relations are satisfied:

(i) *the Hamiltonian inclusion*

$$(-p'(t), \bar{x}'(t)) \in \partial H(t; \bar{x}(t), q(t)) \text{ for a.e. } t \in [t_0, \infty);$$

(ii) *the maximality condition*

$$\langle q(t), \bar{x}'(t) \rangle + L(t, \bar{x}(t), \bar{u}(t)) = H(t, \bar{x}(t), q(t)) \text{ for a.e. } t \in [t_0, \infty);$$

(iii) *the sensitivity relation*

$$(-H(t, \bar{x}(t), q(t)), q(t)) \in \partial^{int} V(t, \bar{x}(t)) \text{ for a.e. } t \in [t_0, \infty);$$

(iv) *the transversality condition at the initial time*

$$p(t_0) \in \partial_x^{L,-} V(t_0, x_0). \tag{18}$$

Furthermore, if H is continuous, then

$$(-H(t, \bar{x}(t), q(t)), q(t)) \in \partial^{int} V(t, \bar{x}(t)) \ \forall t \in (t_0, \infty).$$

Remark 9

(a) Recall that Theorem 8 provides sufficient conditions for the local Lipschitz continuity of V on $\mathbb{R}_+ \times K$. Instead one can also assume that (H2) holds true for $t \in [j, \infty)$ and some integer j (in place of $[0, \infty)$) to deduce Lipschitz continuity of V on $[j, \infty) \times K$.
(b) It is not difficult to realize that (i) contains the maximality condition (ii). We have stated (ii) in order to underline the link with the more familiar maximum principle.

Proof For every sufficiently large integer $j > t_0$, let $g_j : \mathbb{R}^n \to \mathbb{R}$ be a locally Lipschitz function that coincides with $V(j, \cdot)$ on K. Consider the Mayer problem $(\mathcal{M}_j)$

$$V^j(t_0, x_0, z_0) = \sup(g_j(x(j)) + z(j))$$

over all the trajectory-control pairs of the control system

$$\begin{cases} x'(t) = f(t, x(t), u(t)) & \text{a.e. } t \in [t_0, j] \\ z'(t) = L(t, x(t), u(t)) & \text{a.e. } t \in [t_0, j] \\ x(t_0) = x_0, \ z(t_0) = z_0 & \\ u(t) \in U(t) & \text{a.e. } t \in [t_0, j] \\ x(t) \in K & \forall t \in [t_0, j]. \end{cases}$$

From [26, Theorem 5.1] (after rewriting the above Mayer problem as a minimization problem), we deduce that V^j is locally Lipschitz on $[0, j] \times K \times \mathbb{R}$ whenever j is sufficiently large. It is not difficult to realize that $V^j(t_0, x_0, 0) = V^B(t_0, x_0)$ for every $(t_0, x_0) \in [0, j] \times K$, where V^B is the value function of the Bolza problem (introduced in the previous section) for $T = j$. Hence, by Proposition 2, V is locally Lipschitz continuous on $\mathbb{R}_+ \times K$.

To prove the second statement of our theorem, consider an optimal pair $(\bar{x}, \bar{u})$ of the infinite horizon problem at $(t_0, x_0) \in \mathbb{R}_+ \times \text{int}(K)$. Setting

$$\bar{z}(t) = \int_{t_0}^{t} L(s, \bar{x}(s), \bar{u}(s))\, ds,$$

it follows that the restriction of $(\bar{x}, \bar{z}, \bar{u})$ to $[t_0, j]$ is optimal for $(\mathcal{M}_j)$ at $(t_0, x_0, 0)$. By [26, Theorem 5.3] (after rewriting the above Mayer problem as a minimization problem on the interval $[t_0, j]$ instead of $[0, 1]$), for every integer $j > t_0$ there exist an absolutely continuous function p_j, a positive Borel measure μ_j and a Borel measurable selection $\nu_j(s) \in N_K(\bar{x}(s)) \cap B$ μ_j- a.e. defined on the time interval $[t_0, j]$ such that, setting

$$\eta_j(t) := \begin{cases} \int_{[t_0,t]} \nu_j(s)\, d\mu_j(s) & t \in (t_0, j] \\ 0 & t = t_0 \end{cases}$$

and $q_j(t) := p_j(t) + \eta_j(t)$, the following relations hold true for a.e. $t \in [t_0, j]$:

$$(-p_j'(t), \bar{x}'(t)) \in \partial H(t; \bar{x}(t), q_j(t)); \tag{19}$$

$$\langle q_j(t), f(t, \bar{x}(t), \bar{u}(t))\rangle + L(t, \bar{x}(t), \bar{u}(t)) = H(t, \bar{x}(t), q_j(t)); \tag{20}$$

$$(-H(t, \bar{x}(t), q_j(t)), q_j(t)) \in \partial^{int} V(t, \bar{x}(t)); \tag{21}$$

$$p_j(t_0) \in \partial_x^{L,-} V(t_0, x_0). \tag{22}$$

Let $k > t_0$ be a fixed integer.

Step 1 Since $x_0 \in \operatorname{int}(K)$ and V is locally Lipschitz, we deduce from (22) that the sequence $\{p_j(t_0)\}_j$ is bounded. The measure μ_j being regular, we know that q_j is right continuous on (t_0, j). Since V is locally Lipschitz on $\mathbb{R}_+ \times K$ from (21) we deduce that the functions $\{q_j\}_{j>k}$ are equibounded on $[t_0, k]$.

On the other hand, if $R > 0$ is so that $\sup_{t\in[t_0,k]} |\bar{x}(t)| < R$, then for a.e. $t \in [t_0, k]$ and all $(a, b) \in \partial H(t; \bar{x}(t), q_j(t)) \subset \mathbb{R}^n \times \mathbb{R}^n$ we have $|a| \le c_R(t)|q_j(t))| + \bar{c}_R(t)$. This and (19) imply that $|p_j'(t)| \le c_R(t)|q_j(t))| + \bar{c}_R(t)$ a.e. in $[t_0, k]$. Hence also the mappings $\{p_j\}_j$ are equibounded on $[t_0, k]$ and therefore so are $\{\eta_j\}_j$.

For every $j > k$ define the positive measure $\bar{\mu}_j$ on Borel subsets of $[t_0, k]$ by $\bar{\mu}_j(A) = \int_A |\nu_j(t)| d\mu_j(t)$ for any Borel set $A \subset [t_0, k]$. We claim that $\bar{\mu}_j([t_0, k])$ are equibounded. Indeed, the inward pointing condition implies that for every $t \ge t_0$, $\operatorname{int}(C_K(\bar{x}(t))) \ne \emptyset$. Thus the set valued map $t \rightsquigarrow C_K(\bar{x}(t))$ is lower semicontinuous on $[t_0, \infty)$. By [18, proof of Lemma 11] we know that there exists a continuous selection $\psi(t) \in C_K(t)$ defined on $[t_0, k]$ such that

$$\sup_{n\in N_K^1(\bar{x}(t))} \langle n, \psi(t)\rangle \leqslant -2 \quad \forall t \in [t_0, k] \text{ with } \bar{x}(t) \in \operatorname{bd}(K).$$

Consider $\bar{\psi} \in C^\infty([t_0, k]; \mathbb{R}^n)$ such that $\sup_{t\in[t_0,k]} |\psi(s) - \bar{\psi}(s)| \leqslant 1$. Then

$$\sup_{n\in N_K^1(\bar{x}(t))} \langle n, \bar{\psi}(t)\rangle \leqslant -1 \quad \forall t \in [t_0, k] \text{ with } \bar{x}(t) \in \operatorname{bd}(K).$$

Consequently, for every $j > k$,

$$\int_{[t_0,k]} \langle \bar{\psi}(s), \nu_j(s)\rangle \, d\mu_j(s) = \int_{[t_0,k]\cap\{s\,:\,\nu_j(s)\ne 0\}} \left\langle \bar{\psi}(s), \frac{\nu_j(s)}{|\nu_j(s)|} \right\rangle |\nu_j(s)| \, d\mu_j(s)$$

$$\leqslant -\int_{[t_0,k]\cap\{s\,:\,\nu_j(s)\ne 0\}} |\nu_j(s)| \, d\mu_j(s) = -\int_{[t_0,k]} |\nu_j(s)| \, d\mu_j(s).$$

Using that $\{\eta_j\}_j$ are equibounded on $[t_0, k]$ and have bounded total variation, integrating by parts we deduce from the above inequality that for some constant $C \geqslant 0$ and all $j > k$,

$$\int_{[t_0,k]} |\nu_j(s)| \, d\mu_j(s) \leqslant \int_{[t_0,k]} \langle -\bar{\psi}(s), \nu_j(s)\rangle \, d\mu_j(s) = \int_{[t_0,k]} -\bar{\psi}(s) \, d\eta_j(s)$$

$$= -\eta_j(k)\bar{\psi}(k) + \int_{[t_0,k]} \eta_j(s)\bar{\psi}'(s) \, ds \leqslant C\Big(\sup_{s\in[t_0,k]} |\bar{\psi}(s)| + \sup_{s\in[t_0,k]} |\bar{\psi}'(s)|\Big).$$

Since $\bar{\psi}(\cdot)$ does not depend on j our claim follows.

Step 2 By Step 1, the Ascoli–Arzelà theorem and the Dunford–Pettis criterion, taking a subsequence and keeping the same notation, we deduce that there exists an absolutely continuous function $p^k : [t_0, k] \to \mathbb{R}^n$ such that $p_j \to p^k$ uniformly on $[t_0, k]$ and $p'_j \rightharpoonup (p^k)'$ weakly in $L^1([t_0, k]; \mathbb{R}^n)$. Since the total variations of η_j are bounded by $\int_{[t_0,k]} |\nu_j(s)| \, d\mu_j(s)$ and $\eta_j(0) = 0$, from the Helly's selection theorem, taking a subsequence and keeping the same notation, we conclude that there exists a function of bounded variation η^k on $[t_0, k]$ such that $\eta_j \to \eta^k$ pointwise on $[t_0, k]$ and $\eta^k(t_0) = 0$. Furthermore, there exists a nonnegative finite measure μ^k on $[t_0, k]$ such that, by further extraction of a subsequence and preserving the same notation, $\bar{\mu}_j \rightharpoonup^* \mu^k$ (weakly-*) in $C([t_0, k]; \mathbb{R})^*$. Let

$$\gamma_j(t) := \begin{cases} \dfrac{\nu_j(t)}{|\nu_j(t)|} & \nu_j(t) \neq 0 \\ 0 & \text{otherwise.} \end{cases}$$

Since $\gamma_j(t) \in N_K(\bar{x}(t)) \cap B$ $\bar{\mu}_j$–a.e. in $[t_0, k]$ is a Borel measurable selection, applying [45, Proposition 9.2.1], we deduce that, for a subsequence $\{j_i\}_i$, there exists a Borel measurable function ν^k such that $\nu^k(t) \in N_K(\bar{x}(t)) \cap B$ μ^k-a.e. in $[t_0, k]$ and for every $\phi \in C([t_0, k]; \mathbb{R}^n)$

$$\int_{[t_0,k]} \langle \phi(s), \gamma_{j_i}(s) \rangle \, d\bar{\mu}_{j_i}(s) \to \int_{[t_0,k]} \langle \phi(s), \nu^k(s) \rangle \, d\mu^k(s) \text{ as } i \to \infty. \tag{23}$$

Using that for all $t \in (t_0, k]$,

$$\eta_{j_i}(t) = \int_{[t_0,t]} \nu_{j_i}(s) \, d\mu_{j_i}(s) = \int_{[t_0,t]} \gamma_{j_i}(s) \, d\bar{\mu}_{j_i}(s),$$

from (23) and the separation theorem it follows that for every $t \in (t_0, k]$

$$\eta^k(t) = \int_{[t_0,t]} \nu^k(s) \, d\mu^k(s).$$

Recall that $\partial H(t; \cdot, \cdot)$ is upper semicontinuous and has convex compact images and that $\partial_x^{L,-} V(0, x_0)$, $\partial^{int} V(t, \bar{x}(t))$ are closed sets.

Define $q^k = p^k + \eta^k$. Passing to the limits in (20) and (21) a.e. in $[t_0, k]$ for the subsequence q_{j_i} when $i \to \infty$, we deduce that (ii) and (iii) are satisfied on $[t_0, k]$ with q replaced by q^k. From (22) we obtain (18) for $p^k(t_0)$. Taking the weak limit in (19) for the subsequence p'_{j_i} when $i \to \infty$ and using the Mazur lemma and the upper semicontinuity of $\partial H(t; \cdot, \cdot)$ we get (i) on $[t_0, k]$ with p replaced by p^k and q by q^k.

Step 3 Consider now the interval $[t_0, k+1]$. By the same argument as in the second step, taking suitable subsequences $\{p_{j_{i_l}}\}_l \subset \{p_{j_i}\}_i$ and $\{\eta_{j_{i_l}}\}_l \subset \{\eta_{j_i}\}_i$, we deduce that there exist an absolutely continuous function p^{k+1}, a function of bounded variation η^{k+1}, a Borel measurable selection $\nu^{k+1}(t) \in N_K(\bar{x}(t)) \cap B$ on $[t_0, k+1]$ and a positive Borel measure μ^{k+1} on $[t_0, k+1]$ which satisfy $(i) - (iv)$ on $[t_0, k+1]$ with p, q replaced by p^{k+1}, q^{k+1} and moreover, when $\ell \to \infty$

$$\begin{aligned} & p_{j_{i_l}} \to p^{k+1} \text{ uniformly on } [t_0, k+1], \\ & p'_{j_{i_l}} \rightharpoonup (p^{k+1})' \text{ weakly in } L^1([t_0, k+1]; \mathbb{R}^n) \\ & q_{j_{i_l}} \to q^{k+1} \text{ pointwise on } [t_0, k+1], \\ & \mu_{j_{i_l}} \rightharpoonup^* \mu^{k+1} \text{ weakly-* in } C([t_0, k+1]; \mathbb{R})^* \\ & p^{k+1}|_{[t_0,k]} = p^k, \quad q^{k+1}|_{[t_0,k]} = q^k, \end{aligned}$$

and for all $t \in [t_0, k+1]$

$$\eta_{j_{i_l}}(t) \to \eta^{k+1}(t) = \begin{cases} \int_{[t_0,t]} \nu^{k+1}(s)\, d\mu^{k+1}(s) & t \in (t_0, k+1] \\ 0 & t = t_0. \end{cases}$$

Furthermore, since $\eta^{k+1}|_{[t_0,k]} = \eta^k$ and $\mu^{k+1}|_{[t_0,k]} = \mu^k$, we have

$$\nu^{k+1}|_{[t_0,k]} = \nu^k \qquad \mu^k\text{-a.e. on } [t_0, k].$$

The mappings p^{k+1}, η^{k+1}, and ν^{k+1} extend p^k, η^k, and ν^k respectively on the time interval $[t_0, k+1]$, and the measure μ^{k+1} extends the measure μ^k.

Applying the induction argument we get p, η, ν and μ defined on $[t_0, \infty)$ and satisfying the second claim of our theorem.

If H is continuous, then the right continuity of q on (t_0, ∞) and the upper semicontinuity of $\partial^{int} V$ on $\mathbb{R}_+ \times K$ imply the last statement of our theorem.

9 Hamilton–Jacobi Equation

The Hamilton–Jacobi (HJ) equation associated to the non-autonomous infinite horizon optimal control problem is as follows

$$\partial_t W + H(t, x, \partial_x W) = 0 \quad \text{on } (0, \infty) \times K, \qquad (HJ)$$

where ∂_t, ∂_x denote the partial derivatives of W with respect to t and x.

If the value function is differentiable, then it is well known that it satisfies (HJ). If in addition f, U are time independent and $L(t, x, u) = e^{-\lambda t}\ell(x, u)$ for some $\ell : \mathbb{R}^n \times \mathbb{R}^m \to \mathbb{R}$, then $V(t, x) = e^{-\lambda t} V(0, x)$. Setting $W_1(x) := V(0, x)$ we deduce from (HJ) that W_1 is a solution of the following, more familiar in the context of infinite horizon problems, *stationary* Hamilton–Jacobi equation

$$-\lambda W + \mathcal{H}(x, \partial_x W) = 0 \quad \text{on } K,$$

where $\mathcal{H}(x, p) := \sup_{u \in U}(\langle p, f(x, u)\rangle + \ell(x, u))$.

Conversely, if W_1 is a smooth solution of this stationary Hamilton–Jacobi equation, then $W(t, x) := e^{-\lambda t} W_1(x)$ solves (HJ). If f and/or ℓ are time dependent, then, in general, (HJ) can not be replaced by the above stationary equation.

It is well known that (HJ) may not have differentiable solutions. Furthermore, if a smooth solution does exist, it may be not uniquely defined because, in the difference with the finite horizon case, there is no "terminal" condition involved, as in (4). In Sect. 4 we have shown that under some integrable boundedness assumptions for large times, the value function $V(t, \cdot)$ restricted to its domain of definition converges to zero as $t \to \infty$. In such a setting, the natural "terminal" condition should be as follows

$$\lim_{t\to\infty} \sup_{y\in \mathrm{dom}\,(W(t,\cdot))} |W(t, y)| = 0.$$

The value function V being not differentiable, and even possibly discontinuous, solutions of (HJ) have to be understood in a generalized sense. In [9] we have investigated uniqueness of generalized solutions in the class of upper semicontinuous functions satisfying the above "terminal" condition. In this section we restrict our attention to locally Lipschitz continuous solutions on $\mathbb{R}_+ \times K$ only, the case, where results are much simpler to state and to prove. Then the above "terminal" condition becomes

$$\lim_{t\to\infty} \sup_{y\in K} |W(t, y)| = 0. \tag{24}$$

Denote by **(H3)** the following assumptions on f, L:

(i) $\exists\, c \in L^1_{\mathrm{loc}}(\mathbb{R}_+; \mathbb{R}_+)$ such that for a.e. $t \geqslant 0$ and for all $x \in \mathbb{R}^n$, $u \in U(t)$

$$|f(t, x, u)| + |L(t, x, u)| \leqslant c(t)(1 + |x|);$$

(ii) $\exists\, k \in \mathcal{L}_{\mathrm{loc}}$ such that for a.e. $t \geqslant 0$ and for all $x,\ y \in \mathbb{R}^n$, $u \in U(t)$

$$|f(t, x, u) - f(t, y, u)| + |L(t, x, u) - L(t, y, u)| \leqslant k(t)|x - y|;$$

(iii) $\limsup_{t\to\infty} \frac{1}{t} \int_0^t (c(s) + k(s))\, ds < \infty$;

(iv) $\exists\, q \in \mathcal{L}_{\mathrm{loc}}$ such that for a.e. $t \geqslant 0$

$$\sup_{u\in U(t)} (|f(t, x, u)| + |L(t, x, u)|) \leqslant q(t), \ \ \forall x \in \mathrm{bd}\,(K);$$

(v) for a.e. $t \geqslant 0$ and all $x \in \mathbb{R}^n$, $G(t,x)$ is compact;
(vi) There exists $S > 0$ such that $|L(t,x,u)| \leq \alpha(t)$ for a.e. $t \geq S$ and all $x \in K$, $u \in U(t)$, where $\alpha : [S,+\infty) \to \mathbb{R}_+$ is integrable on $[S,+\infty)$.

The *Outward Pointing Condition* (OPC^∞) is the same as (IPC^∞) but with f replaced by $-f$.

Theorem 11 *Assume (H1) (iii), (vi) and (H3). Let $W : \mathbb{R}_+ \times K \to \mathbb{R}$ be a locally Lipschitz function satisfying (24).*

1. If (OPC^∞) holds true, then $W = V$ if and only if W is a bilateral solution of (HJ), that is for a.e. $t > 0$

$$\begin{cases} p_t + H(t,x,p_x) = 0 \ \forall (p_t,p_x) \in \partial_+ W(t,x), \ \ \forall x \in \text{int}\,(K) \\ p_t + H(t,x,p_x) \geq 0 \ \forall (p_t,p_x) \in \partial_+ W(t,x), \ \ \forall x \in \text{bd}\,(K). \end{cases} \tag{25}$$

2. If (IPC^∞) holds true, then $W = V$ if and only if W is a viscosity solution of (HJ), that is for a.e. $t > 0$

$$\begin{cases} p_t + H(t,x,p_x) \leqslant 0 \ \forall (p_t,p_x) \in \partial_- W(t,x), \ \ \forall x \in \text{int}\,(K), \\ p_t + H(t,x,p_x) \geqslant 0 \ \forall (p_t,p_x) \in \partial_+ W(t,x), \ \ \forall x \in K. \end{cases} \tag{26}$$

Now, let $l : \mathbb{R}_+ \times \mathbb{R}^n \times \mathbb{R}^m \to \mathbb{R}_+$, $\lambda > 0$, and

$$L(t,x,u) = e^{-\lambda t} l(t,x,u). \tag{27}$$

From Theorems 7, 11 and 8 we immediately get the following corollaries.

Corollary 1 *Assume (27), (H2), (IPC^∞), (OPC^∞) and that f is bounded. Then for every $\lambda > 0$ sufficiently large, the value function V_λ is the only locally Lipschitz function $W : \mathbb{R}_+ \times K \to \mathbb{R}$ satisfying for a.e. $t > 0$*

$$\begin{cases} p_t + H(t,x,p_x) = 0 \ \forall (p_t,p_x) \in \partial_+ W(t,x), \ \ \forall x \in \text{int}\,(K), \\ p_t + H(t,x,p_x) \geqslant 0 \ \forall (p_t,p_x) \in \partial_+ W(t,x), \ \ \forall x \in \text{bd}\,(K) \\ \lim_{t\to\infty} \sup_{y\in K} |W(t,y)| = 0. \end{cases}$$

Corollary 2 *Assume (27), (H2), (IPC^∞) and that f is bounded. Then for every $\lambda > 0$ sufficiently large, the value function V_λ is the only locally Lipschitz function $W : \mathbb{R}_+ \times K \to \mathbb{R}$ satisfying for a.e. $t > 0$*

$$\begin{cases} p_t + H(t,x,p_x) \leqslant 0 \ \forall (p_t,p_x) \in \partial_- W(t,x), \ \ \forall x \in \text{int}\,(K), \\ p_t + H(t,x,p_x) \geqslant 0 \ \forall (p_t,p_x) \in \partial_+ W(t,x), \ \ \forall x \in K \\ \lim_{t\to\infty} \sup_{y\in K} |W(t,y)| = 0. \end{cases}$$

The proof of Theorem 11 follows the path initiated in [24] that was subsequently applied in many papers, see for instance [25, 28–31].

We first state several auxiliary results.

Proposition 3 *The following relations hold true:*

- *for any* $x \in \mathrm{dom}(V(0,\cdot))$, $\limsup_{s\to 0+,\, y\to_K x} V(s,y) = V(0,x)$;
- *for a.e.* $t > 0$, $\forall x \in \mathrm{dom}(V(t,\cdot))$, $\exists \bar{u} \in U(t)$ *such that*

$$D_{\downarrow}V(t,x)(1, f(t,x,\bar{u})) \geq -L(t,x,\bar{u});$$

- *for a.e.* $t > 0$, $\forall x \in \mathrm{int}\,(K) \cap \mathrm{dom}(V(t,\cdot))$, $\forall u \in U(t)$ *the following two inequalities are satisfied*

$$D_{\downarrow}V(t,x)(-1, -f(t,x,u)) \geq L(t,x,u)$$

and

$$D_{\uparrow}V(t,x)(1, f(t,x,u)) \leq -L(t,x,u).$$

The above proposition follows easily from the dynamic programming principle and [31, Theorems 2.9, 4.2 and Corollary 2.7]. The equivalence results below allow to state the Hamilton–Jacobi inequalities involving sub and superdifferentials in terms of directional derivatives. Such equivalence is important, because it allows to deduce dynamic programming properties of a generalized solution which, in turn, yield uniqueness of solutions to (HJ) with the terminal condition (24).

Proposition 4 *Let* $W : \mathbb{R}_+ \times K \to \mathbb{R}$ *be locally Lipschitz. Then*

1. *the following two statements are equivalent:*

   (a) *for a.e.* $t > 0$, $\forall x \in K$, $\exists \bar{u} \in U(t)$ *such that*

$$D_{\downarrow}W(t,x)(1, f(t,x,\bar{u})) \geq -L(t,x,\bar{u});$$

   (b) $p_t + H(t,x,p_x) \geq 0$ *for a.e.* $t > 0$, $\forall x \in K$ *and every* $(p_t, p_x) \in \partial_+ W(t,x)$.

2. *the following two statements are equivalent:*

   (a)′ *for a.e.* $t > 0$, $\forall x \in \mathrm{int}\,(K)$, $\forall u \in U(t)$,

$$D_{\downarrow}W(t,x)(-1, -f(t,x,u)) \geq L(t,x,u);$$

   (b)′ $p_t + H(t,x,p_x) \leq 0$ *for a.e.* $t > 0$, $\forall x \in \mathrm{int}\,(K)$ *and every* $(p_t, p_x) \in \partial_+ W(t,x)$.

Proof Note that $(a) \Rightarrow (b)$ and $(a)' \Rightarrow (b)'$ by the very definition of superdifferential. From [31, Corollary 2.7 and Theorem 2.9] and [27, Corollary 3.2] we deduce that $(b) \Rightarrow (a)$ and $(b)' \Rightarrow (a)'$.

An analogous proposition can be stated also for the lower directional derivatives of W, we shall not dwell on it however.

Lemma 1 *Let* $W : \mathbb{R}_+ \times K \to \mathbb{R}$ *be locally Lipschitz and satisfying* $(a)'$ *of Proposition 4. Then for all* $0 < \tau_0 < \tau_1$ *and any trajectory-control pair* $(x(\cdot), u(\cdot))$ *of (5) on* $[\tau_0, \tau_1]$, *with* $x([\tau_0, \tau_1]) \subset \mathrm{int}\,(K)$, *the function* $w(\cdot)$ *defined by*

$$w(t) := W(\tau_1, x(\tau_1)) + \int_t^{\tau_1} L(s, x(s), u(s))ds$$

satisfies $(x(t), w(t)) \in \mathrm{hyp}\, W(t, \cdot)$ *for all* $t \in [\tau_0, \tau_1]$.

Consequently, $W(\tau_0, x(\tau_0)) \geq W(\tau_1, x(\tau_1)) + \int_{\tau_0}^{\tau_1} L(t, x(t), u(t))dt$.

Proof Define $\phi(t) = W(t, x(t))$. By the Lipschitz continuity of W for a.e. $t \in [\tau_0, \tau_1]$

$$-\phi'(t) = D_\downarrow W(t, x(t))(-1, -f(t, x(t), u(t))) \geq L(t, x(t), u(t)).$$

Integrating on $[t, \tau_1]$ implies

$$\phi(t) - W(\tau_1, x(\tau_1)) \geq \int_t^{\tau_1} L(s, x(s), u(s))ds = w(t) - W(\tau_1, x(\tau_1)).$$

Hence $W(t, x(t)) \geq w(t)$.

Similarly, we have the following lemma (after stating an analogue of Proposition 4 for the lower directional derivatives).

Lemma 2 *Let* $W : \mathbb{R}_+ \times K \to \mathbb{R}$ *be locally Lipschitz. If for a.e.* $t > 0$ *and all* $x \in \mathrm{int}\,(K)$,

$$p_t + H(t, x, p_x) \leqslant 0 \quad \forall\, (p_t, p_x) \in \partial_- W(t, x),$$

then for all $0 < \tau_0 < \tau_1$ *and any trajectory-control pair* $(x(\cdot), u(\cdot))$ *of (5) on* $[\tau_0, \tau_1]$ *with* $x([\tau_0, \tau_1]) \subset \mathrm{int}\,(K)$, *the function* $w(\cdot)$ *defined by*

$$w(t) := W(\tau_0, x(\tau_0)) - \int_{\tau_0}^{t} L(s, x(s), u(s))ds$$

satisfies $(x(t), w(t)) \in \mathrm{epi}\, W(t, \cdot)$ *for all* $t \in [\tau_0, \tau_1]$.

Consequently, $W(\tau_0, x(\tau_0)) \geq W(\tau_1, x(\tau_1)) + \int_{\tau_0}^{\tau_1} L(t, x(t), u(t))dt$.

Proof of Theorem 11 To show that the locally Lipschitz value function satisfies (25) and (26) it is sufficient to apply Proposition 3 and definitions of super and subdifferentials. Let $(t_0, x_0) \in \mathbb{R}_+ \times K$. To show that $W(t_0, x_0) \geq V(t_0, x_0)$, it is enough to consider the case when $V(t_0, x_0) > -\infty$. Let $(\bar{x}(\cdot), \bar{u}(\cdot))$ be optimal at (t_0, x_0). We apply (NFT) theorem from [10] under either the assumption (OPC^∞) or (IPC^∞) and associate with any $\varepsilon > 0$, the time interval $[t_\varepsilon, T_\varepsilon] \subset \mathbb{R}_+$ and a trajectory-control pair $(x_\varepsilon(\cdot), u_\varepsilon(\cdot))$ defined on $[t_\varepsilon, T_\varepsilon]$ satisfying $x_\varepsilon([t_\varepsilon, T_\varepsilon]) \subset \text{int}(K)$, $\lim_{\varepsilon\to 0+} T_\varepsilon = \infty$, $\lim_{\varepsilon\to 0+}(t_\varepsilon, x_\varepsilon(t_\varepsilon)) = (t_0, x_0)$ and

$$\left|\int_{t_\varepsilon}^{T_\varepsilon} L(t, x_\varepsilon(t), u_\varepsilon(t))dt - \int_{t_\varepsilon}^{T_\varepsilon} L(t, \bar{x}(t), \bar{u}(t))dt\right| \leq \varepsilon.$$

Then Lemmas 1 and 2 yield

$$\begin{aligned} W(\tau_\varepsilon, x_\varepsilon(\tau_\varepsilon)) &\geq W(T_\varepsilon, x_\varepsilon(T_\varepsilon)) + \textstyle\int_{\tau_\varepsilon}^{T_\varepsilon} L(t, x_\varepsilon(t), u_\varepsilon(t))dt \\ &\geq W(T_\varepsilon, x_\varepsilon(T_\varepsilon)) + \textstyle\int_{t_\varepsilon}^{T_\varepsilon} L(t, \bar{x}(t), \bar{u}(t))dt - \varepsilon. \end{aligned}$$

Hence, whenever $T_\varepsilon > S$ we get

$$W(\tau_\varepsilon, x_\varepsilon(\tau_\varepsilon)) \geq W(T_\varepsilon, x_\varepsilon(T_\varepsilon)) + \int_{t_\varepsilon}^{\infty} L(t, \bar{x}(t), \bar{u}(t))dt - \int_{T_\varepsilon}^{\infty} \alpha(t)dt - \varepsilon.$$

Taking the limit when $\varepsilon \to 0+$, using (24) and the integrability of α on $\mathbb{R}_+$ we obtain $W(t_0, x_0) \geq V(t_0, x_0)$. Hence $W \geq V$.

Finally, Proposition 4 (a) and a measurable viability theorem [27, Corollary 3.2] imply that for every $(t_0, x_0) \in \mathbb{R}_+ \times K$ and $T > t_0$ there exists a trajectory-control pair $(\bar{x}(\cdot), \bar{u}(\cdot))$ such that

$$W(t_0, x_0) - \int_{t_0}^{T} L(t, \bar{x}(t), \bar{u}(t))dt \leq W(T, \bar{x}(T)).$$

Hence for any $T > S$, $W(t_0, x_0) \leq W(T, \bar{x}(T)) + V(t_0, x_0) + \int_T^\infty \alpha(t)dt$. This, (24) and the integrability of α on $[S, \infty)$ imply that for every $\varepsilon > 0$ we have $W(t_0, x_0) \leq V(t_0, x_0) + \varepsilon$. Consequently $W \leq V$ and we deduce that (25) yields $W = V$. Similarly, (26) implies $W = V$.

Acknowledgement This material is based upon work supported by the Air Force Office of Scientific Research under award number FA9550-18-1-0254.

References

1. Aseev SM (2013) On some properties of the adjoint variable in the relations of the Pontryagin maximum principle for optimal economic growth problems. Tr Inst Mat Mekh 19:15–24
2. Aseev SM, Kryazhimskiy AV (2004) The Pontryagin maximum principle and transversality conditions for a class of optimal control problems with infinite time horizons. SIAM J Control Optim 43:1094–1119
3. Aseev SM, Veliov VM (2012) Maximum principle for infinite-horizon optimal control problems with dominating discount. Dynam Contin Discrete Impuls Systems Series B: Appl Algorithms 19:43–63
4. Aseev SM, Veliov VM (2015) Maximum principle for infinite-horizon optimal control problems under weak regularity assumptions. Proc Steklov Inst Math 291:S22–S39
5. Aseev SM, Veliov VM (2017) Another view of the maximum principle for infinite-horizon optimal control problems in economics. Technical report. IIASA, Laxenburg
6. Aseev SM, Krastanov M, Veliov VM (2017) Optimality conditions for discrete-time optimal control on infinite horizon. Pure Appl Funct Anal 2:395–409
7. Aubin J-P, Clarke F (1979) Shadow prices and duality for a class of optimal control problems. SIAM J Control Optim 17:567–586
8. Aubin J-P, Frankowska H (1990) Set-valued analysis. Birkhäuser, Basel
9. Basco V, Frankowska H (2019) Hamilton–Jacobi–Bellman equations with time-measurable data and infinite horizon. Nonlinear Differ Equ Appl 26:7. https://doi.org/10.1007/s00030-019-0553-y
10. Basco V, Frankowska H (2019) Lipschitz continuity of the value function for the infinite horizon optimal control problem under state constraints. In: Trends in control theory and partial differential equations, Alabau F, et al. (eds). Springer INdAM series, vol 32. Springer, Cham
11. Basco V, Cannarsa P, Frankowska H (2018) Necessary conditions for infinite horizon optimal control problems with state constraints. Math Control Relat Fields (MCRF) 8:535–555
12. Benveniste LM, Scheinkman JA (1982) Duality theory for dynamic optimization models of economics: the continuous time case. J Econom Theory 27:1–19
13. Bettiol P, Frankowska H, Vinter R (2015) Improved sensitivity relations in state constrained optimal control. Appl Math Optim 71:353–377
14. Cannarsa P, Frankowska H (2018) Value function, relaxation, and transversality conditions in infinite horizon optimal control. J Math Anal Appl 457:1188–1217
15. Cannarsa P, Sinestrari C (2004) Semiconcave functions, Hamilton–Jacobi equations, and optimal control. Birkhäuser, Boston
16. Cannarsa P, Frankowska H, Scarinci T (2015) Second-order sensitivity relations and regularity of the value function for Mayer's problem in optimal control. SIAM J Control Optim 53:3642–3672
17. Carlson DA, Haurie A (1987) Infinite horizon optimal control: theory and applications. Lecture notes in economics and mathematical systems. Springer, Berlin
18. Cernea A, Frankowska H (2006) A connection between the maximum principle and dynamic programming for constrained control problems. SIAM J Control Optim 44:673–703
19. Clarke FH (1983) Optimization and nonsmooth analysis. Wiley, New York
20. Crandall MG, Lions P-L (1983) Viscosity solutions of Hamilton–Jacobi equations. Trans Amer Math Soc 277:1–42
21. Crandall MG, Evans LC, Lions P-L (1984) Some properties of viscosity solutions of Hamilton–Jacobi equation. Trans Amer Math Soc 282:487–502
22. Frankowska H (1989) Optimal trajectories associated to a solution of contingent Hamilton–Jacobi equations. Appl Math Optim 19:291–311
23. Frankowska H (1990) A priori estimates for operational differential inclusions. J Diff Eqs 84:100–128
24. Frankowska H (1993) Lower semicontinuous solutions of Hamilton–Jacobi–Bellman equation. SIAM J Control Optim 31:257–272

25. Frankowska H, Mazzola M (2013) Discontinuous solutions of Hamilton–Jacobi–Bellman equation under state constraints. Calc Var PDEs 46:725–747
26. Frankowska H, Mazzola M (2013) On relations of the co-state to the value function for optimal control problems with state constraints. Nonlinear Differ Equ Appl 20:361–383
27. Frankowska H, Plaskacz S (1996) A measurable upper semicontinuous viability theorem for tubes. J Nonlinear Anal TMA 26:565–582
28. Frankowska H, Plaskacz S (2000) Hamilton–Jacobi equations for infinite horizon control problems with state constraints. In: Proceedings of international conference "Calculus of variations and related topics", Haifa, 1998. Chapman and Hall/CRC research notes in mathematics series, vol 411. pp 97–116
29. Frankowska H, Plaskacz S (2000) Semicontinuous solutions of Hamilton–Jacobi–Bellman equations with degenerate state constraints. J Math Anal Appl 251:818–838
30. Frankowska H, Vinter RB (2000) Existence of neighbouring feasible trajectories: applications to dynamic programming for state constrained optimal control problems. J Optim Theory Appl 104:21–40
31. Frankowska H, Plaskacz S, Rzeżuchowski T (1995) Measurable viability theorems and Hamilton–Jacobi–Bellman equations. J Diff Eqs 116:265–305
32. Halkin H (1974) Necessary conditions for optimal control problems with infinite horizons. Econometrica 42:267–272
33. Khlopin DV (2013) Necessity of vanishing shadow price in infinite horizon control problems. J Dyn Con Sys 19:519–552
34. Lykina V, Pickenhain S (2017) Weighted functional spaces approach in infinite horizon optimal control problems: a systematic analysis of hidden opportunities and advantages. J Math Anal Appl 454:195–218
35. Michel P (1982) On the transversality condition in infinite horizon optimal problems. Econometrica 50:975–984
36. Pickenhain S (2010) On adequate transversality conditions for infinite horizon optimal control problems–a famous example of Halkin. In: Dynamic systems, economic growth, and the environment. Springer, Berlin, pp 3–12
37. Pickenhain S (2015) Infinite horizon optimal control problems in the light of convex analysis in Hilbert spaces. Set-Valued Var Anal 23:169–189
38. Pickenhain S (2019) Infinite horizon problems in the calculus of variations. Set-Valued Var Anal 2:331–354
39. Ramsey FP (1928) A mathematical theory of saving. Economic J 38:543–559
40. Sagara N (2010) Value functions and transversality conditions for infinite-horizon optimal control problems. Set-Valued Var Anal 18:1–28
41. Seierstad A, Sydsaeter K (1987) Optimal control theory with economic applications. North-Holland, Amsterdam
42. Seierstad A, Sydsaeter K (2009) Conditions implying the vanishing of the Hamiltonian at infinity in optimal control problems. Optim Lett 3:507–512
43. Soner HM (1986) Optimal control with state-space constraints. SIAM J Control Optim 24:552–561
44. Tauchnitz N (2015) The Pontryagin maximum principle for nonlinear optimal control problems with infinite horizon. J Optim Theory Appl 167:27–48
45. Vinter RB (2000) Optimal control. Birkhäuser, Boston
46. Von Weizsacker CC (1965) Existence of optimal programs of accumulation for an infinite time horizon. Rev Econ Stud 32:85–92
47. Ye J (1993) Nonsmooth maximum principle for infinite-horizon problems. J Optim Theory Appl 76:485–500

The Nonemptiness of the Inner Core

Tomoki Inoue

Abstract We prove that if a non-transferable utility (NTU) game is cardinally balanced and if, at every individually rational and efficient payoff vector, every non-zero normal vector to the set of payoff vectors feasible for the grand coalition is strictly positive, then the inner core is nonempty. The condition on normal vectors is satisfied if the set of payoff vectors feasible for the grand coalition is non-leveled. An NTU game generated by an exchange economy where every consumer has a continuous, concave, and strongly monotone utility function satisfies our sufficient condition. Our proof relies on Qin's theorem on the nonemptiness of the inner core.

Keywords Inner core · Inhibitive set · Cardinal balancedness · NTU game

Article type: Research Article
Received: February 27, 2019
Revised: July 24, 2019

1 Introduction

The inner core is a solution concept for a non-transferable utility (NTU) game. Players' utilities are not transferable, but an inner core payoff vector is stable even if players can transfer their utilities with some fictitious rates λ. Namely, at an inner core payoff vector, any coalition cannot improve upon the λ-weighted sum

JEL Classification: C62, C71
Mathematics Subject Classification (2010): 91A12, 91B50

T. Inoue (✉)
School of Political Science and Economics, Meiji University, Tokyo, Japan

T. Maruyama (ed.), *Advances in Mathematical Economics*, Advances in Mathematical Economics 23, https://doi.org/10.1007/978-981-15-0713-7_3

of utilities. This stability is stronger than the one required for the core. Thus, the inner core is a refinement of the core.

The inner core is of significance in the relations to (1) Walrasian equilibria for some economies, and to (2) the strictly inhibitive set of payoff vectors that are stable against randomized blocking plans. An exchange economy or a production economy generates a unique NTU game, but an NTU game may be generated by multiple economies. Billera [3] proposed a method how to induce a production economy from a totally cardinally balanced NTU game so that the economy actually generates the given NTU game. Qin [14] proved that the inner core of a totally cardinally balanced NTU game coincides with the set of utility vectors at Walrasian equilibria for Billera's induced production economy.[1] Inoue [9] constructed a coalition production economy generating a given NTU game and proved that the inner core coincides with the set of utility vectors at Walrasian equilibria for his induced coalition production economy.

When a payoff vector can be improved upon by multiple coalitions, it is not clear which coalition actually improves upon the payoff vector. Myerson [12, Section 9.8] considered the following scenario. A mediator who is not a player of a given game randomly chooses a coalition and payoffs to the members of the coalition. Every player accepts the mediator's randomized blocking plan if, conditional on being a member of a coalition, the expected payoff to him is better than the given payoff to him. The strictly inhibitive set is the set of payoff vectors that are feasible for the grand coalition and that are stable against randomized blocking plans. Qin [13] proved that the inner core is always a subset of the strictly inhibitive set and, in some classes of NTU games, these two sets coincide.

A sufficient condition for the inner core to be nonempty was given by Qin [15, Theorem 1]. Although his sufficient condition is mathematically general, it is not easy to check whether a given NTU game satisfies the condition. Accordingly, it is useful to give classes of NTU games satisfying Qin's sufficient condition. Qin [15, Corollary 1] proved that every cardinally balanced-with-slack NTU game satisfies his own sufficient condition. We give another class of NTU games satisfying Qin's sufficient condition: If an NTU game is cardinally balanced and if, at every individually rational and efficient payoff vector, every non-zero normal vector to the set of payoff vectors feasible for the grand coalition is strictly positive, then the NTU game satisfies Qin's sufficient condition for the inner core to be nonempty. The condition on normal vectors is met if the set of payoff vectors feasible for the grand coalition is non-leveled. Our class of NTU games with the nonempty inner core contains NTU games generated by exchange economies where every consumer has a continuous, concave, and strongly monotone utility function. de Clippel and Minelli [8] proved that the utility vector at a Walrasian equilibrium for such an exchange

[1]Furthermore, given an NTU game and given any payoff vector in the inner core, Qin [14] constructed a production economy generating the given NTU game such that the given payoff vector in the inner core is the utility vector at the unique Walrasian equilibrium for his induced production economy. Brangewitz and Gamp [6] extended this result from one point in the inner core to a closed subset with certain properties.

economy is always in the inner core and, therefore, the nonemptiness of the inner core follows from the well-known existence theorem of a Walrasian equilibrium. Hence, our theorem extends de Clippel and Minelli's class of NTU games with the nonempty inner core.

In the present paper, we prove our theorem on the nonemptiness of the inner core by applying Qin's [15] theorem. There are two other methods of proof of our theorem. The first way is due to Aubin [2, Theorem 2.1, Proposition 2.3]. Aubin adopted an abstract description of an NTU game. If we adopt a specified "representative function" of an NTU game, what Aubin called "an equilibrium for a representative function" turns out to be an inner core payoff vector. Since Aubin proved the existence of an equilibrium for a representative function, he indeed proved the nonemptiness of the inner core. At an inner core payoff vector, fictitious transfer rates λ of utilities must be strictly positive. Aubin first finds nonnegative fictitious transfer rates λ with certain properties and then provides a sufficient condition (condition (c) of Proposition 2 in Sect. 3) for λ to be strictly positive.[2] As discussed in Sect. 3, Aubin's sufficient condition for λ to be strictly positive can be slightly weakened. The second way is due to Inoue [11] who took two steps as well as Aubin [2]. Inoue [11] first gives a new coincidence theorem, a synthesis of Brouwer's fixed point theorem and a stronger separation theorem for convex sets due to Debreu and Schmeidler [7], and then, by applying the coincidence theorem, he obtains nonnegative fictitious transfer rates of utilities with certain properties. Inoue's [11] second step of the proof is the same as Aubin's [2].

The present paper is organized as follows. In Sect. 2, we give a precise description of an NTU game and the definition of the inner core. Also, we give Qin's sufficient condition for the inner core to be nonempty (Theorem 1). In Sect. 3, we provide characterizations of the efficient surface with strictly positive normal vectors. In Sect. 4, we prove the nonemptiness of the inner core (Theorem 2). In Sect. 5, we prove that our class of NTU games with the nonempty inner core contains de Clippel and Minelli's class of NTU games generated by exchange economies with continuous, concave, and strongly monotone utility functions. In Sect. 6, we give some remarks.

2 NTU Games and the Inner Core

We begin with some notation. We follow the notation of Qin [15]. Let $N = \{1, \ldots, n\}$ be a set with $n \geq 2$ elements and let $\mathbb{R}^N$ be the n-dimensional Euclidean space of vectors x with coordinates x_i indexed by $i \in N$. The inner product of x and y in $\mathbb{R}^N$ is denoted by $x \cdot y$, i.e., $x \cdot y = \sum_{i \in N} x_i y_i$. For $x, y \in \mathbb{R}^N$, we write $x \geq y$ if $x_i \geq y_i$ for every $i \in N$; $x \gg y$ if $x_i > y_i$ for every $i \in N$. The symbol 0 denotes

[2] Inoue [10, Appendix] summarizes Aubin's description of an NTU game and reproduces Aubin's method of proof.

the origin in $\mathbb{R}^N$ as well as the real number zero. Let $\mathbb{R}^N_{++} = \{x \in \mathbb{R}^N \mid x \gg 0\}$. For a nonempty subset S of N, let $\mathbb{R}^S = \{x \in \mathbb{R}^N \mid x_i = 0 \text{ for every } i \in N \setminus S\}$, let $\mathbb{R}^S_+ = \{x \in \mathbb{R}^S \mid x_i \geq 0 \text{ for every } i \in S\}$, and let $e^S \in \mathbb{R}^N$ be the characteristic vector of S, i.e., $e^S_i = 1$ if $i \in S$ and 0 otherwise. For $x \in \mathbb{R}^N$, x^S denotes the projection of x to $\mathbb{R}^S$. Let $\Delta = \{\lambda \in \mathbb{R}^N_+ \mid \sum_{i \in N} \lambda_i = 1\}$, $\Delta^\circ = \{\lambda \in \Delta \mid \lambda \gg 0\}$, and, for every $m \geq n$, $\Delta_{1/m} = \{\lambda \in \Delta \mid \lambda_i \geq 1/m \text{ for every } i \in N\}$.

We regard N as the set of n players. Let $\mathcal{N}$ be the family of all nonempty subsets of N, i.e., $\mathcal{N} = \{S \subseteq N \mid S \neq \emptyset\}$. Elements in $\mathcal{N}$ are called *coalitions*. A *non-transferable utility game* (NTU game, for short) with n players is a correspondence $V : \mathcal{N} \twoheadrightarrow \mathbb{R}^N$ such that, for every $S \in \mathcal{N}$, $V(S)$ is a nonempty subset of $\mathbb{R}^S$ and satisfies $V(S) - \mathbb{R}^S_+ = V(S)$. An NTU game is *compactly generated* if, for every $S \in \mathcal{N}$, there exists a nonempty compact subset C_S of $\mathbb{R}^S$ with $V(S) = C_S - \mathbb{R}^S_+$. In the present paper, we consider only compactly generated NTU games V with $V(N)$ convex.

The core is the set of payoff vectors which is feasible for the grand coalition N and which cannot be improved upon by any coalition. By adopting a different notion of improvement by a coalition, we can define the inner core.

Definition 1

(1) The *core* $C(V)$ of NTU game V is the set of payoff vectors $u \in \mathbb{R}^N$ such that $u \in V(N)$ and there exists no $S \in \mathcal{N}$ and $u' \in V(S)$ with $u'_i > u_i$ for every $i \in S$.
(2) The *inner core* $IC(V)$ of NTU game V is the set of payoff vectors $u \in \mathbb{R}^N$ such that $u \in V(N)$ and there exists $\lambda \in \mathbb{R}^N_{++}$ such that, for every $S \in \mathcal{N}$ and every $u' \in V(S)$, $\lambda^S \cdot u \geq \lambda^S \cdot u'$ holds.

By definition, we have $IC(V) \subseteq C(V)$. The vector $\lambda \in \mathbb{R}^N_{++}$ in the definition of the inner core represents fictitious transfer rates of utilities among players. Note that $u \in IC(V)$ if and only if $u \in V(N) \cap C(V_\lambda)$ for some $\lambda \in \mathbb{R}^N_{++}$, where $V_\lambda : \mathcal{N} \twoheadrightarrow \mathbb{R}^N$ is the λ-transfer game defined by

$$V_\lambda(S) = \left\{u \in \mathbb{R}^S \mid \lambda \cdot u \leq \lambda \cdot u' \text{ for some } u' \in V(S)\right\} \quad \text{for every } S \in \mathcal{N}.$$

Note also that we can restrict the space of fictitious transfer rates to Δ°.

For $\beta \in \mathbb{R}^N_+$, let

$$\Gamma(\beta) = \left\{ \gamma = (\gamma_S)_{S \in \mathcal{N}} \;\middle|\; \gamma_S \geq 0 \text{ for every } S \in \mathcal{N} \text{ and } \sum_{S \in \mathcal{N}} \gamma_S e^S = \beta \right\}$$

and let $\widehat{\Gamma}(\beta) = \{\gamma \in \Gamma(\beta) \mid \gamma_N = 0\}$. Note that $\Gamma(e^N)$ is the set of balancing vectors of weights. An NTU game V is *cardinally balanced* if, for every $\gamma \in \Gamma(e^N)$,

$$\sum_{S \in \mathcal{N}} \gamma_S V(S) \subseteq V(N).$$

This notion of balancedness is stronger than the balancedness due to Scarf [16]. Scarf's ordinal balancedness is sufficient for the nonemptiness of the core. For our theorem (Theorem 2) on the nonemptiness of the inner core, this stronger balancedness is assumed.

Before we give a condition equivalent to the cardinal balancedness, we introduce some notation. Let $V : \mathcal{N} \twoheadrightarrow \mathbb{R}^N$ be a compactly generated NTU game with $V(N)$ convex. For every $S \in \mathcal{N}$, let C_S be a compact subset of $\mathbb{R}^S$ generating $V(S)$, i.e., $V(S) = C_S - \mathbb{R}^S_+$, and let C_N be also convex. For every $\lambda \in \mathbb{R}^N_+$ and every $S \in \mathcal{N}$, define

$$v_\lambda(S) = \max\{\lambda \cdot u \mid u \in V(S)\} = \max\{\lambda \cdot u \mid u \in C_S\}\,.$$

Note that, by Berge's maximum theorem, $\mathbb{R}^N_+ \ni \lambda \mapsto v_\lambda(S) \in \mathbb{R}$ is continuous. For every $i \in N$, let $b_i \in \mathbb{R}$ be the utility level that player i can achieve by himself, i.e,

$$b_i = \max\{u_i \in \mathbb{R} \mid u \in V(\{i\})\}\,.$$

Define a correspondence $B : \mathbb{R}^N_{++} \twoheadrightarrow \mathbb{R}^N$ by

$$B(\lambda) = \{u \in V(N) \mid \lambda \cdot u = v_\lambda(N)\} = \{u \in C_N \mid \lambda \cdot u = v_\lambda(N)\}\,.$$

Note that $B(\lambda)$ is positively homogeneous of degree zero.

The following proposition gives a condition equivalent to the cardinal balancedness.

Proposition 1 *Let $V : \mathcal{N} \twoheadrightarrow \mathbb{R}^N$ be a compactly generated NTU game with $V(N)$ convex. Then, V is cardinally balanced if and only if, for every $\lambda \in \Delta^\circ$ and every $\gamma \in \widehat{\Gamma}(e^N)$, $\sum_{S \in \mathcal{N}} \gamma_S\, v_\lambda(S) \le v_\lambda(N)$.*

This equivalence is due to Shapley (see Qin [15, Proposition 1]).

For $m \ge n$, define continuous functions $p^m : \Delta \to \mathbb{R}^N_{++}$ and $\beta^m : \Delta \to \mathbb{R}^N_+ \setminus \{0\}$ by

$$p^m_j(\lambda) = \begin{cases} \lambda_j & \text{if } \lambda_j \ge 1/m, \\ 1/m & \text{if } \lambda_j < 1/m, \end{cases}$$

and

$$\beta^m_j(\lambda) = \frac{\lambda_j}{p^m_j(\lambda)} = \begin{cases} 1 & \text{if } \lambda_j \ge 1/m, \\ m\lambda_j & \text{if } \lambda_j < 1/m. \end{cases}$$

We are now ready to give Qin's theorem [15, Theorem 1] on the nonemptiness of the inner core. Inoue [11] gives another proof to the theorem.

Theorem 1 (Qin) *Let $V : \mathcal{N} \twoheadrightarrow \mathbb{R}^N$ be a compactly generated NTU game with $V(N)$ convex. If there exists $m \geq n$ such that*

(i) *for every $\lambda \in \Delta_{1/m}$ and every $\gamma \in \widehat{\Gamma}(e^N)$, $\sum_{S \in \mathcal{N}} \gamma_S v_\lambda(S) \leq v_\lambda(N)$, and*
(ii) *for every $\lambda \in \Delta^\circ \setminus \Delta_{1/m}$ and every $\gamma \in \widehat{\Gamma}(\beta^m(\lambda))$, there exists $u \in B(p^m(\lambda))$ such that $\sum_{S \in \mathcal{N}} \gamma_S v_{p^m(\lambda)}(S) \leq \lambda \cdot u$,*

then the inner core $IC(V)$ of V is nonempty.

By Proposition 1, condition (i) is weaker than the cardinal balancedness. Qin [15, Corollary 1] gives a class of NTU games that satisfy both conditions (i) and (ii) for some m: If a compactly generated NTU game V with $V(N)$ convex is *cardinally balanced with slack*, i.e., for every $\gamma \in \widehat{\Gamma}(e^N)$, $\sum_{S \in \mathcal{N}} \gamma_S V(S)$ is a subset of the interior of $V(N)$, then V satisfies conditions (i) and (ii) for some $m \geq n$, and, therefore, the inner core of V is nonempty.

Theorem 2 in Sect. 4 gives another class of NTU games with the nonempty inner core.

3 Characterization of Efficient Surface with Strictly Positive Normal Vectors

In the next section, we prove the nonemptiness of the inner core for an NTU game V where the normal cone to $V(N)$ at any individually rational and efficient payoff vector is a subset of $\mathbb{R}^N_{++} \cup \{0\}$. In this section, we provide characterizations of the set of individually rational and efficient payoff vectors with the above mentioned property.

Let $V(N)$ be a nonempty, closed, convex subset of $\mathbb{R}^N$ generated by a compact set $C_N \subseteq \mathbb{R}^N$, i.e., $V(N) = C_N - \mathbb{R}^N_+$. Let $b \in \mathbb{R}^N$ be such that $\{x \in V(N) \mid x \geq b\} \neq \emptyset$. Define

$$\begin{aligned}&\mathrm{Eff}(V(N), b)\\ &= \left\{x \in V(N) \;\middle|\; x \geq b,\ \text{there exists no } x' \in V(N) \text{ with } x' \geq x \text{ and } x' \neq x\right\}\end{aligned}$$

and

$$\mathrm{Eff}_{\mathrm{w}}(V(N), b) = \left\{x \in V(N) \;\middle|\; x \geq b,\ \text{there exists no } x' \in V(N) \text{ with } x' \gg x\right\}.$$

The set $\mathrm{Eff}(V(N), b)$ (resp. $\mathrm{Eff}_{\mathrm{w}}(V(N), b)$) is the set of individually rational and efficient payoff vectors (resp. individually rational and weakly efficient payoff vectors).

Remark 1

(1) $\emptyset \neq \mathrm{Eff}(V(N), b) \subseteq \mathrm{Eff}_{\mathrm{w}}(V(N), b)$.
(2) $\mathrm{Eff}_{\mathrm{w}}(V(N), b)$ is compact.

The nonemptiness of $\mathrm{Eff}(V(N), b)$ in statement (1) follows from the assumption that $V(N)$ is compactly generated and $\{x \in V(N) \mid x \geq b\} \neq \emptyset$. Statement (2) can be easily shown.

The following lemma gives a characterization of $\mathrm{Eff}_{\mathrm{w}}(V(N), b)$ by normal vectors to $V(N)$.[3]

Lemma 1

$$\begin{aligned} &\mathrm{Eff}_{\mathrm{w}}(V(N), b) \\ &= \{x \in V(N) \mid x \geq b, \\ &\qquad \text{there exists } \lambda \in \Delta \text{ such that, for every } y \in V(N),\ \lambda \cdot x \geq \lambda \cdot y\}. \end{aligned}$$

Proof Let $x \in \mathrm{Eff}_{\mathrm{w}}(V(N), b)$. Then, $x \in V(N)$ and $x \geq b$. Furthermore, $V(N) \cap (\{x\} + \mathbb{R}^N_{++}) = \emptyset$. By the separation theorem for convex sets, there exists $\lambda \in \mathbb{R}^N \setminus \{0\}$ such that, for every $y \in V(N)$ and every $z \in \{x\} + \mathbb{R}^N_{++}$, $\lambda \cdot z \geq \lambda \cdot y$. Then, we have $\lambda \in \mathbb{R}^N_+ \setminus \{0\}$. By normalization, we may assume that $\lambda \in \Delta$. Since the inner product is continuous, we have $\lambda \cdot x \geq \lambda \cdot y$ for every $y \in V(N)$.

We next prove the inverse inclusion. Let $x \in V(N)$ be such that $x \geq b$ and there exists $\lambda \in \Delta$ such that, for every $y \in V(N)$, $\lambda \cdot x \geq \lambda \cdot y$. Suppose, to the contrary, that $x \notin \mathrm{Eff}_{\mathrm{w}}(V(N), b)$. Then, there exists $x' \in V(N)$ with $x' \gg x$. Thus, we have $\lambda \cdot x' > \lambda \cdot x$, a contradiction. Hence, we have $x \in \mathrm{Eff}_{\mathrm{w}}(V(N), b)$. □

The following proposition gives a characterization of the individually rational and efficient surface such that every non-zero normal vector to $V(N)$ at every point of the surface is strictly positive.

Proposition 2 *The following three conditions are equivalent.*

(a) *Let* $x \in \mathrm{Eff}(V(N), b)$ *and* $\lambda \in \mathbb{R}^N \setminus \{0\}$ *be such that* $\lambda \cdot x = \max_{y \in V(N)} \lambda \cdot y$. *Then,* $\lambda \gg 0$. *Namely, for every* $x \in \mathrm{Eff}(V(N), b)$, *the normal cone to* $V(N)$ *at* x *is a subset of* $\mathbb{R}^N_{++} \cup \{0\}$.[4]
(b) *There exists* $b' \in \mathbb{R}^N$ *such that* $b' \ll b$ *and* $\mathrm{Eff}(V(N), b') = \mathrm{Eff}_{\mathrm{w}}(V(N), b')$.
(c) *For every* $x \in V(N)$ *with* $x \geq b$ *and every* $S \in \mathcal{N}$ *with* $S \subsetneq N$, *there exist* $\varepsilon, \delta > 0$ *such that* $x - \varepsilon\, e^S + \delta\, e^{N \setminus S} \in V(N)$.

Condition (b) holds if $V(N)$ is *non-leveled*, i.e., $x = y$ whenever x and y are on the boundary of $V(N)$ and $x \geq y$. Condition (c) is due to Aubin [2, Proposition 2.3]. Although $\mathrm{Eff}(V(N), b)$ need not be closed (see Arrow et al. [1] for such an example), condition (a) implies that $\mathrm{Eff}(V(N), b)$ is closed as shown by the next lemma.

[3]For a convex subset A of $\mathbb{R}^N$ and $x \in A$, $\lambda \in \mathbb{R}^N$ is *normal* to A at x if $\lambda \cdot x \geq \lambda \cdot y$ for every $y \in A$.

[4]For a convex subset A of $\mathbb{R}^N$ and $x \in A$, the *normal cone* to A at x is the set of all vectors $\lambda \in \mathbb{R}^N$ normal to A at x.

Lemma 2 *Condition* (a) *of Proposition* 2 *implies that*

$$\mathrm{Eff}(V(N), b) = \mathrm{Eff_w}(V(N), b).$$

Therefore, under condition (a), $\mathrm{Eff}(V(N), b)$ *is compact.*

***Proof of Lemma* 2** Suppose, to the contrary, that $\mathrm{Eff}(V(N), b) \subsetneq \mathrm{Eff_w}(V(N), b)$. Then, there exists $x \in \mathrm{Eff_w}(V(N), b)$ with $x \notin \mathrm{Eff}(V(N), b)$. By Lemma 1, there exists $\lambda \in \Delta$ such that, for every $y \in V(N)$, $\lambda \cdot x \geq \lambda \cdot y$.

Claim 1 $\lambda_i = 0$ for some $i \in N$.

***Proof of Claim* 1** Suppose, to the contrary, that $\lambda \gg 0$. Since $x \in V(N)$, $x \geq b$, and $x \notin \mathrm{Eff}(V(N), b)$, there exists $x' \in V(N)$ with $x' \geq x$ and $x' \neq x$. Therefore, we have $\lambda \cdot x' > \lambda \cdot x$, a contradiction. Thus, $\lambda_i = 0$ for some $i \in N$. □

Define $S = \{i \in N \mid \lambda_i = 0\}$. By Claim 1 above, we have $S \neq \emptyset$. Define

$$A = \left\{ y \in V(N) \;\middle|\; y^{N\setminus S} = x^{N\setminus S},\; y^S \geq x^S \right\}.$$

Since A is nonempty and compact, A has a maximal element with respect to $\geq$, i.e., there exists $\bar{y} \in A$ such that there exists no $y' \in A$ with $y' \geq \bar{y}$ and $y' \neq \bar{y}$. Note that, for every $x' \in V(N)$ with $x' \geq x$, we have $x' \in A$, since $\lambda \cdot x = \max_{y \in V(N)} \lambda \cdot y$.

Claim 2 $\bar{y} \in \mathrm{Eff}(V(N), b)$.

***Proof of Claim* 2** Suppose, to the contrary, that $\bar{y} \notin \mathrm{Eff}(V(N), b)$. Then, there exists $y' \in V(N)$ with $y' \geq \bar{y}$ and $y' \neq \bar{y}$. Since $y' \geq \bar{y} \geq x$, we have $y' \in A$. This contradicts that $\bar{y}$ is a maximal element in A. Thus, $\bar{y} \in \mathrm{Eff}(V(N), b)$. □

Since $\bar{y}^{N\setminus S} = x^{N\setminus S}$, we have $\lambda \cdot \bar{y} = \lambda \cdot x = \max_{y \in V(N)} \lambda \cdot y$. Thus, by condition (a), we have $\lambda \gg 0$, which contradicts that $\lambda_i = 0$ for every $i \in S$. Therefore, $\mathrm{Eff}(V(N), b) = \mathrm{Eff_w}(V(N), b)$. The compactness of $\mathrm{Eff}(V(N), b)$ follows from Remark 1. This completes the proof of Lemma 2. □

Remark 2 By Lemma 2, we can replace $\mathrm{Eff}(V(N), b)$ by $\mathrm{Eff_w}(V(N), b)$ in condition (a) of Proposition 2.

We are now ready to prove Proposition 2.

***Proof of Proposition* 2** (a) ⇒ (b): Suppose, to the contrary, that for every $r \in \mathbb{N}$,

$$\mathrm{Eff}(V(N), b - 1/r\, e^N) \subsetneq \mathrm{Eff_w}(V(N), b - 1/r\, e^N).$$

Then, for every $r \in \mathbb{N}$, there exists $x^r \in \mathrm{Eff_w}(V(N), b - 1/r\, e^N)$ such that $x^r \notin \mathrm{Eff}(V(N), b - 1/r\, e^N)$. By Lemma 1, for every $r \in \mathbb{N}$, there exists $\lambda^r \in \Delta$ with $\lambda^r \cdot x^r = \max_{y \in V(N)} \lambda^r \cdot y$. For every $r \in \mathbb{N}$, define $I_r = \{i \in N \mid \lambda_i^r = 0\}$. Since

$x^r \notin \mathrm{Eff}(V(N), b - 1/r\, e^N)$, we have $I_r \neq \emptyset$ for every $r \in \mathbb{N}$.[5] Since N has finitely many nonempty subsets, there exists $S \in \mathcal{N}$ such that $I_r = S$ for infinitely many r. By passing to a subsequence if necessary, we may assume that $I_r = S$ for every r. Since both sequences $(x^r)_r$ and $(\lambda^r)_r$ are bounded, by passing to further subsequences if necessary, we have

$$x^r \to x \quad \text{and} \quad \lambda^r \to \lambda \in \Delta.$$

Then, $x \in V(N)$ and $x \geq b$. Thus, we have also that $x \in \mathrm{Eff}_{\mathrm{w}}(V(N), b)$. By Lemma 2, $x \in \mathrm{Eff}(V(N), b)$. Since $\lambda_i^r = 0$ for every $i \in S$ and every r, we have $\lambda_i = 0$ for every $i \in S$. Since $\tilde{\lambda} \mapsto \max_{y \in C_N} \tilde{\lambda} \cdot y$ is continuous, from

$$\lambda^r \cdot x^r = \max_{y \in V(N)} \lambda^r \cdot y = \max_{y \in C_N} \lambda^r \cdot y,$$

it follows that

$$\lambda \cdot x = \max_{y \in C_N} \lambda \cdot y = \max_{y \in V(N)} \lambda \cdot y.$$

Since $\lambda_i = 0$ for every $i \in S$, we have a contradiction.
(b) $\Rightarrow$ (c): Let $x \in V(N)$ with $x \geq b$ and let $\emptyset \neq S \subsetneq N$. Since $b' \ll b$, there exists $\varepsilon > 0$ such that $x - \varepsilon\, e^S \geq b'$. Since $x - \varepsilon\, e^S \notin \mathrm{Eff}(V(N), b') = \mathrm{Eff}_{\mathrm{w}}(V(N), b')$ and $V(N) - \mathbb{R}^N_+ = V(N)$, there exists $\delta > 0$ such that $x - \varepsilon\, e^S + \delta\, e^N \in V(N)$. Since $x - \varepsilon\, e^S + \delta\, e^{N \setminus S} \leq x - \varepsilon\, e^S + \delta\, e^N$, we have $x - \varepsilon\, e^S + \delta\, e^{N \setminus S} \in V(N)$.
(c) $\Rightarrow$ (a): Let $x \in \mathrm{Eff}(V(N), b)$ and $\lambda \in \mathbb{R}^N \setminus \{0\}$ be such that $\lambda \cdot x = \max_{y \in V(N)} \lambda \cdot y$. Since $V(N) = C_N - \mathbb{R}^N_+$, we have $\lambda \in \mathbb{R}^N_+$. Suppose, to the contrary, that $\lambda_i = 0$ for some $i \in N$. By condition (c), there exist $\varepsilon, \delta > 0$ such that $x - \varepsilon\, e^{\{i\}} + \delta\, e^{N \setminus \{i\}} \in V(N)$. Since

$$\lambda \cdot \left(x - \varepsilon\, e^{\{i\}} + \delta\, e^{N \setminus \{i\}}\right) = \lambda \cdot x + \delta \sum_{j \in N \setminus \{i\}} \lambda_j > \lambda \cdot x,$$

we have a contradiction. Thus, $\lambda \gg 0$. □

Remark 3 If a compactly generated NTU game V satisfies one of the conditions of Proposition 2, its inner core $IC(V)$ is closed.

Proof Let $b \in \mathbb{R}^N$ be such that $b_i = \max\{u_i \in \mathbb{R} \mid u \in V(\{i\})\}$ for every $i \in N$. Let $(x^r)_r$ be a sequence in $IC(V)$ such that $x^r \to x$. Since $x^r \in \mathrm{Eff}(V(N), b)$ for every r and since, by Lemma 2, $\mathrm{Eff}(V(N), b)$ is closed, we have $x \in \mathrm{Eff}(V(N), b)$. For every r, let $\lambda^r \in \Delta^\circ$ be such that, for every $S \in \mathcal{N}$, $\lambda^{r,S} \cdot x^r \geq v_{\lambda^r}(S)$. Since $(\lambda^r)_r$ is a bounded sequence, by passing to a subsequence if necessary, we have $\lambda^r \to \lambda \in \Delta$. Since $\Delta \ni \tilde{\lambda} \mapsto v_{\tilde{\lambda}}(S) \in \mathbb{R}$ is continuous, we have $\lambda^S \cdot x \geq v_\lambda(S)$

[5]This can be shown by the same argument as Claim 1 in the proof of Lemma 2.

for every $S \in \mathcal{N}$. Thus, in particular, $\lambda \cdot x = \max_{y \in V(N)} \lambda \cdot y$. By condition (a) of Proposition 2, we have $\lambda \in \Delta^{\circ}$. Therefore, we have $x \in IC(V)$. Hence, $IC(V)$ is closed. □

The following example illustrates that condition (a), (b), or (c) of Proposition 2 can be weakened for the nonemptiness of the inner core.

Example 1 Let $N = \{1, 2\}$ and let $V : \mathcal{N} \twoheadrightarrow \mathbb{R}^N$ be such that $b_i = \max\{u_i \in \mathbb{R} \mid u \in V(\{i\})\} = 1$ for $i \in N$ and

$$V(N) = \left\{u \in \mathbb{R}^N \mid u_1 + u_2 \leq 3,\ u_1 \leq 2,\ u_2 \leq 3\right\}.$$

Then, V is compactly generated and $V(N)$ is convex. Note that $x := (2, 1) \in \mathrm{Eff}(V(N), b)$ (see Fig. 1). The vector $\lambda := (1, 0)$ is normal to $V(N)$ at x. Thus, condition (a) of Proposition 2 is violated. In this example, however, $V(N)$ can be extended to $V'(N)$ such that the pair $(V'(N), b)$ satisfies condition (a) of Proposition 2 and the inner core of the extended NTU game V' is *not* larger than the inner core of V. Therefore, we can weaken the conditions of Proposition 2 sufficient for the nonemptiness of the inner core of V.

For example, define

$$V'(N) = \left\{u \in \mathbb{R}^N \mid u_1 + u_2 \leq 3,\ u_1 \leq 3,\ u_2 \leq 3\right\}.$$

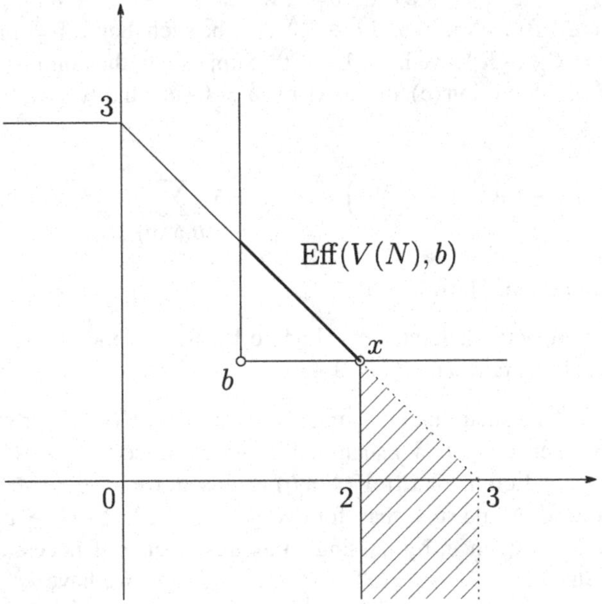

Fig. 1 V of Example 1

The shaded region in Fig. 1 is extended. The pair $(V'(N), b)$ satisfies condition (a) of Proposition 2. Any payoff vector y in the extended region is not in the inner core of the new game V', because it violates the individual rationality. Thus, the inner core $IC(V')$ of V' is not larger than the inner core $IC(V)$ of V. By the extension of $V(N)$, since the normal cone to the set of payoff vectors feasible for N at x becomes smaller,[6] $IC(V')$ can be smaller than $IC(V)$.[7] In this example, however, both inner cores are the same.

Condition (d) of the following proposition represents that, for some extension $V'(N)$ of $V(N)$, every non-zero normal vector to $V'(N)$ at every payoff vector in $\mathrm{Eff}(V'(N), b)$ is strictly positive.

Proposition 3 *The following two conditions are equivalent.*

(d) *There exists a nonempty, closed, convex subset $V'(N)$ of $\mathbb{R}^N$ such that $V(N) \subseteq V'(N)$, $V'(N)$ is generated by a compact set C'_N, $\{x \in V(N) \mid x \geq b\} = \{x \in V'(N) \mid x \geq b\}$, and condition* (a) *of Proposition* 2 *holds for $(V'(N), b)$, i.e.,*

$$\left[x \in \mathrm{Eff}(V'(N), b),\ \lambda \in \mathbb{R}^N \setminus \{0\},\ \text{and } \lambda \cdot x = \max_{y \in V'(N)} \lambda \cdot y\right] \text{implies } \lambda \gg 0.$$

(e) *There exists a compact subset K of Δ° such that for every $x \in \mathrm{Eff_w}(V(N), b)$, there exists $\lambda \in K$ with $\lambda \cdot x = \max_{y \in V(N)} \lambda \cdot y$.*

Clearly, condition (a) of Proposition 2 implies condition (d). Thus, by Propositions 2 and 3, we have

$$\text{(a)} \Leftrightarrow \text{(b)} \Leftrightarrow \text{(c)} \Rightarrow \text{(d)} \Leftrightarrow \text{(e)}.$$

***Proof of Proposition* 3** (d) $\Rightarrow$ (e): Define a correspondence $\xi : \mathrm{Eff}(V'(N), b) \twoheadrightarrow \Delta$ by

$$\xi(x) = \left\{\lambda \in \Delta \,\middle|\, \lambda \cdot x = \max_{y \in V'(N)} \lambda \cdot y\right\}.$$

Then, by Lemma 1, ξ is nonempty-valued. By Lemma 2, condition (d) implies that $\mathrm{Eff}(V'(N), b) = \mathrm{Eff_w}(V'(N), b)$ and this common set is compact. It is clear that ξ is compact-valued and upper hemi-continuous. Thus, $\xi(\mathrm{Eff}(V'(N), b))$ is compact.

Let $K = \xi(\mathrm{Eff}(V'(N), b))$. By condition (d), we have $K \subseteq \Delta^\circ$. From $\{x \in V(N) \mid x \geq b\} = \{x \in V'(N) \mid x \geq b\}$, it follows that

$$\mathrm{Eff_w}(V(N), b) = \mathrm{Eff_w}(V'(N), b) = \mathrm{Eff}(V'(N), b).$$

[6]In this example, $(2/3, 1/3)$ is normal to $V(N)$ at x, but this vector is not normal to $V'(N)$ at x.
[7]For an example of $IC(V') \subsetneq IC(V)$, see Example 2.

Therefore, for every $x \in \mathrm{Eff_w}(V(N), b)$, there exists $\lambda \in K$ with

$$\lambda \cdot x = \max_{y \in V'(N)} \lambda \cdot y \geq \max_{y \in V(N)} \lambda \cdot y.$$

Since $x \in V(N)$, we have $\lambda \cdot x = \max_{y \in V(N)} \lambda \cdot y$.
(e) $\Rightarrow$ (d): By condition (e), for every $x \in \mathrm{Eff_w}(V(N), b)$, we choose and fix a $\lambda^x \in K$ with $\lambda^x \cdot x = \max_{y \in V(N)} \lambda^x \cdot y$. Let $b' \in \mathbb{R}^N$ be such that $b' \ll b$ and $b' \leq y$ for every $y \in C_N$, where C_N is a compact set generating $V(N)$. Define

$$C'_N = \left\{ y \in \mathbb{R}^N \mid y \geq b' \text{ and, for every } x \in \mathrm{Eff_w}(V(N), b),\ \lambda^x \cdot y \leq \lambda^x \cdot x \right\}.$$

Then, $C_N \subseteq C'_N$ and thus C'_N is nonempty. We have also that C'_N is compact and convex. Define $V'(N) = C'_N - \mathbb{R}^N_+$. Then, $V'(N)$ is nonempty, closed, convex, and satisfies $V(N) \subseteq V'(N)$. Therefore, $\{y \in V(N) \mid y \geq b\} \subseteq \{y \in V'(N) \mid y \geq b\}$.

Claim 3 $\{y \in V'(N) \mid y \geq b\} \subseteq \{y \in V(N) \mid y \geq b\}$.

***Proof of Claim* 3** Let $\bar{y} \in V'(N)$ be such that $\bar{y} \geq b$. Suppose, to the contrary, that $\bar{y} \notin V(N)$. Let $\bar{z} \in \mathbb{R}^N$ be the closest point in $\{y \in V(N) \mid y \geq b\}$ from $\bar{y}$, i.e., $\bar{z} \in V(N)$, $\bar{z} \geq b$, and

$$\|\bar{y} - \bar{z}\| = \min\left\{\|\bar{y} - z\| \mid z \in V(N),\ z \geq b\right\},$$

where $\|\cdot\|$ stands for the Euclidean norm. Since $\{y \in V(N) \mid y \geq b\}$ is nonempty, compact, and convex, $\bar{z}$ is uniquely determined.

We prove that $\bar{y} - \bar{z} \in \mathbb{R}^N_+ \setminus \{0\}$. Since $\bar{y} \notin V(N)$ and $\bar{z} \in V(N)$, we have $\bar{y} - \bar{z} \neq 0$. Suppose, to the contrary, that $\bar{y}_i < \bar{z}_i$ for some $i \in N$. Define $\widehat{z} \in \mathbb{R}^N$ by

$$\widehat{z}_j = \begin{cases} \bar{z}_j & \text{if } j \neq i, \\ \bar{y}_i & \text{if } j = i. \end{cases}$$

Since $\bar{z} \geq b$ and $\bar{y} \geq b$, we have $\widehat{z} \geq b$. We have also that $\widehat{z} \in \{\bar{z}\} - \mathbb{R}^N_+ \subseteq V(N) - \mathbb{R}^N_+ = V(N)$. Since $\|\bar{y} - \widehat{z}\| < \|\bar{y} - \bar{z}\|$, we have a contradiction. Hence, $\bar{y} - \bar{z} \in \mathbb{R}^N_+ \setminus \{0\}$.

We next prove that $\bar{z} \in \mathrm{Eff_w}(V(N), b)$. Suppose, to the contrary, that $\bar{z} \notin \mathrm{Eff_w}(V(N), b)$. Then, there exists $z' \in V(N)$ with $z' \gg \bar{z}$. Since $\bar{y} - \bar{z} \in \mathbb{R}^N_+ \setminus \{0\}$, $\bar{y}_k > \bar{z}_k$ for some $k \in N$. Define $\tilde{z} \in \mathbb{R}^N$ by

$$\tilde{z}_j = \begin{cases} \bar{z}_j & \text{if } j \neq k, \\ \min\{\bar{y}_k, z'_k\} & \text{if } j = k. \end{cases}$$

Since $z' \gg \bar{z} \geq b$ and $\bar{y} \geq b$, we have $\tilde{z} \geq b$. We have also that $\tilde{z} \in \{z'\} - \mathbb{R}^N_+ \subseteq V(N) - \mathbb{R}^N_+ = V(N)$. Since $\|\bar{y} - \tilde{z}\| < \|\bar{y} - \bar{z}\|$, we have a contradiction. Hence, $\bar{z} \in \mathrm{Eff_w}(V(N), b)$.

From $\bar{z} \in \mathrm{Eff_w}(V(N), b)$, it follows that $\lambda^{\bar{z}} \in K \subseteq \Delta^\circ$ and $\lambda^{\bar{z}} \cdot \bar{z} = \max_{y \in V(N)} \lambda^{\bar{z}} \cdot y$. Since $\lambda^{\bar{z}} \gg 0$ and $\bar{y} - \bar{z} \in \mathbb{R}^N_+ \setminus \{0\}$, we have $\lambda^{\bar{z}} \cdot \bar{y} > \lambda^{\bar{z}} \cdot \bar{z}$. Thus, by the definition of $V'(N)$, we have $\bar{y} \notin V'(N)$, which is a contradiction. Therefore, we have proven that $\{y \in V'(N) \mid y \geq b\} \subseteq \{y \in V(N) \mid y \geq b\}$. □

It remains to prove that condition (a) of Proposition 2 holds for $(V'(N), b)$. Let $x \in \mathrm{Eff}(V'(N), b)$ and $\lambda \in \mathbb{R}^N \setminus \{0\}$ be such that $\lambda \cdot x = \max_{y \in V'(N)} \lambda \cdot y$. Since $V'(N) - \mathbb{R}^N_+ = V'(N)$, we have $\lambda \in \mathbb{R}^N_+$. Suppose, to the contrary, that $\lambda_i = 0$ for some $i \in N$. Since $b' \ll b$, we have $x - \alpha\, e^{\{i\}} \geq b'$ for sufficiently small $\alpha > 0$. Since $K \subseteq \Delta^\circ$ is compact, there exists $\delta > 0$ such that for every $\lambda' \in K$ and every $j \in N$, $\lambda'_j > \delta$ holds. Thus, for every $z \in \mathrm{Eff_w}(V(N), b)$, we have $\lambda^z \cdot \left(\alpha\, e^{\{i\}}\right) = \alpha\lambda^z_i > \alpha\delta$. For every $z \in \mathrm{Eff_w}(V(N), b)$, since $x \in V'(N)$, we have $\lambda^z \cdot z \geq \lambda^z \cdot x$. Let $l \in N$ be such that $\lambda_l > 0$. Then, for every $z \in \mathrm{Eff_w}(V(N), b)$,

$$\begin{aligned}\lambda^z \cdot \left(x - \alpha\, e^{\{i\}} + \alpha\delta\, e^{\{l\}}\right) &\leq \lambda^z \cdot z + \lambda^z \cdot \left(-\alpha\, e^{\{i\}} + \alpha\delta\, e^{\{l\}}\right) \\ &< \lambda^z \cdot z - \alpha\delta + \alpha\delta = \lambda^z \cdot z.\end{aligned}$$

Since $x - \alpha\, e^{\{i\}} + \alpha\delta\, e^{\{l\}} \geq b'$, we have

$$x - \alpha\, e^{\{i\}} + \alpha\delta\, e^{\{l\}} \in C'_N \subseteq V'(N).$$

We have also that

$$\lambda \cdot \left(x - \alpha\, e^{\{i\}} + \alpha\delta\, e^{\{l\}}\right) = \lambda \cdot x + \alpha\delta\lambda_l > \lambda \cdot x,$$

which contradicts that $\lambda \cdot x = \max_{y \in V'(N)} \lambda \cdot y$. Therefore, $\lambda \gg 0$. This completes the proof of Proposition 3. □

The following example inspired by Qin [13, Example 2] illustrates that (1) $IC(V')$ can be strictly smaller than $IC(V)$ and (2) the inner core $IC(V)$ of V need not be closed when V satisfies condition (d) or (e) of Proposition 3. Recall that if V satisfies one of the conditions of Proposition 2, $IC(V)$ is closed (see Remark 3).

Example 2 Let $N = \{1, 2, 3\}$. An NTU game $V : \mathscr{N} \twoheadrightarrow \mathbb{R}^N$ is given by

$$\begin{aligned} V(\{1,2\}) &= \{u \in \mathbb{R}^{\{1,2\}}_+ \mid u_1^2 + u_2^2 \leq 4/25\} - \mathbb{R}^{\{1,2\}}_+, \\ V(N) &= \{u \in \mathbb{R}^N \mid u \cdot e^N = 1 \text{ and } u \geq 0\} - \mathbb{R}^N_+, \text{ and} \\ V(S) &= \{(0,0,0)\} - \mathbb{R}^S_+ \quad \text{for any other coalition } S. \end{aligned}$$

Then, V is compactly generated, $V(N)$ is convex, and $b = 0 \in \mathbb{R}^N$.

Note that, for every $t \in (2/5, 1]$, we have $x(t) := (0, t, 1-t) \in IC(V)$ as shown below. Let $t \in (2/5, 1]$. Then, $x(t) \in V(N)$. For $\lambda := (\lambda_1, (1-\lambda_1)/2, (1-\lambda_1)/2) \in$

$\mathbb{R}^N$ with $\lambda_1 \in (0, 1/3)$, let

$$z = \left(\frac{4\lambda_1}{5\sqrt{1-2\lambda_1+5\lambda_1^2}}, \frac{2(1-\lambda_1)}{5\sqrt{1-2\lambda_1+5\lambda_1^2}}, 0 \right) \in V(\{1,2\}).$$

Then, $v_\lambda(\{1,2\}) = \lambda^{\{1,2\}} \cdot z$. Thus,

$$\lambda^{\{1,2\}} \cdot x(t) - v_\lambda(\{1,2\}) = \lambda^{\{1,2\}} \cdot x(t) - \lambda^{\{1,2\}} \cdot z = \frac{1-\lambda_1}{2} \cdot t - \frac{\sqrt{1-2\lambda_1+5\lambda_1^2}}{5}.$$

For sufficiently small $\lambda_1 > 0$, e.g., $\lambda_1 = (t - 2/5)/2$, we have

$$\frac{1-\lambda_1}{2} \cdot t - \frac{\sqrt{1-2\lambda_1+5\lambda_1^2}}{5} > 0.$$

Thus, $v_\lambda(\{1,2\}) \le \lambda^{\{1,2\}} \cdot x(t)$ for sufficiently small $\lambda_1 > 0$. Since $0 < \lambda_1 < 1/3$, we have $v_\lambda(N) = (1-\lambda_1)/2 = \lambda \cdot x(t)$. Since $v_\lambda(S) = 0$ for any other coalition S, it is clear that $v_\lambda(S) = 0 \le \lambda^S \cdot x(t)$. Thus, for every $t \in (2/5, 1]$, $x(t) \in IC(V)$ holds.

Let $x^* := (0, 1/2, 1/2) \in IC(V)$. Note that $x^* \in \mathrm{Eff}(V(N), 0)$ and the vector $(0, 1/2, 1/2) \in \Delta$ is normal to $V(N)$ at x^*. Thus, the pair $(V(N), 0)$ does not satisfy condition (a) of Proposition 2. Define $V' : \mathcal{N} \twoheadrightarrow \mathbb{R}^N$ by

$$V'(N) = \left\{ u \in \mathbb{R}^N \;\middle|\; u \cdot e^N = 1 \quad \text{and} \quad u \ge -1/5\, e^N \right\} - \mathbb{R}^N_+$$

and $V'(S) = V(S)$ for every $S \in \mathcal{N} \setminus \{N\}$. Then, the pair $(V'(N), 0)$ satisfies condition (d) of Proposition 3.

We prove that $x^* \notin IC(V')$. Since $\mu := (1/3, 1/3, 1/3) \in \Delta$ is a unique normal vector to $V'(N)$ at x^* and since

$$\max_{u \in V'(\{1,2\})} \mu^{\{1,2\}} \cdot u = \frac{1}{3} \cdot \frac{\sqrt{2}}{5} + \frac{1}{3} \cdot \frac{\sqrt{2}}{5} = \frac{2\sqrt{2}}{15} > \frac{1}{6} = \mu^{\{1,2\}} \cdot x^*,$$

we have $x^* \notin IC(V')$. Hence, $IC(V') \subsetneq IC(V)$.

We finally prove that $IC(V)$ is not closed. Let $y := (0, 2/5, 3/5)$. For every $\lambda \in \Delta^\circ$, we have $v_\lambda(\{1,2\}) > \lambda^{\{1,2\}} \cdot y$. Thus, $y \notin IC(V)$. Since $x(t) \in IC(V)$ for every $t \in (2/5, 1]$ and $x(t) \to y$ as $t \to 2/5$, $IC(V)$ is not closed. Therefore, even if an NTU game satisfies condition (d) or (e) of Proposition 3, its inner core need not be closed.

4 Nonemptiness of the Inner Core

We give the main result.

Theorem 2 *Let* $V : \mathcal{N} \twoheadrightarrow \mathbb{R}^N$ *be a compactly generated NTU game with* $V(N)$ *convex. If* V *is cardinally balanced and if* V *satisfies condition* (d) *or* (e) *of Proposition* 3, *then the inner core* $IC(V)$ *of* V *is nonempty.*

We prove this theorem by applying Theorem 1.

***Proof of Theorem* 2** Note first that the inner core satisfies the following covariance property.

Lemma 3 *Let* $V : \mathcal{N} \twoheadrightarrow \mathbb{R}^N$ *be an NTU game and let* $a \in \mathbb{R}^N$. *Define an NTU game* $V + a : \mathcal{N} \twoheadrightarrow \mathbb{R}^N$ *by*

$$(V + a)(S) = V(S) + \{a^S\} \quad \text{for every } S \in \mathcal{N}.$$

Then, $IC(V + a) = IC(V) + \{a\}$.

This can be easily shown, so we omit the proof.

Since, by Proposition 3, conditions (d) and (e) are equivalent, we may assume that V satisfies condition (e). Let $b' \in \mathbb{R}^N$ be such that $b' \ll b$ and, for every $S \in \mathcal{N}$ and every $y \in C_S$, $b' \leq y$, where C_S is a compact subset of $\mathbb{R}^S$ with $V(S) = C_S - \mathbb{R}^S_+$. We define V' as in the proof of (e) $\Rightarrow$ (d). By condition (e), for every $x \in \text{Eff}_w(V(N), b)$, we choose and fix a $\lambda^x \in K$ with $\lambda^x \cdot x = \max_{y \in V(N)} \lambda^x \cdot y$. Define

$$C'_N = \{y \in \mathbb{R}^N \mid y \geq b' \text{ and, for every } x \in \text{Eff}_w(V(N), b), \lambda^x \cdot y \leq \lambda^x \cdot x\},$$

$V'(N) = C'_N - \mathbb{R}^N_+$, and $V'(S) = V(S)$ for every $S \in \mathcal{N}$ with $S \subsetneq N$. By the proof of (e) $\Rightarrow$ (d) of Proposition 3, $V'(N)$ satisfies all the properties of condition (d). Define an NTU game $\widehat{V}$ by $\widehat{V} = V' - b'$. Then, $\widehat{V}(N)$ is convex, $\widehat{V}$ is cardinally balanced,[8] and, for every $S \in \mathcal{N}$, $\widehat{V}(S)$ is generated by a compact subset of $\mathbb{R}^S_+$. Since $IC(\widehat{V} + b') = IC(V') \subseteq IC(V)$, by Lemma 3, it suffices to prove that $IC(\widehat{V}) \neq \emptyset$. We will prove that $\widehat{V}$ satisfies all the conditions of Theorem 1. Since $\widehat{V}$ is cardinally balanced, by Proposition 1, condition (i) of Theorem 1 is met for $\widehat{V}$ and for every $m \geq n$. It remains to prove that condition (ii) is met for $\widehat{V}$ and for some $m \geq n$, i.e., there exists $m \geq n$ such that, for every $\lambda \in \Delta^\circ \setminus \Delta_{1/m}$ and every $\gamma \in \widehat{\Gamma}(\beta^m(\lambda))$, there exists $y \in B(p^m(\lambda))$ with $\sum_{S \in \mathcal{N}} \gamma_S v_{p^m(\lambda)}(S) \leq \lambda \cdot y$, where $B(p^m(\lambda))$ and $v_{p^m(\lambda)}$ are defined for NTU game $\widehat{V}$.

Since, by the definition of b', for every $S \in \mathcal{N} \setminus \{N\}$ and every $y \in C_S$, $y \geq b'$ holds and since, for every $y \in C'_N$, $y \geq b'$ holds, for every $m \geq n$, every $\lambda \in$

[8] Since V is cardinally balanced and $V(N) \subset V'(N)$, V' is cardinally balanced. Thus, $\widehat{V} = V' - b'$ also is cardinally balanced.

$\Delta^\circ \setminus \Delta_{1/m}$, and every $S \in \mathcal{N}$, we have $v_{p^m(\lambda)}(S) \geq 0$. Since $0 \leq \beta^m(\lambda) \leq e^N$ for every $\lambda \in \Delta$, $\widehat{V}$ is cardinally balanced, and $v_{p^m(\lambda)}(S) \geq 0$ for every $S \in \mathcal{N}$, we have, for every $m \geq n$, every $\lambda \in \Delta^\circ \setminus \Delta_{1/m}$, every $\gamma \in \widehat{\Gamma}(\beta^m(\lambda))$, and every $y \in B(p^m(\lambda))$,

$$\sum_{S \in \mathcal{N}} \gamma_S \, v_{p^m(\lambda)}(S) \leq v_{p^m(\lambda)}(N) = p^m(\lambda) \cdot y.$$

Thus, it suffices to prove that there exists $m \geq n$ such that, for every $\lambda \in \Delta^\circ \setminus \Delta_{1/m}$ and every $y \in B(p^m(\lambda))$, $p^m(\lambda) \cdot y \leq \lambda \cdot y$ holds.

Since $K \subseteq \Delta^\circ$ is compact, there exists $k \geq n$ with $K \subseteq \Delta_{1/k}$. Let $m \in \mathbb{N}$ be such that $m > (n-1)(k-n+1)+1$.

Claim 4 $m > k$.

***Proof of Claim* 4** Since $k \geq n \geq 2$, we have

$$\begin{aligned} m - k &> (n-1)(k-n+1) + 1 - k = k(n-2) - (n-1)^2 + 1 \\ &\geq n(n-2) - (n-1)^2 + 1 = 0. \end{aligned}$$

Hence, $m > k$. □

Claim 5 Let $\lambda \in \Delta^\circ \setminus \Delta_{1/m}$ and $x \in \mathrm{Eff_w}(\widehat{V}(N), 0)$ be such that

$$\lambda \cdot x = \max_{y \in \widehat{V}(N)} \lambda \cdot y.$$

Then, for every $i \in N$ with $\lambda_i < 1/m$, $x_i = 0$ holds.

***Proof of Claim* 5** Suppose, to the contrary, that $x_i > 0$ for some $i \in N$ with $\lambda_i < 1/m$. Since $\lambda \in \Delta$ and $\lambda_i < 1/m$, there exists $j \in N \setminus \{i\}$ such that

$$\lambda_j > \frac{1 - \frac{1}{m}}{n-1} = \frac{m-1}{m(n-1)}.$$

Define

$$E = \left\{ y \in \mathbb{R}^N_+ \;\middle|\; y^{N \setminus \{i,j\}} = x^{N \setminus \{i,j\}}, \; \mu \cdot y \leq \mu \cdot x \quad \text{for every } \mu \in \Delta_{1/k} \right\}.$$

We first prove that $E \subseteq C'_N - \{b'\}$. Let $y \in E$. Since $x \in \widehat{V}(N) = C'_N - \{b'\} - \mathbb{R}^N_+$, we have $x + b' \in C'_N - \mathbb{R}^N_+$. Hence, by the definition of C'_N, for every $z \in \mathrm{Eff_w}(V(N), b)$, $\lambda^z \cdot (x + b') \leq \lambda^z \cdot z$ holds. Since $\lambda^z \in K \subseteq \Delta_{1/k}$,

$$\lambda^z \cdot (y + b') \leq \lambda^z \cdot (x + b') \leq \lambda^z \cdot z.$$

Since $y \geq 0$, we have $y + b' \geq b'$. Thus, $y + b' \in C'_N$. Hence, $y \in C'_N - \{b'\}$ and $E \subseteq C'_N - \{b'\}$.

For every $l \in N$, define $\mu^{k,l} \in \Delta_{1/k}$ by $\mu_t^{k,l} = 1/k$ for $t \in N \setminus \{l\}$ and $\mu_l^{k,l} = 1 - (n-1)/k$. Then, $\{\mu^{k,1}, \ldots, \mu^{k,n}\}$ is the set of all extreme points of $\Delta_{1/k}$. Therefore,

$$\begin{aligned} E &= \left\{ y \in \mathbb{R}^N_+ \;\middle|\; y^{N\setminus\{i,j\}} = x^{N\setminus\{i,j\}},\ \mu^{k,l} \cdot y \leq \mu^{k,l} \cdot x \quad \text{for every } l \in N \right\} \\ &= \left\{ y \in \mathbb{R}^N_+ \;\middle|\; y^{N\setminus\{i,j\}} = x^{N\setminus\{i,j\}},\ \mu^{k,i} \cdot y \leq \mu^{k,i} \cdot x,\ \mu^{k,j} \cdot y \leq \mu^{k,j} \cdot x \right\} \\ &= \{ y \in \mathbb{R}^N_+ \mid y^{N\setminus\{i,j\}} = x^{N\setminus\{i,j\}},\ (k-n+1)y_i + y_j \leq (k-n+1)x_i + x_j, \\ &\qquad y_i + (k-n+1)y_j \leq x_i + (k-n+1)x_j \}. \end{aligned}$$

Define $z^* \in \mathbb{R}^N$ by

$$\begin{aligned} z^{*N\setminus\{i,j\}} &= x^{N\setminus\{i,j\}}, \\ z^*_i &= 0, \\ z^*_j &= x_j + \tfrac{1}{k-n+1} x_i \end{aligned}$$

(See Fig. 2). Then, $z^* \in E \subseteq C'_N - \{b'\} \subseteq \widehat{V}(N)$. Since $\lambda_j > \frac{m-1}{m(n-1)}$ and $m > (n-1)(k-n+1)+1$, we have

$$\frac{\lambda_j}{k-n+1} > \frac{m-1}{m(n-1)(k-n+1)} > \frac{m-1}{m(m-1)} = \frac{1}{m} > \lambda_i.$$

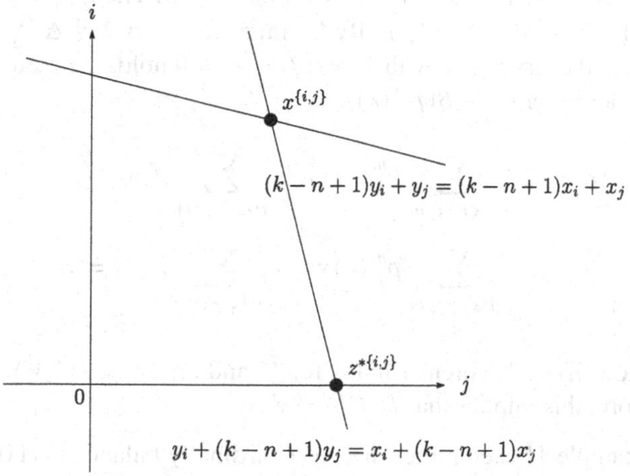

Fig. 2 Definition of $z^{*\{i,j\}}$

Therefore, since $x_i > 0$, we have

$$\begin{aligned}\lambda \cdot z^* &= \lambda \cdot x^{N\setminus\{i,j\}} + \lambda_j x_j + \frac{\lambda_j}{k-n+1} x_i \\ &> \lambda \cdot x^{N\setminus\{i,j\}} + \lambda_j x_j + \lambda_i x_i \\ &= \lambda \cdot x = \max_{y \in \widehat{V}(N)} \lambda \cdot y \geq \lambda \cdot z^*,\end{aligned}$$

a contradiction. Therefore, $x_i = 0$ for every $i \in N$ with $\lambda_i < 1/m$. □

Define $\sigma^m : \Delta \to \Delta_{1/(m+n-1)}$ by

$$\sigma_j^m(\lambda) = \frac{p_j^m(\lambda)}{p^m(\lambda) \cdot e^N}.$$

Since

$$\begin{aligned}B(p^m(\lambda)) &= \left\{ y \in \widehat{V}(N) \;\middle|\; p^m(\lambda) \cdot y = \max_{z \in \widehat{V}(N)} p^m(\lambda) \cdot z \right\} \\ &= \left\{ y \in \widehat{V}(N) \;\middle|\; \sigma^m(\lambda) \cdot y = \max_{z \in \widehat{V}(N)} \sigma^m(\lambda) \cdot z \right\} \\ &= \left\{ y \in C'_N - \{b'\} \;\middle|\; \sigma^m(\lambda) \cdot y = \max_{z \in \widehat{V}(N)} \sigma^m(\lambda) \cdot z \right\},\end{aligned}$$

by Lemma 1 and the definition of b', we have $B(p^m(\lambda)) \subseteq \mathrm{Eff}_w(\widehat{V}(N), 0)$ for every $\lambda \in \Delta$. Let $\lambda \in \Delta^\circ \setminus \Delta_{1/m}$ and let $i \in N$ with $\lambda_i < 1/m$. Then, $p_i^m(\lambda) = 1/m$ and $p^m(\lambda) \cdot e^N > 1$. Thus, $\sigma_i^m(\lambda) < 1/m$. By Claim 5, for every $\lambda \in \Delta^\circ \setminus \Delta_{1/m}$, every $y \in B(p^m(\lambda))$, and every $i \in N$ with $\lambda_i < 1/m$, $y_i = 0$ holds. Therefore, for every $\lambda \in \Delta^\circ \setminus \Delta_{1/m}$ and every $y \in B(p^m(\lambda))$,

$$\begin{aligned}p^m(\lambda) \cdot y &= \sum_{j:\lambda_j \geq 1/m} p_j^m(\lambda) y_j + \sum_{j:\lambda_j < 1/m} p_j^m(\lambda) y_j \\ &= \sum_{j:\lambda_j \geq 1/m} p_j^m(\lambda) y_j = \sum_{j:\lambda_j \geq 1/m} \lambda_j y_j = \lambda \cdot y.\end{aligned}$$

Hence, condition (ii) of Theorem 1 holds for $\widehat{V}$ and m. Thus, $IC(\widehat{V}) \neq \emptyset$. As we mentioned before, this implies that $IC(V) \neq \emptyset$. □

Qin [15, Example 1] exemplifies that, if a cardinally balanced NTU game does not satisfy condition (d) or (e) of Proposition 3, then its inner core can be empty. We give another example of a totally cardinally balanced NTU game V with every $V(S)$

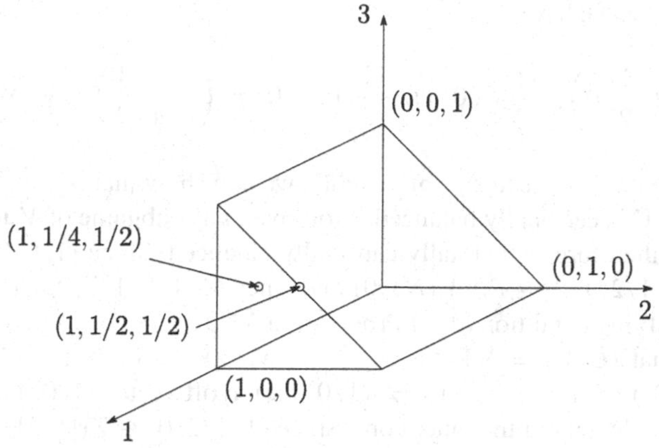

Fig. 3 $V(N)$ of Example 3

polyhedral such that the inner core $IC(V)$ of V is empty.[9] Billera and Bixby [4] proved that any totally cardinally balanced NTU game with every $V(S)$ polyhedral can be generated by an exchange economy where agents' consumption sets have the form $[0, 1]^l$ and their utility functions are concave and continuous. Therefore, by the following example, the nonemptiness of the inner core is irrelevant to such representation of NTU games. It should be worth mentioning that in the first step of the proof of Billera and Bixby's representation, the inner core plays an essential role. Actually, $\bar{x}$ in the proof of Lemma 3.2 of Billera and Bixby [4] is an inner core payoff vector.

Example 3 Let $N = \{1, 2, 3\}$. An NTU game $V : \mathcal{N} \twoheadrightarrow \mathbb{R}^N$ is given by

$$\begin{aligned} V(\{i\}) &= \{(0,0,0)\} - \mathbb{R}_+^{\{i\}} \quad \text{for every } i \in N, \\ V(\{1,2\}) &= \{(1,1/2,0)\} - \mathbb{R}_+^{\{1,2\}}, \\ V(\{1,3\}) &= \{(1,0,1)\} - \mathbb{R}_+^{\{1,3\}}, \\ V(\{2,3\}) &= \{(0,0,0)\} - \mathbb{R}_+^{\{2,3\}}, \text{ and} \\ V(N) &= \{x \in \mathbb{R}^N \mid x_i \le 1 \text{ for every } i \in N \text{ and } x_2 + x_3 \le 1\}. \end{aligned}$$

Then, $b = 0 \in \mathbb{R}^N$. Figure 3 depicts $V(N) \cap \mathbb{R}_+^N$. Note that every $V(S)$ is a polyhedron. For a balancing vector γ with $\gamma_{\{1,2\}} = \gamma_{\{1,3\}} = \gamma_{\{2,3\}} = 1/2$ and

[9] An NTU game V is *totally cardinally balanced* if every subgame of V is cardinally balanced, i.e. for every $S \in \mathcal{N}$ and every $\gamma \in \Gamma(e^S)$, $\sum_{T \in \mathcal{N}} \gamma_T V(T) \subseteq V(S)$ holds.

$\gamma_S = 0$ otherwise, we have

$$\frac{1}{2}\left(1, \frac{1}{2}, 0\right) + \frac{1}{2}(1, 0, 1) + \frac{1}{2}(0, 0, 0) = \left(1, \frac{1}{4}, \frac{1}{2}\right) \in V(N).$$

For any other balancing vector γ' of weights, we can show that $\sum_{S \in \mathcal{N}} \gamma'_S V(S) \subseteq V(N)$. Hence, V is cardinally balanced. Moreover, any subgame of V is cardinally balanced and, therefore, V is totally cardinally balanced. Since $(1, 0, 0)$ is normal to $V(N)$ at $(1, 1/2, 1/2) \in \mathrm{Eff}(V(N), 0)$ and since $(1, 1/2, 1/2) \gg b$, there exists no $V'(N)$ satisfying condition (d) of Proposition 3.

We prove that $IC(V) = \emptyset$. Let $\lambda \in \mathbb{R}^N_{++}$. Since $\lambda^{\{1,3\}} \cdot (1, 0, 1) > \lambda^{\{1,3\}} \cdot x$ for every $x \in V(N)$ with $x \geq 0$ and $x \neq (1, 0, 1)$, payoff vector $(1, 0, 1)$ is a unique candidate of an element of the inner core. Since $(1, 1/2, 0) \in V(\{1, 2\})$ and

$$\lambda^{\{1,2\}} \cdot \left(1, \frac{1}{2}, 0\right) = \lambda_1 + \frac{1}{2}\lambda_2 > \lambda_1 = \lambda^{\{1,2\}} \cdot (1, 0, 1),$$

payoff vector $(1, 0, 1)$ is not in the inner core. Therefore, $IC(V) = \emptyset$.

5 NTU Games Generated by Exchange Economies

de Clippel and Minelli [8, Proposition 1] proved that, in an exchange economy where every agent has a continuous, concave, and strongly monotone utility function, the utility vector at any Walrasian allocation is always in the inner core. Since there exists a Walrasian equilibrium for such an exchange economy, the inner core is nonempty. In this section, we prove that an NTU game generated by such an exchange economy satisfies condition (a) of Proposition 2. Thus, by Theorem 2, its inner core is nonempty. Therefore, our class of NTU games with the nonempty inner core contains de Clippel and Minelli's [8] class.

An *exchange economy* with n consumers is a list of the commodity space $\mathbb{R}^L$, where L is a finite set of commodities, and consumers' characteristics $(u^i, \omega^i)_{i \in N}$ such that, for every consumer i, utility function $u^i : \mathbb{R}^L_+ \to \mathbb{R}$ is continuous, concave, and strongly monotone, and endowment vector ω^i is in $\mathbb{R}^L_+ \setminus \{0\}$. An exchange economy is denoted by $\mathcal{E} = (\mathbb{R}^L, (u^i, \omega^i)_{i \in N})$. For every coalition $S \in \mathcal{N}$, let $F_{\mathcal{E}}(S)$ be the set of feasible S-allocations, i.e.,

$$F_{\mathcal{E}}(S) = \left\{ (z^i)_{i \in S} \;\middle|\; z^i \in \mathbb{R}^L_+ \quad \text{for every } i \in S \text{ and } \quad \sum_{i \in S} \left(z^i - \omega^i\right) = 0 \right\}.$$

A feasible N-allocation $(z^i)_{i \in N}$ is a *Walrasian allocation* for $\mathcal{E}$ if there exists a price vector $p \in \mathbb{R}^L$ such that, for every $i \in N$, $p \cdot z^i \leq p \cdot \omega^i$ and $u^i(z^i) \geq u^i(y)$ for every $y \in \mathbb{R}^L_+$ with $p \cdot y \leq p \cdot \omega^i$. Let $W(\mathcal{E})$ be the set of all Walrasian allocations for $\mathcal{E}$.

A feasible N-allocation $(z^i)_{i \in N}$ is an *inner core allocation* for $\mathscr{E}$ if there exists $\lambda \in \mathbb{R}^N_{++}$ such that, for every $S \in \mathscr{N}$ and every $(y^i)_{i \in S} \in F_{\mathscr{E}}(S)$, $\sum_{i \in S} \lambda_i u^i(z^i) \geq \sum_{i \in S} \lambda_i u^i(y^i)$ holds. Let $IC(\mathscr{E})$ be the set of all inner core allocations for $\mathscr{E}$.

An exchange economy $\mathscr{E} = (\mathbb{R}^L, (u^i, \omega^i)_{i \in N})$ generates an NTU game $V_{\mathscr{E}} : \mathscr{N} \twoheadrightarrow \mathbb{R}^N$ by defining

$$V_{\mathscr{E}}(S) = \left\{ v \in \mathbb{R}^S \;\middle|\; \text{there exists } (z^i)_{i \in S} \in F_{\mathscr{E}}(S) \text{ such that, for every } i \in S,\, v_i \leq u^i(z^i) \right\}.$$

Since $F_{\mathscr{E}}(S)$ is nonempty, compact, and convex for every $S \in \mathscr{N}$, NTU game $V_{\mathscr{E}}$ is compactly convexly generated. Note that, for every $i \in N$,

$$b_i = \max\{v_i \in \mathbb{R} \mid v \in V_{\mathscr{E}}(\{i\})\} = u^i(\omega^i),$$

and $IC(V_{\mathscr{E}}) = \{(u^i(z^i))_{i \in N} \mid (z^i)_{i \in N} \in IC(\mathscr{E})\}$.

Qin [13, Remark 2] states the following without proof. For completeness, we give its proof.

Proposition 4 *Let $\mathscr{E} = (\mathbb{R}^L, (u^i, \omega^i)_{i \in N})$ be an exchange economy where, for every $i \in N$, $u^i : \mathbb{R}^L_+ \twoheadrightarrow \mathbb{R}$ is continuous, concave, and strongly monotone, and $\omega^i \in \mathbb{R}^L_+ \setminus \{0\}$. Then, NTU game $V_{\mathscr{E}}$ generated by $\mathscr{E}$ satisfies condition* (a) *of Proposition 2, i.e., if $x \in \mathrm{Eff}(V_{\mathscr{E}}(N), b)$ and $\lambda \in \mathbb{R}^N \setminus \{0\}$ satisfy $\lambda \cdot x = \max_{y \in V_{\mathscr{E}}(N)} \lambda \cdot y$, then $\lambda \gg 0$.*

Proof Let $x \in \mathrm{Eff}(V_{\mathscr{E}}(N), b)$ and $\lambda \in \mathbb{R}^N \setminus \{0\}$ be such that $\lambda \cdot x = \max_{y \in V_{\mathscr{E}}(N)} \lambda \cdot y$. Since $V_{\mathscr{E}}(N) - \mathbb{R}^N_+ = V_{\mathscr{E}}(N)$, we have $\lambda \in \mathbb{R}^N_+$. Suppose, to the contrary, that there exists $j \in N$ with $\lambda_j = 0$. Since $\lambda \in \mathbb{R}^N_+ \setminus \{0\}$, there exists $k \in N$ with $\lambda_k > 0$. From $x \in V_{\mathscr{E}}(N)$, it follows that there exists a feasible N-allocation $(z^i)_{i \in N}$ such that $u^i(z^i) \geq x_i$ for every $i \in N$. Since $u^j(z^j) \geq x_j \geq b_j = u^j(\omega^j)$, $\omega^j \in \mathbb{R}^L_+ \setminus \{0\}$, and u^j is strongly monotone, we have $z^j \in \mathbb{R}^L_+ \setminus \{0\}$. Define an allocation $(y^i)_{i \in N}$ by

$$y^i = \begin{cases} z^i & \text{if } i \in N \setminus \{j, k\}, \\ z^k + z^j & \text{if } i = k, \\ 0 & \text{if } i = j. \end{cases}$$

Then, $(y^i)_{i \in N}$ is a feasible N-allocation. Therefore, $u := (u^i(y^i))_{i \in N} \in V_{\mathscr{E}}(N)$. Since $\lambda_j = 0$, $\lambda_k > 0$, and $u^k(z^k + z^j) > u^k(z^k)$, we have

$$\begin{aligned} \lambda \cdot u = \sum_{i \in N} \lambda_i u^i(y^i) &= \sum_{i \in N \setminus \{j,k\}} \lambda_i u^i(z^i) + \lambda_k u^k(z^k + z^j) + \lambda_j u^j(0) \\ &> \sum_{i \in N \setminus \{j,k\}} \lambda_i u^i(z^i) + \lambda_k u^k(z^k) + \lambda_j u^j(z^j) \geq \lambda \cdot x. \end{aligned}$$

This contradicts that $\lambda \cdot x = \max_{y \in V_{\mathscr{E}}(N)} \lambda \cdot y$. Therefore, $\lambda \gg 0$ and hence $V_{\mathscr{E}}$ satisfies condition (a). □

Since an exchange economy with the properties in Proposition 4 generates a cardinally balanced NTU game,[10] by Theorem 2, we have the following.

Corollary 1 *Let $\mathscr{E} = (\mathbb{R}^L, (u^i, \omega^i)_{i \in N})$ be an exchange economy where, for every $i \in N$, $u^i : \mathbb{R}^L_+ \to \mathbb{R}$ is continuous, concave, and strongly monotone, and $\omega^i \in \mathbb{R}^L_+ \setminus \{0\}$. Then, the inner core $IC(V_{\mathscr{E}})$ of NTU game $V_{\mathscr{E}}$ generated by $\mathscr{E}$ is nonempty.*

We can show this corollary without relying on Theorem 2. de Clippel and Minelli [8, Proposition 1] proved the inclusion $W(\mathscr{E}) \subseteq IC(\mathscr{E})$ for an exchange economy $\mathscr{E}$ satisfying the properties in Corollary 1. Since there exists a Walrasian equilibrium for $\mathscr{E}$, the inner core of $V_{\mathscr{E}}$ is nonempty.

6 Concluding Remarks

The inner core has a relation to the *strictly inhibitive set*, the set of payoff vectors stable against randomized blocking plans (see Myerson [12, Section 9.8] and Qin [13]). For a compactly generated NTU game, its inner core is a subset of the strictly inhibitive set (Qin [13, Theorem 2]). Thus, our Theorem 2 gives a sufficient condition for the nonemptiness of the strictly inhibitive set. Furthermore, if $V(S)$ is convex for every $S \in \mathscr{N}$ and if V satisfies one of the conditions of Proposition 2, then the inner core coincides with the strictly inhibitive set (Qin [13, Theorem 4]).

By Proposition 4, an exchange economy with continuous, concave, and strongly monotone utility functions generates an NTU game satisfying the cardinal balancedness and condition (a) of Proposition 2. Since different economies can generate the same NTU game, exchange economies without the properties in Proposition 4 or production economies may generate NTU games satisfying condition (d) or (e) of Proposition 3. Billera [3] proved that every compactly generated, totally cardinally balanced NTU game V with every $V(S)$ convex can be generated by a production economy where every consumer has a upper semi-continuous and concave utility function on a compact convex consumption set and has his own compact convex production set. Billera's induced production economy can be converted to an exchange economy (see Billera and Bixby [5]). Inoue [9] proved that every compactly generated NTU game can be generated by a coalition production economy. In both induced economies due to Billera [3] and due to Inoue [9], the inner core coincides with the set of utility vectors at Walrasian allocations (see Qin [14] and Inoue [9], respectively). Thus, if an NTU game satisfies all the assumptions of Theorem 2, then there exists a Walrasian equilibrium for both Billera's and Inoue's induced economies.

---

[10]This can be shown by the same method as Billera and Bixby [4, Theorem 2.1].

Acknowledgements The main part of the present paper was written while the author was a member of Institut für Mathematische Wirtschaftsforschung (IMW), Universität Bielefeld. The author is grateful to Sonja Brangewitz, Jan-Philip Gamp, and Walter Trockel for stimulating discussions on this research. The author is also grateful to Jean-Marc Bonnisseau and Tadashi Sekiguchi for helpful comments, as well as participants at the 6th EBIM Workshop held at Université Paris 1 Panthéon-Sorbonne, at the 19th European Workshop on General Equilibrium Theory held at Cracow University of Economics, and at the 2011 RCGEB Workshop on Markets and Games held at Shandong University, and seminar participants at Kyoto University and Hitotsubashi University.

References

1. Arrow KJ, Barankin EW, Blackwell D (1953) Admissible points of convex sets. In: Kuhn HW, Tucker AW (eds) Contributions to the theory of games, vol. 2. Princeton University Press, Princeton, pp 87–91
2. Aubin JP (1973) Equilibrium of a convex cooperative game, MRC Technical Summary Report #1279, Mathematics Research Center, University of Wisconsin-Madison
3. Billera LJ (1974) On games without side payments arising from a general class of market. J Math Econ 1:129–139
4. Billera LJ, Bixby RE (1973) A characterization of polyhedral market games. Int J Game Theory 2:253–261
5. Billera LJ, Bixby RE (1974) Market representation of n-person games. Bull Am Math Soc 80:522–526
6. Brangewitz S, Gamp J-P (2014) Competitive outcomes and the inner core of NTU market games. Econ Theory 57:529–554
7. Debreu G, Schmeidler D (1972) The Radon-Nikodým derivative of a correspondence. In: Le Cam LM, Neyman J, Scott EL (eds) Proceedings of the sixth berkeley symposium on mathematical statistics and probability, vol. 2. University of California Press, Berkeley, pp 41–56
8. de Clippel G, Minelli E (2005) Two remarks on the inner core. Games Econ Behav 50:143–154
9. Inoue T (2013) Representation of non-transferable utility games by coalition production economies. J Math Econ 49:141–149
10. Inoue T (2019) Coincidence theorem and the inner core. Available at SSRN: https://ssrn.com/abstract=1954547
11. Inoue T, Coincidence theorem and the inner core. to appear in Pure Appl Funct Anal
12. Myerson RB (1991) Game theory: analysis of conflict. Harvard University Press, Cambridge
13. Qin C-Z (1993) The inner core and the strictly inhibitive set. J Econ Theory 59:96–106
14. Qin C-Z (1993) A conjecture of Shapley and Shubik on competitive outcomes in the cores of NTU market games. Int J Game Theory 22:335–344
15. Qin C-Z (1994) The inner core of an n-person game. Games Econ Behav 6:431–444
16. Scarf HE (1967) The core of an n person game. Econometrica 35:50–69

Complicated Dynamics and Parametric Restrictions in the Robinson-Solow-Srinivasan (RSS) Model

M. Ali Khan and Tapan Mitra

Abstract The delineation of the optimal policy function (OPF) in the discounted setting has remained an open question since the 2005 demonstration of optimal topological chaos (OTC) in a particular instance of the 2-sector RSS model. This paper provides an explicit solution of the OPF when the discount factor is less than the labor/capital-output ratio a. With OTC conceived both as period-three cycles and turbulence, it establishes the existence of OTC for non-negligible parametric ranges of the model, shows the identified ranges also to be necessary, presents exact restrictions on a, and extends the 1996 Mitra-Nishumura-Yano theorems on discount-factor restrictions.

Keywords Two-sector model · Optimal policy function · Period-three cycles · Turbulence · Optimal topological chaos · Parametric restrictions

Article type: Research Article
Received: August 10, 2019
Revised: September 1, 2019

A preliminary version of this paper was first circulated on October 15, 2009. It was presented at the *Mini Conference on Economic Theory* held at the University of Illinois at Urbana-Champaign, November 5–7, 2010, to honor the retirement of Professor Nicholas Yannelis. It was subsequently presented at Professor Maruyama's seminar at Keio University on January 28, 2011. Tapan Mitra, the Goldwin Smith Professor of Economics at Cornell University, passed away February 3, 2019 after a seven-year illness. During the last two years of his life, he worked on chaotic dynamics in active collaboration with Liuchun Deng of NUS-Yale and the author. The surviving author is submitting this manuscript for publication in its original form (with references updated) not

---

M. Ali Khan (✉)
Department of Economics, The Johns Hopkins University, Baltimore, MD, USA
e-mail: akhan@jhu.edu

T. Mitra
Department of Economics, Cornell University, Ithaca, NY, USA

T. Maruyama (ed.), *Advances in Mathematical Economics*, Advances in Mathematical Economics 23, https://doi.org/10.1007/978-981-15-0713-7_4

> The Golden Ratio's attractiveness seems first and foremost from the fact that it has an almost uncanny way of popping up where it is least expected.[1]
>
> Livio (1970, p. 7)

1 Introduction

It has been well-understood, at least since the early nineties, that solutions to single-agent intertemporal optimization problems can exhibit complicated dynamics in the form of *topological* and *ergodic* chaos precisely defined; see [2] and [23] for anthologies of the pioneering papers. A recent survey [31] of this work delineates how endogenous sources of chaos revolve around a variety of considerations: "upward inertia" as a consequence of zero consumption levels, "downward inertia" as a consequence of depreciating capital, supermodularity of the felicity functions, factor-intensity reversals in a two-sector technology, and high levels of impatience have all been given salience.[2] In [15], optimal topological chaos has been shown in a particularly parsimonious instance of a stripped-down version of the two-sector model, the so-called *borderline* case of the two-sector RSS model, one that involves a specific relationship between only two parameters: ξ, the marginal rate of transformation of capital today into that of tomorrow, given full employment of both factors; and d, the rate of depreciation.[3] The result is executed with linear felicities, a polar form of the factor intensity assumption, and a positive rate of

only because of its continuing analytical interest, but also in the hope that it'll facilitate the understanding and reception of the Deng-Khan-Mitra results as and when they are written up. In this connection, he thanks Professors Mukul Majumdar, Toru Maruyama, Debraj Ray and Santanu Roy for their kind encouragement. He also thanks Mordecai Kurz, Chris Metcalf and Paulo Sousa for correspondence and conversation on the original draft presented at Urbana-Champaign.

JEL Classification: D90, C62, O21
Mathematics Subject Classification (2010): 91B62, 49J45, 49O2

[1]The reader uninitiated into the mysteries of the "golden-ratio" may want to check out [13, pp. 25–27] or [22, pp. 78–86]. A case could be made for singling out [28, 36] as the pioneering applications of the number in economic theory.

[2]A narrative is laid out in [31, Section 6] and it revolves around (1) capital depreciation and linear felicities but with factor intensity reversals in a two-sector model with Cobb-Douglas and Leontief technologies, as in the numerical results in [6], (2) fully circulating capital but with zero consumption levels on the optimal path, as in [33], (3) the inclusion of depreciation with Leontief technologies in both sectors, the so-called Leontief-Shinkai model, and supermodular felicities, as in [34], and with linear felicities, as in [35, 37], (4) non-zero optimal consumption levels in the extension in [40] of [35], and finally, (5) the establishment of ergodic chaos and geometric sensitivity in [39]. Also see the early attempt in [32].

[3]See [3] where the principal result involves eight parameters, and [40], where the result is whittled down to a simpler setting, but still with four parameters. This footnote is an obvious subscription to the simplicity imperative in [24, 45] and others. With Saari [45], it is also a resigned acceptance.

capital depreciation. It is striking on two counts. First, it goes against the continuous-time intuition, rather well-established in the early seventies, that optimal programs, even for a considerably generalized setting of such a model, exhibit saddle-point stability.[4] Second, relative to instances of the more recent literature, it overcomes the disadvantage of zero-consumption levels[5] (in two of the three periods in the period-three cycle established in [33]) without the tagging of somewhat ad-hoc felicity functions, as in [35] and [40].[6]

All this being said, it is important to be clear that the methodological import of the result in [15] lies not so much in the existence of optimal topological chaos (work done more than two decades earlier had already established this), but rather in the fact that complicated dynamics could be shown without any knowledge of the shape of the OPF other than its continuity,[7] and to bring out the power of some sufficient tests for topological chaos, specifically those guaranteeing turbulence in the resulting system. However, as interest deepens in the two-sector RSS model, it is natural to ask whether the OPF can be pinned down for a non-negligible range of parametric values; and if so, what additional light can be thrown on the question of the robustness of optimal topological chaos in the model, and in particular, its demonstration as a consequence of conditions that are easy to check. This question was explicitly left open in [19], where the authors asked for the optimal policy correspondence when the discount factor ρ was less than the inverse of the marginal rate of transformation of capital $(1/\xi)$, and conjectured that "it is possible that ... the graph of the optimal policy correspondence is the lower boundary of the graph of G, which, following the terminology in [8], can be referred to as the "check-map" policy function since its graph resembles the standard check mark.[8] The first, and perhaps primary, contribution of this paper is to answer this question.

---

[4]See [50, 51] and [14] for further genealogical details; also [31, 161–162].

[5]A point of view insisted on in Joan Robinson's response to Stiglitz; see [41] and [51]. In [15], despite a linear utility function, consumption is *never* zero along any optimal path, except one starting at zero capital stock, and then only in the initial period. Put differently, Joan Robinson's criticism does not apply to the OPF reported in this paper, and seems to be purely an artifact of the continuous-time formulation.

[6]In [35], the authors work with the Weitzman-Samuelson reduced-form utility function, and in [40], with a constant elasticity of substitution felicity function. The substantive motivation for either specification is not fully apparent.

[7]For complicated dynamics, see the textbook [7] and the pioneering papers [46, 47]. For the economic literature, see [29, 38]. Also note that Khan and Mitra [15] shares a similarity with [40] as regards this feature of working without a specific analytical form of the OPF.

[8]The concluding remarks in [19] conjecture the shape of the optimal policy correspondence at the threshold discount factor $\rho = (1/\xi)$, to be G, described in their Eq. (23), and seen to be a composite of the pan- and check-maps, and everything in between; also see Figs. 1 and 2 in this paper.

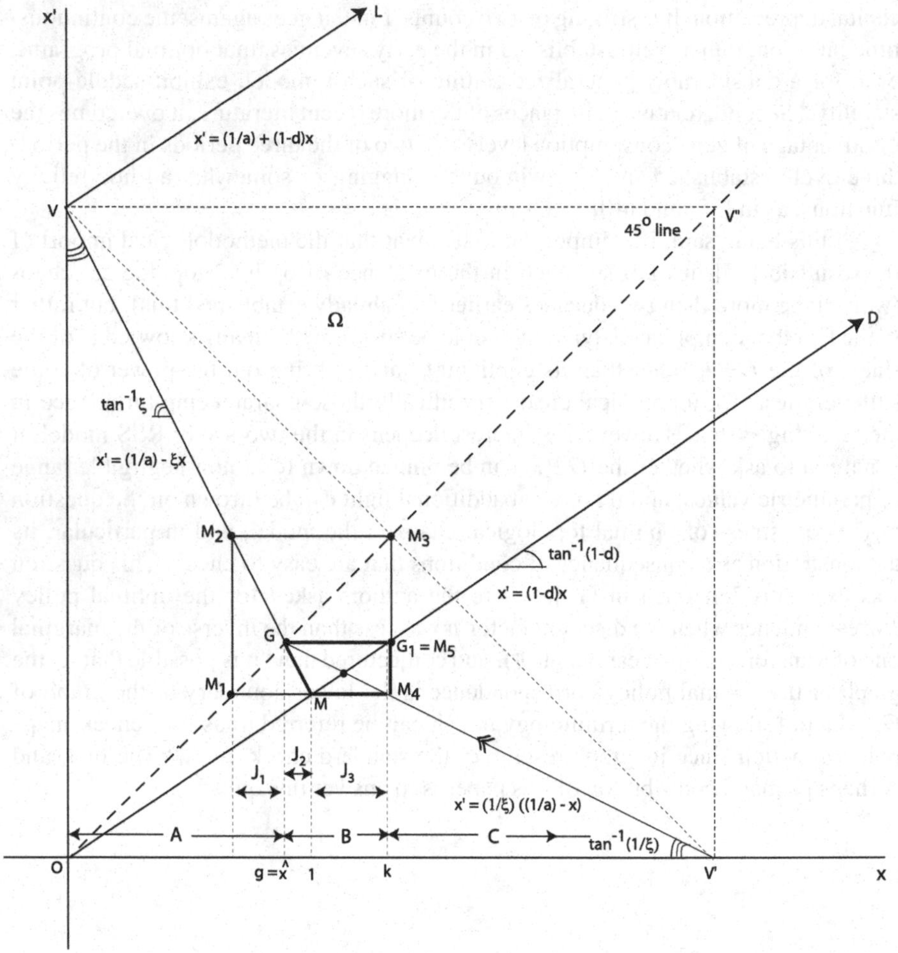

Fig. 1 The 2-sector RSS model in the borderline case $(\xi-(1/\xi))(1-d)=1$ or $a\xi^3 = (\xi^2-1)$ or $(1-d)((d/a)+(1-d)^2)=(1/(1+ad))$

In the first substantive section of the paper, Sect. 3, we show that for all values of the discount factor less than the labor-output ratio[9] a in the investment goods sector,

$$\rho < a = (1/(\xi + 1 - d)) < (1/\xi), \quad \xi > 1, \quad 0 < d < 1, \tag{1}$$

[9]Note that the use of the abbreviation "labor-output ratio" is ambiguous since there are two outputs in the model; in this paper, we shall use it to refer to the labor required to produce a machine, the labor/capital-output ratio, so to speak.

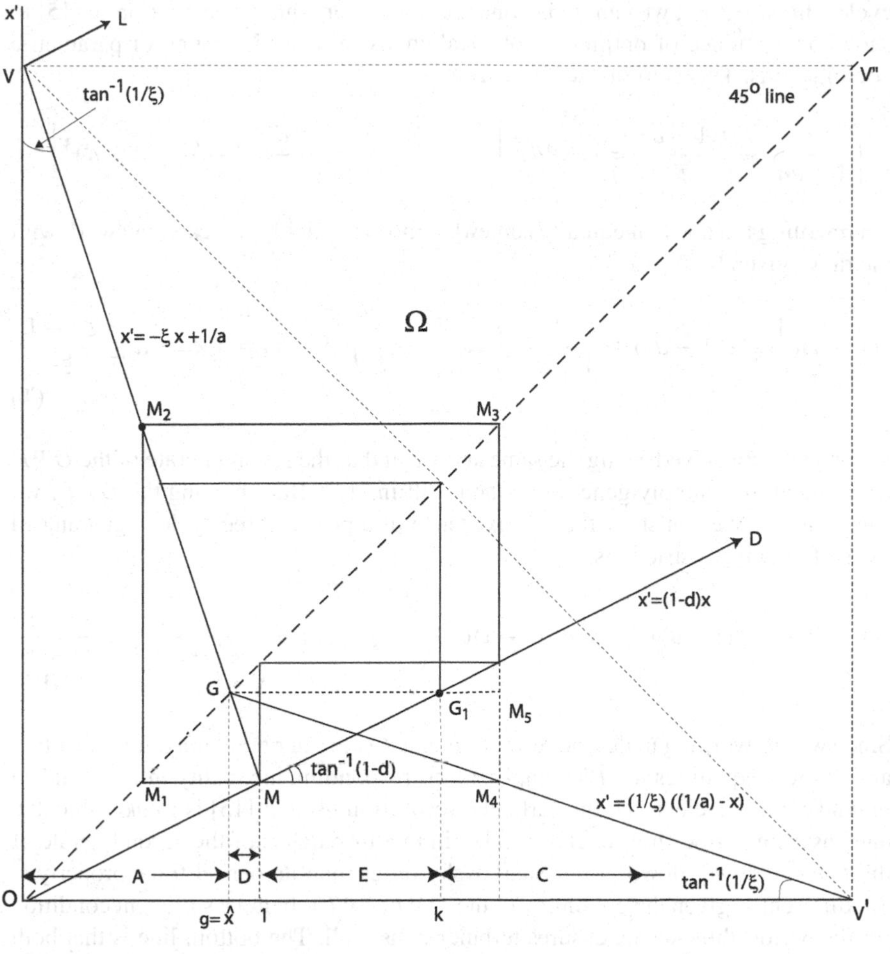

Fig. 2 The 2-sector RSS model in the outside case $(\xi - 1)(1{-}d){=}1$ or $a(\xi^2 - \xi + 1) = (\xi - 1)$ or $(1/a){-}1 = (1{-}d){+}(1{-}d)^{-1}$

the OPF is indeed given by the check-map.[10] And once this information is factored into the equation, we move, in the second substantive section of the paper, beyond the specific instance of the two-sector RSS model analyzed in [15] for turbulence, and give a rather complete treatment of both turbulence and period-three

[10] See [19] and earlier references for the straightforward details of the case when $-1 < \xi \leq 1$. The terminology *check-map* appears in [8, p. 46], but a detailed analysis goes back to [9], and subsequently, in an optimal intertemporal context, to [11]. For a more recent numerical attempt rooted in the RSS setting, but again without giving it an optimality underpinning, see [27]. It is of interest that of the five figures in [31], none concern the check-map, though Figure 6.5 comes closest to it.

cycles. In particular, we can go beyond the parameter equality restriction in [15] to show the existence of optimal topological chaos in an entire range of parametric combinations. This is to say, to move from

$$\left[\frac{1}{1+ad} - \frac{d(1-d)}{a} = (1-d)^3\right] \text{ to } \left[\frac{1}{1+ad} - \frac{d(1-d)}{a} \leq (1-d)^3\right].$$

On rewriting the above inequality somewhat more explicitly, we can now work with the range given by[11]

$$(1-d)(\frac{1}{a} - \xi(1-d)) \geq \frac{1}{1+ad} \Longleftrightarrow \left(\xi - \frac{1}{\xi}\right)(1-d) \geq 1 \Longleftrightarrow a \leq \frac{\xi^2 - 1}{\xi^3} \tag{T}$$

Technically conceived, using the same argument that the second iterate of the OPF is turbulent, this simply generalizes the result in [15]. But knowing the OPF, we can do more. We can show that the existence of a period-three cycle is guaranteed by the following restrictions.

$$(1-d)(\frac{1}{a} - \xi(1-d)) \geq 1 \Longleftrightarrow (\xi - 1)(1-d) \geq 1 \Longleftrightarrow a \leq \frac{\xi - 1}{\xi^2 - \xi + 1} = \frac{\xi^2 - 1}{\xi^3 + 1} \tag{PT}$$

Since we are working under the restriction that ξ is greater than unity, it is clear that any of the inequalities in (PT) imply the corresponding inequality in (T). And so in hindsight, it is clear that the particular restriction used in [15] is *weaker* than the one ensuring period-three cycles; weaker in that turbulence of the second iterate of the OPF is lower down in the Sharkovsky order than its period-three property.[12] Put differently, given the continuity of the OPF, (PT), being a sufficient condition for the period-three cycle, ensures turbulence as well. The bottom line is that both cases (turbulence and the period-three property) are very easy to describe once one has the explicit form of the OPF. These results are presented in Sect. 4.

While these results provide robust parameter configurations for which the second iterate of the OPF is turbulent, or the OPF satisfies the Li-Yorke condition, they

[11]The derivation of these formulae is relegated to the Appendix 8.1. The numbering (T) and (PT) is dictated by the words *turbulence* and *period-three:* as we shall see in the sequel, the specification (T), and guaranteeing turbulence actually places a weaker restriction on the parameters than the specification (PT), guaranteeing and optimal period-three cycle. It may also be worth pointing out that already in 2005, the authors had shown the existence of an optimal program with period-three cycles in another instance of the two-sector RSS model, in which the inequality is replaced by an equality in (PT).

[12]It comes after all the odd-period cycles greater than a single period, but before those of period-six cycles; see [30] for extended discussion. It is also worth emphasizing here that (T) is sufficient and not a necessary condition for turbulence.

also show that concrete restrictions on the parameters of the RSS model appear to be involved in exhibiting such phenomena. The third substantive section of this paper takes this as a point of departure, and turns to what, following the *anything-goes* theorems of Sonnenschein-Mantel-Debreu, has come to be referred to as the *rationalizability* theory.[13] The basic rationale of this theory in terms of the problematic at hand is a simple one: unlike the situation for the Arrow-Debreu-McKenzie (ADM) model that involves the construction of an economy with a *given* excess demand function, here one constructs here a single-agent, Gale-McKenzie (GM) intertemporal optimization model with a policy function identical to a *given* one; see [12, 25]. However, in the context of the GM model, one can ask sharper questions, and furnish sharper answers. In particular, given that the discount factor stands on its own in the GM model, one can focus on it, and ask for necessary and sufficient conditions on its magnitude under which a given policy function (the tent- or logistic map, for example) exhibiting period-three cycles, turbulence of the second iterate of the policy function, or more generally positive topological entropy, can be rationalized. In the literature, *discount-factor restrictions* associated with such phenomena have been obtained in a variety of intertemporal allocation models: these are the so-called "exact MNY discount-factor restrictions" of [28, 36] which involve, in the context of the GM model, the golden number.[14] The three results reported in [48, Theorems 4.4–4.6] summarize the state of the art, though substantial ongoing work continues; see [49] and his references. A key element of the class of intertemporal allocation models studied in this context is that the reduced form utility function exhibits (beyond the standard concavity assumption) some form of strict concavity on its domain. The function might be required to be strictly concave in both arguments, or at least in one of them (that is, either in the initial stock or the terminal stock). The strict concavity requirement plays two roles. It ensures the existence of an optimal policy function. But, in addition, it is seen to be indispensable to the methods used to derive the discount-factor restrictions for complicated optimal behavior.

With the OPC determined for a non-negligible range of discount factors, and with topological chaos in the form of turbulence and period-three cycles robustly identified, this work then prompts two sets of questions in the context of the two-sector RSS model. First, do the existing theorems apply to the restricted setting of the two-sector RSS model? and if not, are there reformulated counterparts of these theorems that can be proved? The first question is easily answered. The point is that even strict concavity of the felicity function does imply strict concavity of the reduced-form utility function of the RSS model, and therefore discount-factor restrictions for complicated optimal behavior in the RSS model, if any, will have to be established by methods different from those employed in the literature. As

---

[13]For references to the Sonnenschein-Mantel-Debreu theorem, and for a recent survey of the available theory that gives references to the pioneering papers, see [48].

[14]See for example [13, 22] for the fascinating career of the golden number; also the footnoted epigraph.

regards the second, one has to take into account the fact that the two-sector RSS model cannot generate the maps (for example, the usual tent maps)that have been used in the derivation of the (MNY) bound, and consider the bounds on the discount factor that arise from the check-map.[15] In particular, we show that $\rho < (1/3)$ implies that there exist (a, d) such that the RSS model (a, d, ρ) has an optimal policy function which generates a period three cycle. Conversely, if the RSS model (a, d, ρ) has an optimal policy function which generates a period three cycle, then $\rho < (1/2)$. Although not *exact* restrictions, it is quite remarkable how closely the restriction on ρ on the sufficiency side compares with that on the necessity side. Further, the restriction of $\rho < (1/2)$ on the necessity side is really a very strong discount-factor restriction for period-three optimal cycles to occur, since it involves a discount rate of a 100%. Thus, "period three implies heavy discounting" turns out to be a robust conclusion, valid for a broad class of intertemporal allocation models, *including* the RSS model. In addition, we offer discount-factor restrictions arising from optimal turbulence. We show that ρ being less than $(\sqrt{\mu})^3$, where μ is the golden-number $(\sqrt{5} - 1)/2$, implies that there exist (a, d) such that the RSS model (a, d, ρ) has an optimal policy function whose second iterate exhibits turbulence. Conversely, if the RSS model (a, d, ρ) has an optimal policy function whose second iterate exhibits turbulence, then $\rho < (1/2)$. These results are new, and even though irreversible investment has been shown to bring in the role of depreciation of capital, as in [35], we have been able to go further by exploiting the special structure of the two-sector RSS model. These results are an important third contribution of the paper, and constitute Sect. 6.

Indeed, given the rather specific technological structure—the toy-nature of the two-sector RSS setting, so to speak—there is the obvious suggestion that it might be possible to exploit this structure to make even further progress on this topic. The fact that the model can be completely summarized by the two parameters (a, d), makes it possible to address the problem of identifying *technological restrictions* involved when the OPF generates, for instance, optimal turbulence or period-three cycles. Because intertemporal allocation models are phrased in terms of a general convex technology set, a similar exercise with respect to technological parameters has not been attempted before, to the best of our knowledge. In any case, the results here are especially satisfying. We provide an exact labor-output ratio (in the investment good sector) restriction for period-three cycles in the two-sector RSS model of optimal growth in the following sense. We show that if the labor-output ratio $a < (1/3)$, then there exist $\rho \in (0, 1)$ and $d \in (0, 1)$, such that the RSS model defined by (ρ, a, d) has an optimal policy function, h, which generates a period-three cycle. Conversely, we show that if there is an RSS model, defined by parameters (ρ, a, d),which has an optimal policy function that generates a period-three cycle, then $a < (1/3)$. It is useful to note in this context that regardless of the value of the depreciation

[15] It is interesting that despite being sidelined by the tent-map and the logistic function in the earliest economic applications, the check-map has been resiliently present from the very inception of the work. Also see Footnote 8 and its references.

factor, d, (and the value of the discount factor ρ) our result indicates that it is *not* possible to have an optimal policy function generating period three cycles when $a \geq (1/3)$. However, more to the point, we also supplement this result by showing that condition (T) above is necessary and sufficient for the optimal policy function of the RSS model to exhibit optimal turbulence. These results are the important fourth contribution of the paper, and constitute Sect. 5.

This extended introduction to the problem and the results obtained has already outlined the paper to a substantial extent. In terms of a summary, after a brief introduction to the two-sector RSS model in Sects. 2 and 3 establishes the OPF under the restriction $\rho < a$, and with this pinned down, Sect. 4 turns to optimal topological chaos formalized in terms of both turbulence and period-three cycles. Subsequent parts of the paper turn to the parametric restrictions: Sect. 5 to *exact* restrictions on the labor-output coefficient a, for optimal period-three cycles and to (T) for optimal turbulence; and Sect. 6 to discount-factor restrictions, again in the context of both optimal turbulence and period-three cycles. Section 7 concludes the paper with open questions and complementary analyses that will be reported elsewhere. The substantial technicalities of the paper lie in the proofs of the necessity results, and their details are relegated to an Appendix so that they may not interfere with the reader primarily interested in the substantive contribution of this work.

2 The Two-Sector RSS Model

A single consumption good is produced by infinitely divisible labor and machines with the further Leontief specification that a unit of labor and a unit of a machine produce a unit of the consumption good. In the investment-goods sector, only labor is required to produce machines, with $a > 0$ units of labor producing a single machine. Machines depreciate at the rate $0 < d < 1$. A constant amount of labor, normalized to unity, is available in each time period $t \in \mathbb{N}$, where $\mathbb{N}$ is the set of non-negative integers. Thus, in the canonical formulation surveyed in [25, 26], the collection of production plans (x, x'), the amount x' of machines in the next period (tomorrow) from the amount x available in the current period (today), is given by the *transition possibility set.* Here it takes the specific form

$$\Omega = \{(x, x') \in \mathbb{R}^2_+ : x' - (1-d)x \geq 0 \text{ and } a(x' - (1-d)x) \leq 1\},$$

where $z \equiv (x' - (1-d)x)$ is the number of machines that are produced, and $z \geq 0$ and $az \leq 1$ respectively formalize constraints on the irreversibility of investment and the use of labor. Associated with Ω is the transition correspondence, $\Gamma : \mathbb{R}_+ \to \mathbb{R}_+$, given by $\Gamma(x) = \{x' \in \mathbb{R}_+ : (x, x') \in \Omega\}$. For any $(x, x') \in \Omega$, one can also consider the amount y of the machines available for the production of the consumption good, leading to a correspondence

$\Lambda : \Omega \longrightarrow \mathbb{R}_+$ with $\Lambda(x, x') = \{y \in \mathbb{R}_+ : 0 \leq y \leq x$ and $y \leq 1 - a(x' - (1-d)x)\}$.

Welfare is derived only from the consumption good and is represented by a linear function, normalized so that y units of the consumption good yields a welfare level y. A *reduced form utility function*

$$u : \Omega \rightarrow \mathbb{R}_+ \text{ with } u(x, x') = \max\{y \in \Lambda(x, x')\}$$

indicates the maximum welfare level that can be obtained today, if one starts with x of machines today, and ends up with x' of machines tomorrow, where $(x, x') \in \Omega$. Intertemporal preferences are represented by the present value of the stream of welfare levels, using a discount factor $\rho \in (0, 1)$.

A *2-sector RSS model* $\mathcal{G}$ consists of a triple (a, d, ρ), and the following concepts apply to it. A *program from* x_o is a sequence $\{x(t), y(t)\}$ such that $x(0) = x_o$, and for all $t \in \mathbb{N}$, $(x(t), x(t+1)) \in \Omega$ and $y(t) = \max \Lambda((x(t), x(t+1))$. A *program* $\{x(t), y(t)\}$ is simply a program from $x(0)$, and associated with it is a *gross investment sequence* $\{z(t+1)\}$, defined by $z(t+1) = (x(t+1) - (1-d)x(t))$ for all $t \in \mathbb{N}$. It is easy to check that every program $\{x(t), y(t)\}$ is bounded by $\max\{x(0), 1/ad\} \equiv M(x(0))$, and so $\sum_{t=0}^{\infty} \rho^t u(x(t), x(t+1)) < \infty$. A program $\{\bar{x}(t), \bar{y}(t)\}$ from x_o is called *optimal* if

$$\sum_{t=0}^{\infty} \rho^t u(x(t), x(t+1)) \leq \sum_{t=0}^{\infty} \rho^t u(\bar{x}(t), \bar{x}(t+1))$$

for every program $\{x(t), y(t)\}$ from x_o. A program $\{x(t), y(t)\}$ is called *stationary* if for all $t \in \mathbb{N}$, we have $(x(t), y(t)) = (x(t+1), y(t+1))$. A *stationary optimal program* is a program that is stationary and optimal.

The parameter $\xi = (1/a) - (1-d)$ plays an important role in the subsequent analysis. It represents the marginal rate of transformation of capital today into that of tomorrow, given full employment of both factors. In what follows, and without further mention, we always assume that the parameters (a, d) of the RSS model are such that

$$\xi > 1 \Longrightarrow a \in (0, 1). \tag{2}$$

For more details, technical and bibliographic, the reader is referred to Khan-Mitra [14] and its further elaboration in [15, 19]. For the basic geometric representation of the model, see Figs. 1 and 2 also detailed in [16, 17] and their references.

2.1 Constructs from Dynamic Programming

Using standard methods of dynamic programming, one can establish that there exists an optimal program from every $x \in X \equiv [0, \infty)$, and then use it to define a *value function*, $V : X \to \mathbb{R}$ by:

$$V(x) = \sum_{t=0}^{\infty} \rho^t u(\bar{x}(t), \bar{x}(t+1)), \tag{3}$$

where $\{\bar{x}(t), \bar{y}(t)\}$ is an optimal program from x. Then, it is straightforward to check that V is concave, non-decreasing and continuous on X. Further, it can be verified that V is, in fact, increasing on X; see [14] for the verification.

It can be shown that for each $x \in X$, the Bellman equation

$$V(x) = \max_{x' \in \Gamma(x)} \{u(x, x') + \rho V(x')\} \tag{4}$$

holds. For each $x \in X$, we denote by $h(x)$ the set of $x' \in \Gamma(x)$ which maximize $\{u(x, x') + \delta V(x')\}$ among all $x' \in \Gamma(x)$. That is, for each $x \in X$,

$$h(x) = \arg \max_{x' \in \Gamma(x)} \{u(x, x') + \rho V(x')\}.$$

Then, a program $\{x(t), y(t)\}$ from $x \in X$ is an optimal program from x if and only if it satisfies the equation

$$V(x(t)) = u(x(t), x(t+1)) + \delta V(x(t+1) \text{ for } t \geq 0;$$

that is, if and only if $x(t+1) \in h(x(t))$ for $t \geq 0$. We call h the *optimal policy correspondence (OPC)*. When this correspondence is a function, we refer to it as the *optimal policy function (OPF)*.

It is easy to verify, using $\rho \in (0, 1)$, that the function V, defined by (3), is the *unique* continuous function on $Z \equiv [0, (1/ad)]$ which satisfies the functional equation of dynamic programming, given by (4).

2.2 The Modified Golden Rule

A *modified golden rule* is a pair $(\hat{x}, \hat{p}) \in \mathbb{R}^2_+$ such that $(\hat{x}, \hat{x}) \in \Omega$ and

$$u(\hat{x}, \hat{x}) + (\rho - 1)\hat{p}\hat{x} \geq u(x, x') + \hat{p}(\rho x' - x) \text{ for all } (x, x') \in \Omega. \tag{MGR}$$

The existence of a modified golden-rule has already been established in [16, 19]. We reproduce that result here (without proof) for ready reference. A distinctive feature of our model is that we can describe the modified golden-rule stock explicitly in terms of the parameters of the model, and that it is independent of the discount factor.

Proposition 1 *Define* $(\hat{x}, \hat{p}) = (1/(1+ad),\ 1/(1+\rho\xi))$. *Then* $(\hat{x}, \hat{x}) \in \Omega$, *where* $\hat{x}$ *is independent of* ρ, *and satisfies (MGR).*

The connection between the value function and the modified golden-rule may be noted as follows. Given a modified golden-rule $(\hat{x}, \hat{p}) \in \mathbb{R}^2_+$, we know that $\hat{x}$ is a stationary optimal stock (see, for example, [25, p. 1305]. Consequently, it is easy to verify that $V(\hat{x}) = \hat{x}/(1-\rho)$ and that

$$V(x) - \hat{p}x \leq V(\hat{x}) - \hat{p}\hat{x} \text{ for all } x \geq 0. \tag{5}$$

On choosing $x = \hat{x}+\varepsilon$, with $\varepsilon > 0$, in (5), and letting $\varepsilon \to 0$, we obtain $V'_+(\hat{x}) \leq \hat{p}$, and hence from (MGR),

$$V'_+(\hat{x}) \leq \hat{p} = 1/(1+\rho\xi) < (a/\rho). \tag{6}$$

2.3 A Failure of Strict Concavity

As emphasized in the introduction, a key element of the class of intertemporal allocation models studied in the literature in the substantive context of this work is that the reduced form utility function u exhibits some form of strict concavity on its domain. As has been well-understood, this assumption fails in the RSS model that we study here. We provide a formal argument for the reader new to the model. Consider $x, \bar{x}$ with $1 < x < \bar{x} < k$, and $(x', \bar{x}') = (1-d)(x, \bar{x})$. Then, $(x, x') \in \Omega$, and $(\bar{x}, \bar{x}') \in \Omega$, and $u(x, x') = 1 = u(\bar{x}, \bar{x}')$. One can now choose $\tilde{x} = \lambda x + (1-\lambda)\bar{x}$ and $\tilde{x}' = \lambda x' + (1-\lambda)\bar{x}'$ with any $\lambda \in (0, 1)$. Then, it is easy to check that $\tilde{x}' = (1-d)\tilde{x}$, and $(\tilde{x}, \tilde{x}') \in \Omega$, and $u(\tilde{x}, \tilde{x}') = 1$. Thus, while u is concave on Ω, as noted above, it is not strictly concave in either the first or the second argument.

2.4 Basic Properties of the OPC

The basic properties of the OPC, with no additional restrictions on the parameters of our model, have already been described in [19]. We summarize these properties below. This helps us to present an explicit solution of the optimal policy correspondence in the next section.

To this end, we describe three regions of the state space; see Fig. 1.

$$A = [0, \hat{x}],\ B = (\hat{x}, k),\ C = [k, \infty)$$

where $k = \hat{x}/(1-d)$. In addition, we define a function, $g : X \to X$, by:

$$g(x) = \begin{cases} (1-d)x & \text{for } x \in C \\ \hat{x} & \text{for } x \in B \\ (1/a) - \xi x & \text{for } x \in A \end{cases}$$

We refer to g as the pan-map, in view of the fact that its graph resembles a pan. In Figs. 1 and 2, it is given by VGG_1D.

We further subdivide the region B into two regions as follows:

$$D = (\hat{x}, 1), \;\; E = [1, k)$$

and define a correspondence, $G : X \to X$, by:

$$G(x) = \begin{cases} \{(1-d)x\} & \text{for } x \in C \\ [(1-d)x, \hat{x}] & \text{for } x \in E \\ [(1/a) - \xi x, \hat{x}] & \text{for } x \in D \\ \{(1/a) - \xi x\} & \text{for } x \in A \end{cases} \tag{7}$$

Proposition 2 *The optimal policy correspondence, h, satisfies:*

$$h(x) \subset \begin{cases} \{g(x)\} & \textit{for all } x \in A \cup C \\ G(x) & \textit{for all } x \in B \end{cases} \tag{8}$$

It should be clear from this result that the only part of the optimal policy correspondence for which we do *not* have an explicit solution is for the middle region of stocks, given by $B = (\hat{x}, k) = D \cup E$; see Fig. 2.

Two useful implications of Proposition 2 are that (i) one must have positive optimal consumption levels in all programs that start from positive capital stocks, and (ii) the slope of the value function cannot exceed unity.

Corollary 1

(i) If $\{x(t), y(t)\}$ *is an optimal program from* $x_o > 0$, *then* $y(0) > 0$.
(ii) If $0 < z' < z < \infty$, *then*

$$\frac{V(z) - V(z')}{z - z'} \leq 1 \tag{9}$$

Proof To see (i), note that for $x_o \in (0, \hat{x}]$, (8) implies that $y(0) = x_o > 0$, while for $x_o \in [k, \infty)$, (8) implies that $y(0) = 1$. For $x_o \in (\hat{x}, k)$, (8) implies that if $x' \in h(x_o)$, then $x' \leq \hat{x}$, and this means that $\hat{x} \in \Lambda(x_o, x')$. Thus $y(0) \geq \hat{x} > 0$.

To see (ii), pick any $x > 0$, and let $\{x(t), y(t)\}$ be an optimal program from $x > 0$. Since $y(0) > 0$ by (i), we can choose $0 < x' < x$, so that $\varepsilon \equiv [x - x'] < y(0)$, and define $\tilde{y}(0) = y(0) - \varepsilon > 0$.

Note that $x(1) \geq (1-d)x(0) \geq (1-d)x'$, and:

$$\begin{aligned}\tilde{y}(0) + a[x(1) - (1-d)x'] &= \tilde{y}(0) + a[x(1) - (1-d)x(0)] + a(1-d)\varepsilon \\ &= \tilde{y}(0) + 1 - y(0) + a(1-d)\varepsilon \\ &= 1 - \varepsilon[1 - a(1-d)] < 1,\end{aligned}$$

so that $(x', x(1)) \in \Omega$ and $\tilde{y}(0) \in \Lambda(x', x(1))$. Since $V(x')$ is at least as large as the sum of discounted utilities generated by the program $(x', x(1), x(2), \ldots)$, we have:

$$V(x) - V(x') \leq [y(0) - \tilde{y}(0)] = \varepsilon = (x - x').$$

This yields the desired bound on the slope of the value function, namely $(V(x) - V(x')/(x - x') \leq 1$. Since V is concave on $\mathbb{R}_+$, (9) follows. ■

3 An Explicit Solution of the OPF

In this section we present an explicit solution of the optimal policy function when the discount factor is smaller than the labor-output ratio in the investment good sector. Specifically, we show that in this case, the map

$$H(x) = \begin{cases} (1/a) - \xi x & \text{for } x \in [0, 1] \\ (1-d)x & \text{for } x \in (1, \infty) \end{cases} \tag{10}$$

is the OPF. We refer to the map H as a check-map.[16]

Proposition 3 *Suppose the RSS model (a, d, ρ) satisfies $\rho < a$. Then, its optimal policy correspondence, h, is the function given by H in (10).*

Proof Using Proposition 2, it is clear that we only need to show that H, given by (10), is the OPF for $x \in (\hat{x}, k)$. To this end, let us define $c : X \to X$ by:

$$c(x) = \begin{cases} x & \text{for } x \in (\hat{x}, 1) \\ 1 & \text{for } x \in [1, k) \end{cases}$$

Note that $H(x) \geq (1-d)x > 0$ and $c(x) > 0$ for all $x \in (\hat{x}, k)$. Also, for $x \in (\hat{x}, 1)$, we have $c(x) = x$, and so

$$c(x) + a[H(x) - (1-d)x] = x + a[(1/a) - \xi x - (1-d)x] = x + 1 - a(1/a)x = 1.$$

[16] See Footnote 8, and the text it footnotes.

Thus, $(x, H(x)) \in \Omega$ and $c(x) \in \Lambda(x, H(x))$. For $x \in [1, k)$, we have $c(x) = 1$, and so

$$c(x) + a[H(x) - (1-d)x] = 1 + a[(1-d)x - (1-d)x] = 1.$$

Thus, we have again $(x, H(x)) \in \Omega$ and $c(x) \in \Lambda(x, H(x))$.

Let $\{x(t), y(t)\}$ be an optimal program from $x > 0$. We establish that $x(1) = H(x)$ for $x \in (\hat{x}, k)$. We know from Proposition 2 that $x(1) \geq H(x)$. So, it remains to rule out $x(1) > H(x)$. To this end we break up the verification into two parts corresponding to the two ranges of x, namely, (i) $x \in (\hat{x}, 1)$ and (ii) $x \in [1, k)$.

We begin with case (i). Suppose $x(1) = H(x) + \varepsilon$, where $\varepsilon > 0$. Note that $y(0) + a[x(1) - (1-d)x] \leq 1 = x + a[g(x) - (1-d)x]$ so that $y(0) \leq x + a[g(x) - x(1)] = x - a\varepsilon$. Using the optimality principle, we obtain

$$\begin{aligned} V(x) = y(0) + \rho V(x(1)) &\leq x - a\varepsilon + \rho[V(x(1)) - V(H(x))] + \rho V(H(x)) \\ &\leq x - a\varepsilon + \rho\varepsilon + \rho V(H(x)) < x + \rho V(H(x)), \quad (11) \end{aligned}$$

the second inequality following from Corollary 1, and the last inequality following from the fact that $\rho < a$ and $\varepsilon > 0$. But, since $(x, H(x)) \in \Omega$ and $c(x) = x \in \Lambda(x, H(x))$, we must have $V(x) \geq x + \rho V(H(x))$, which contradicts (11).

Next we turn to case (ii). Suppose $x(1) = H(x) + \varepsilon$, where $\varepsilon > 0$. Note that

$$\begin{aligned} y(0) + a[x(1) - (1-d)x] &= [y(0) - 1] + 1 + a[H(x) + \varepsilon - (1-d)x] \\ &= [y(0) - 1 + a\varepsilon] + 1, \end{aligned}$$

so that $y(0) \leq 1 - a\varepsilon$. Using the optimality principle,

$$\begin{aligned} V(x) = y(0) + \rho V(x(1)) &\leq 1 - a\varepsilon + \rho[V(x(1)) - V(H(x))] + \rho V(H(x)) \\ &\leq 1 - a\varepsilon + \rho\varepsilon + \rho V(H(x)) < 1 + \rho V(H(x)), \quad (12) \end{aligned}$$

the second inequality following from an analogue of Corollary 1, and the last inequality following from the fact that $\rho < a$ and $\varepsilon > 0$. But, since $(x, H(x)) \in \Omega$ and $c(x) = 1 \in \Lambda(x, H(x))$, we must have $V(x) \geq 1 + \rho V(H(x))$, which contradicts (12). ■

Remark Our sufficient condition ($\rho < a$) for an explicit solution of the OPF as the check map (given by (10)) does not directly involve the depreciation factor, d. In view of this, one should not expect this sufficient condition to be a sharp one, even for the instances delineated in (T) and (PT) above. In particular, it has already been established in [16, 17] that for the case $\xi \leq 1/(1-d)$, the optimal policy function is the check map whenever $\rho < (1/\xi)$. Since $(1/\xi) = (a/(1 - a(1-d)) > a$, this shows that when $\xi \leq 1/(1-d)$, the OPF is the check-map even for $\rho \in (a, (1/\xi))$.

4 Optimal Topological Chaos

With the optimal policy function explicitly determined (albeit in the specific case of the discount factor ρ being less than the labor-output coefficient a), we can provide robust sets of parameter configurations for which (1) the second iterate of the optimal policy function exhibits turbulence, and (2) the optimal policy function satisfies the Li-Yorke condition. The set of parameter configurations for which (1) holds (stated in (T) above) generalizes the result obtained in [15]. The set of parameter configurations for which (2) holds (stated in (PT) above is clearly stronger than (T). Both sets of parameter configurations ensure that the optimal policy function exhibits topological chaos.

We recall a few definitions relating to the concepts appearing in the previous paragraph. Let X be a compact interval of the reals $\mathbb{R}$, and f a continuous function from X to itself. The pair (X, f) is said to be a *dynamical system* with *state space* X and *law of motion* f. A dynamical system (X, f) is said to be *turbulent* if there exist points a, b, c in X such that

$$f(b) = f(a) = a, \ f(c) = b, \text{ and either } a < c < b \text{ or } a > c > b$$

(see Fig. 3). It satisfies the *Li-Yorke condition* if there exists $x^* \in X$ such that

$$f^3(x^*) \leq x^* < f(x^*) < f^2(x^*) \text{ or } f^3(x^*) \geq x^* > f(x^*) > f^2(x^*).$$

The *topological entropy* of a dynamical system (X, f) is denoted by $\psi(X, f)$, and the dynamical system itself is said to exhibit *topological chaos* if its topological entropy is positive.

Proposition 4 *Suppose the RSS model (a, d, ρ) satisfies $\rho < a$, and (T) above. Then, the optimal policy correspondence, h, is the function given by H in (10), and h^2 is turbulent.*

Proof The proof naturally splits up into three parts. The first part involves verifying that

$$H^2(1) \geq k \Longleftrightarrow \left[\xi - \frac{1}{\xi}\right](1 - d) \geq 1$$

where H is the check map, given by (10), and the right hand side of the implication is (T). The second part involves showing that, when (T) is satisfied, f is turbulent, where $f(x) = H^2(x)$ for all $x \in \mathbb{R}_+$. The third part is to observe that when we combine these two parts with Proposition 3 we can conclude that when $\rho < a$, and (T) holds, then the optimal policy correspondence h is a function, given by the check map H, and h^2 is turbulent.

Fig. 3 Turbulence of f(.)

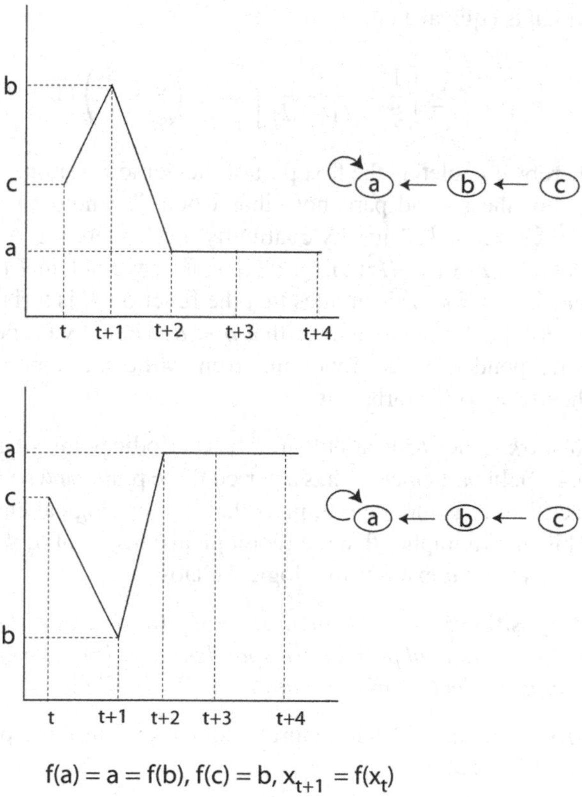

For the first part, let us define the closed intervals (see Fig. 1),

$$J_1 = [1 - d, \hat{x}];\; J_2 = [\hat{x}, 1];\; J_3 = [1, k],$$

and denote $H^2(1)$ by k'. Denote the length of the interval J_2 by θ. Notice that H maps J_2 onto J_1, and the relevant slope for this domain is $(-\xi)$, so that the length of J_1 is $\xi\theta$. Further, H maps J_3 onto J_1, and the relevant slope for this domain is $(1 - d)$, so that the length of $J_3 = \xi\theta/(1 - d)$. Thus, the length of $J_2 \cup J_3 = [\hat{x}, k]$ is $\{\theta + [\xi\theta/(1 - d)]\}$. On the other hand, H maps J_1 onto $[\hat{x}, k']$, and the relevant slope for this domain is $(-\xi)$, so that $k' > \hat{x}$, and $[k' - \hat{x}] = \xi^2\theta$. Thus, we obtain

$$k' \geq k \Longleftrightarrow \xi^2 \geq 1 + \frac{\xi}{(1 - d)}. \tag{13}$$

One can rewrite the right-hand inequality in (13) as

$$1 \geq \frac{1}{\xi^2} + \frac{1}{\xi(1 - d)} = \frac{1}{\xi}\left[\frac{1}{\xi} + \frac{1}{(1 - d)}\right],$$

which is equivalent to

$$\xi \geq \left[\frac{1}{\xi} + \frac{1}{(1-d)}\right] \Longleftrightarrow \left(\xi - \frac{1}{\xi}\right)(1-d) \geq 1 \Longleftrightarrow (T),$$

thereby completing the first part of the demonstration.

For the second part, note that when (T) holds, we have $H^2(1) \geq k$, while $H^2(\hat{x}) = \hat{x} < k$. Thus, by continuity of H, there is $z \in (\hat{x}, 1]$ such that $H^2(z) = k$. Defining $F(x) = H^2(x)$ for all $x \in \mathbb{R}_+$, we obtain $F(z) = k$, $F(k) = \hat{x} = F(\hat{x})$ and $\hat{x} < z < k$. This implies that the function F is turbulent (see [4, p. 25]).

For the third part, assume that $\rho < a$. Then, by Proposition 3, the optimal policy correspondence h is a function, given by the check map H. If in addition (T) holds, then $h^2 = H^2$ is turbulent. ■

Remark When h^2 is turbulent h^2 has periodic points of all periods (see [4, Lemma 3, p. 26]). In particular, h^2 has a period three point, and so h has a period six point. The fact that h^2 is turbulent implies that the topological entropy of h^2 , $\psi(h^2) \geq \ln 2$. This in turn implies that the topological entropy of h, $\psi(h) = (1/2)\psi(h^2) \geq \ln\sqrt{2}$ > 0, so that h exhibits topological chaos.

Proposition 5 *Suppose the RSS model (a, d, ρ) satisfies $\rho < a$, and (PT) above. Then, the optimal policy correspondence, h, is the function given by H in (10), and h satisfies the Li-Yorke condition.*

Proof The proof again naturally splits up into three parts. The first part involves verifying that

$$H^2(1) \geq \frac{1}{1-d} \Longleftrightarrow [\xi - 1](1-d) \geq 1,$$

where H is the check map, given by (10), and the right hand side of the implication is (PT). The second part involves showing that, when (PT) is satisfied, H satisfies the Li-Yorke condition. The third part is to observe that when we combine these two parts with Proposition 3 we can conclude that when $\rho < a$, and (PT) holds, then the optimal policy correspondence h is a function, given by the check map H, and h satisfies the Li-Yorke condition.

For the first part, let us define the closed intervals (mark on Fig. 2),

$$I_1 = [1-d, \hat{x}];\ I_2 = [\hat{x}, 1];\ I_3 = [1, \tilde{k}],$$

where $\tilde{k} = [1/(1-d)] > k$,and denote $H^2(1)$ by k''. Denote the length of the interval I_2 by θ. Notice that H maps I_2 onto I_1, and the relevant slope for this domain is $(-\xi)$, so that the length of I_1 is $\xi\theta$. Further, H maps I_3 onto $[1-d, 1] = I_1 \cup I_2$, and the relevant slope for this domain is $(1-d)$, so that the length of $I_3 = (\xi+1)\theta/(1-d)$. On the other hand, H maps I_1 onto $[\hat{x}, k'']$, and the relevant slope for this domain is $(-\xi)$, so that $k'' > \hat{x}$, and $[k'' - \hat{x}] = \xi^2\theta$. Thus, we obtain

$$k'' \geq \tilde{k} \Longleftrightarrow \xi^2 \geq 1 + \frac{\xi + 1}{(1-d)} \tag{14}$$

One can rewrite the right-hand inequality in (14) as

$$\xi^2 - 1 \geq \frac{\xi + 1}{(1-d)} \Longleftrightarrow (\xi - 1)(1-d) \geq 1 \Longleftrightarrow (PT),$$

thereby completing the first part of our demonstration.

For the second part, note that we have $H(\tilde{k}) = (1-d)\tilde{k} = 1 < \tilde{k}$, and $H^2(\tilde{k}) = H(1) = (1-d) < H(\tilde{k})$. Thus, when (PT) holds, $H^3(\tilde{k}) = H^2(1) \geq [1/(1-d)] = \tilde{k}$, and Thus, we obtain

$$H^3(\tilde{k}) \geq \tilde{k} > H(\tilde{k}) > H^2(\tilde{k}),$$

which clearly means that the Li-Yorke condition is satisfied.

For the third part, assume that $\rho < a$.Then, by Proposition 3, the optimal policy correspondence h is a function, given by the check map H. If in addition (PT) holds, then $h = H$ satisfies the Li-Yorke condition. ■

Remark When h satisfies the Li-Yorke condition, h has periodic points of all periods; see [21]. In particular, h has a period three point. The fact that h has a period three point implies by a result of [5] that the topological entropy of h, $\psi(h) \geq \ln[(\sqrt{5}+1)/2] > 0$, so that h exhibits topological chaos.

5 Technological Restrictions for Optimal Topological Chaos

We present "necessary and sufficient" conditions on technology, the so-called *technological restrictions,* for optimal topological chaos conceived as optimal period-three cycles and as optimal turbulence.

5.1 Technological Restrictions for Optimal Period-Three Cycles

We begin with the sufficiency result.

Proposition 6 *Let* $0 < a < (1/3)$. *Then, there exist* $\rho \in (0, 1)$ *and* $d \in (0, 1)$ *such that the two-sector RSS model with parameters* (ρ, a, d) *has an optimal policy function,* h, *which generates a period-three cycle.*

Proof Given $0 < a < (1/3)$, choose any $\rho \in (0, a)$. Then, since $\rho < a$, the analysis presented in Sect. 3 implies that for every $d \in (0, 1)$, the two-sector RSS model with parameters (ρ, a, d) has an optimal policy function, h, given by the check map, H.

We now proceed to choose $d \in (0, 1)$ such that the optimal policy function h exhibits a period-three cycle. Towards this end, define a function $f : [0, 1] \to \mathbb{R}$ by

$$f(z) = (1 - z)^3 + [z(1 - z)/a],$$

and observe the following:

$$f(0) = 1,\ f(1) = 0 \text{ and } f'(z) = -3(1 - z)^2 + (1/a)(1 - 2z).$$

Since $a \in (0, (1/3))$ guarantees that $f'(0) = -3 + (1/a) > 0$, we have, for z positive and close enough to 0, $f(z) > 1$. Since $f(1) < 1$, we can appeal to the intermediate value theorem, to assert the existence of $d \in (0, 1)$ for which $f(d) = 1$. This means that the RSS model with parameters (ρ, a, d) satisfies:

$$(1 - d)^3 + [d(1 - d)/a] = 1. \tag{15}$$

Since the RSS model (ρ, a, d) has the optimal policy function $h = H$, (15) implies

$$h^2(1) = H^2(1) \equiv (1 - d)^2 + (d/a) = [1/(1 - d)].$$

Since $H[1/(1 - d)] = 1$, we have $h^3(1) = h(h^2(1)) = 1$, and we obtain the period-three cycle

$$h(1) = 1 - d,\ h^2(1) = [1/(1 - d)],\ h^3(1) = 1. \qquad \blacksquare$$

Next, we turn to the necessity result

Proposition 7 *Let (ρ, a, d) be the parameters of a two-sector RSS model such that there is an optimal policy function h which generates a period-three cycle from some initial stock. Then $a < (1/3)$.*

Proof Denote the optimal policy function by h, and the period-three cycle stocks by α, β, γ. Without loss of generality we may suppose that $\alpha < \beta < \gamma$. There are then two possibilities to consider: (1) $\beta = h(\alpha)$, (2) $\gamma = h(\alpha)$.

In case (i), we must have $\alpha \in A$, and $\alpha \neq \hat{x}$ since $\beta > \alpha$. Consequently, $\beta = (1/a) - \xi\alpha$, and $\gamma \neq h(\alpha)$. Thus, we must have $\gamma = h(\beta)$, and since $\gamma > \beta$, we must have $\beta \in A$. But, since $\beta = (1/a) - \xi\alpha$ with $\alpha \in A$, $\alpha \neq \hat{x}$, we must have $\beta \in B \cup C$, a contradiction. Thus, case (i) cannot occur.

Thus case (ii) must occur. In this case, since $\gamma > \alpha$, we must have $\alpha \in A$, and $\alpha \neq \hat{x}$. Consequently, $\gamma = (1/a) - \xi\alpha$, and $\beta \neq h(\alpha)$. Thus, we must have $\beta = h(\gamma)$; it also follows that we must have $\alpha = h(\beta)$. Since $\beta < \gamma$, we must have

$\gamma \in B \cup C$; similarly, since $\alpha < \beta$, we must have $\beta \in B \cup C$. Also, note that β cannot be $\hat{x}$ or k, and similarly γ cannot be $\hat{x}$ or k.

We claim now that $\gamma > k$. For if $\gamma \leq k$, then, we must have $\hat{x} < \beta < \gamma < k$. But, then, $\gamma \in (\hat{x}, k)$, and so $\beta = h(\gamma)$ implies by (8) that $\beta \leq \hat{x}$. Since $\beta \neq \hat{x}$, we must have $\beta \in A$, a contradiction. Thus, the claim that $\gamma > k$ is established. But, then, by (8), we can infer that $\beta = (1-d)\gamma$.

We claim, next, that $\beta \in (\hat{x}, k)$. Since $\beta \in B \cup C$, and β cannot be $\hat{x}$ or k, we must have $\beta > k$ if the claim is false. But, then, by (8), we can infer that $\alpha = (1-d)\beta > \hat{x}$, a contradiction, since $\alpha \in A$. Thus, our claim that $\beta \in (\hat{x}, k)$ is established.

Since $\beta \in (\hat{x}, k)$ and $\alpha = h(\beta)$, we can infer from (8) that $\alpha \geq (1-d)$. To summarize our findings so far, we have:

$$(i)\ \gamma > k > \beta > \hat{x} > \alpha \geq (1-d) \text{ and } (ii)\ \gamma = (1/a) - \xi\alpha, \beta = (1-d)\gamma, \tag{16}$$

from which we can infer that

$$\gamma = (1/a) - \xi\alpha \leq (1/a) - \xi(1-d) \text{ and } \beta = (1-d)\gamma \leq [(1-d)/a] - \xi(1-d)^2.$$

On simplifying the right-hand side of the inequality for β, we obtain the important inequality

$$\beta \leq [d(1-d)/a] + (1-d)^3. \tag{17}$$

Now suppose that contrary to the assertion of the Proposition, we have $a \geq (1/3)$. Then we can appeal to Lemma 1 in the Appendix to conclude that

$$\beta \leq [d(1-d)/a] + (1-d)^3 < 1. \tag{18}$$

Clearly, (18) implies that $\beta \in (\hat{x}, 1)$. Thus, by (8), $\alpha = h(\beta) \geq (1/a) - \xi\beta$, while $\hat{x} = (1/a) - \xi\hat{x}$, so that

$$(\hat{x} - \alpha) \leq \xi(\beta - \hat{x}). \tag{19}$$

Using (16)(ii), we have $(\beta - \hat{x}) = (1-d)(\gamma - k) \leq (1-d)(\gamma - \hat{x})$. Also, using (16)(i), we have $(\gamma - \hat{x}) = \xi(\hat{x} - \alpha)$, so that

$$(\beta - \hat{x}) \leq \xi(1-d)(\hat{x} - \alpha). \tag{20}$$

Combining (19) and (20) yields

$$\xi^2(1-d) \geq 1. \tag{21}$$

We now use (18) in conjunction with (21) to complete the argument. Since $\beta \in (\hat{x}, 1)$, (8) yields $\alpha = h(\beta) \geq [(1/a) - \xi\beta]$ and

$$\gamma = (1/a) - \xi\alpha \Longrightarrow \gamma \leq (1-\xi)(1/a) + \xi^2\beta.$$

Finally, since $\beta = (1-d)\gamma$, the above inequality for γ implies

$$\beta \leq (1-d)(1-\xi)(1/a) + (1-d)\xi^2\beta \Longrightarrow (\xi^2(1-d)-1)\beta \geq (\xi-1)(1-d)(1/a).$$

By (1) and the specification of the two-sector RSS model, the left hand side cannot be zero. And so by (21), it is positive. On appealing to the identities presented as Lemmas 2 and 3, we conclude that $\beta \geq 1$, and contradict (18), establishing the Proposition. ■

5.2 *Technological Restrictions for Optimal Turbulence*

We look at technological restrictions on the RSS model when complicated behavior takes the form of the second iterate of the optimal policy function exhibiting turbulence.

Proposition 8 *(i) Let (a, d) be such that (T) holds. Then, there exist $\rho \in (0, 1)$ such that the two-sector RSS model with parameters (ρ, a, d) has an optimal policy function, h, whose second iterate exhibits turbulence. (ii) Let (ρ, a, d) be the parameters of a two-sector RSS model such that there is an optimal policy function h whose second iterate exhibits turbulence. Then (T) holds.*

Proof We provide a proof of part (i) of the proposition and relegate the proof of part (ii) to the Appendix. Towards this end, given (a, d) satisfying (T), choose any $\rho \in (0, a)$. Then, since $\rho < a$, the analysis presented in Sect. 3 implies that the two-sector RSS model with parameters (ρ, a, d) has an optimal policy function, h, given by the check map, H. Then, by Proposition 4, h^2 exhibits turbulence. ■

6 Discount-Factor Restrictions for Optimal Topological Chaos

We present "necessary and sufficient" conditions on the discount factor, the so-called *discount-factor restrictions*, for optimal topological chaos conceived as optimal period-three cycles and as optimal turbulence. These conditions are not exact in the sense that they are for the technological restrictions presented in Sect. 5.

6.1 Discount-Factor Restrictions for Optimal Period-Three Cycles

As in Sect. 5, we begin with period-three cycles and then turn to turbulence.

Proposition 9 *(i) Let* $0 < \rho < (1/3)$. *Then, there exist* $a \in (0, 1)$ *and* $d \in (0, 1)$ *such that the two-sector RSS model with parameters* (ρ, a, d) *has an optimal policy function,* h, *which generates a period-three cycle. (ii) Let* (ρ, a, d) *be the parameters of a two-sector RSS model such that there is an optimal policy function* h *which generates a period-three cycle from some initial stock. Then* $\rho < (1/2)$.

Proof We begin with the proof of (i). Given $\rho \in (0, 1/3)$, pick $a \in (\rho, 1/3)$, and choose $d \in (0, 1)$ such that condition (PT) is satisfied. Towards this end, define a function $f : [0, 1] \to \mathbb{R}$ by:

$$f(z) = (1 - z)^3 + [z(1 - z)/a],$$

and observe the following

$$f(0) = 1, \;\; f(1) = 0 \text{ and } f'(z) = -3(1 - z)^2 + (1/a)(1 - 2z).$$

Since $a \in (0, (1/3))$ guarantees that $f'(0) = -3 + (1/a) > 0$, we have, for z positive and close enough to 0, $f(z) > 1$. Since $f(1) < 1$, we can appeal to the intermediate value theorem, to assert the existence of $d \in (0, 1)$ for which $f(d) = 1$. This means that the RSS model with parameters (a, d) satisfies

$$(1 - d)^3 + [d(1 - d)/a] = 1 \tag{22}$$

Using (22), we obtain:

$$(d/a) + (1 - d)^2 = H^2(1) = [1/(1 - d)]$$

so that by the equivalence in the proof of Proposition 5, we obtain $(\xi - 1)(1 - d) = 1$, and ensure that (PT) is satisfied. Since $\rho < a$, the analysis presented in Sect. 3 implies that the two-sector RSS model with parameters (ρ, a, d) has an optimal policy function, h, given by the check map, H. Then, by Proposition 5, h satisfies the Li-Yorke Condition, and therefore h has a period-three cycle.

Next we turn to the proof of (ii). Towards this end, we claim that $\rho\xi \leq 1$. Suppose to the contrary, we have $\rho\xi > 1$. Then, the RSS model (a, d, ρ) has an optimal policy function, h, given by the pan map. Since the OPF generates a period-three cycle, let us denote the cycle by α, β, γ, and without loss of generality suppose that $\alpha < \beta < \gamma$. Clearly, none of these values can be equal to $\hat{x}$.

We have either (a) $h(\alpha) = \beta$, or (b) $h(\alpha) = \gamma$. In case (a), noting that $\beta > \alpha$, we must have $\alpha \in A$. In this case, since $\alpha \neq \hat{x}$, $\beta \in B \cup C$. Since $h(\alpha) = \beta$, we must have $h(\beta) = \gamma$; and, since $\gamma > \beta$, we must have $\beta \in A$, a contradiction.

Thus, (a) cannot occur. In case (b), we must have $h(\gamma) = \beta$ and so $h(\beta) = \alpha$. Since $h(\alpha) = \gamma$ and $\gamma > \alpha$, we must have $\alpha \in A$, and since $\alpha \neq \hat{x}$, $\gamma \in B \cup C$. But since h is a pan map, we must have $h(\gamma) \geq \hat{x}$; that is $\beta \geq \hat{x}$. Thus, since $\beta \neq \hat{x}$, $\beta \in B \cup C$ as well, and $h(\beta) \geq \hat{x}$; that is, $\alpha \geq \hat{x}$. Since $\alpha \in A$, we must then have $\alpha = \hat{x}$, a contradiction. Thus, case (b) cannot occur, and this establishes our claim.

Proposition 7 guarantees that (PT) holds, thereby implying

$$(\xi - 1)(1 - d) \geq 1 \Longleftrightarrow \xi \geq 1 + \frac{1}{(1 - d)} > 2,$$

and furnishing the required conclusion that $\rho < (1/2)$. ■

6.2 Discount-Factor Restrictions for Optimal Turbulence

We show that if $\rho < \mu^{3/2}$, then there exist (a, d), such that the RSS model (a, d, ρ) has an optimal policy function whose second iterate exhibits turbulence. Conversely, if the RSS model (a, d, ρ) has an optimal policy function whose second iterate exhibits turbulence, then $\rho < \mu$. Here μ is given by:

$$\mu = \frac{\sqrt{5} - 1}{2}$$

Proposition 10 *(i) Let* $0 < \rho < \mu^{3/2}$. *Then, there exist* $a \in (0, 1)$ *and* $d \in (0, 1)$ *such that the two-sector RSS model with parameters* (ρ, a, d) *has an optimal policy function,* h*, whose second iterate exhibits turbulence. (ii) Let* (ρ, a, d) *be the parameters of a two-sector RSS model such that there is an optimal policy function* h *whose second iterate exhibits turbulence. Then,* $\rho < \mu$.

Proof We begin with the proof of (i). Towards this end, consider the quadratic equation

$$g(x) \equiv x^2 - x - 1 = 0,$$

with its two roots given by $x = (1 \pm \sqrt{5})/2$, and the positive root denoted by R. Since $g(0) = -1,\ g(R) = 0$, we obtain for all $x > R$, $x^2 - 1 > x$ which implies $(x - (1/x)) > 1$. Hence for any $\xi > R$, we can find $d(\xi) \in (0, 1)$, and subsequently $a(\xi) \in (0, 1)$ such that

$$(\xi - \frac{1}{\xi})(1 - d(\xi)) = 1 \text{ and } \frac{1}{a(\xi)} = \xi + (1 - d(\xi)). \tag{23}$$

We now simplify the notation by setting $(a(\xi), d(\xi)) \equiv (a, d)$, and obtain

$$(i)\ \xi = (1/a) - (1-d) \text{ and } (ii)\ (\xi - \frac{1}{\xi})(1-d) = 1.$$

and, furthermore, $\xi > R$.

Since $g(R) = 0$, we obtain $R(R^2 - R - 1) = 0$, and hence $R^3 = R^2 + R = R(R+1)$. Since $\mu R = [(\sqrt{5}-1)/2][(\sqrt{5}+1)/2] = 1$, we obtain

$$\mu^{3/2} = \sqrt{\mu^3} = \sqrt{\frac{1}{R^3}} = \sqrt{\frac{1}{R(R+1)}}.$$

Now for $\xi > R$, as $\xi \to R$, we have by (23) that $d(\xi) \to 0$, and $a(\xi) \to 1/(1+R)$. Thus, given $0 < \rho < \mu^{3/2}$, we can choose $\xi > R$ with ξ sufficiently close to R, so that

$$\rho < \sqrt{\frac{a(\xi)}{\xi}}$$

Then, for the economy $(a, d, \rho) \equiv (a(\xi), d(\xi), \rho)$, we have by the above construction and Lemma 1 in the Appendix that the optimal policy function, h, coincides with the check map H. Further, by the Proposition in Sect. 5, h^2 exhibits turbulence.

Next, we turn to the proof of (ii). Towards this end, we claim that $\rho\xi \leq 1$. Suppose to the contrary that $\rho\xi > 1$. Then, the RSS model (a, d, ρ) has an optimal policy function, h, given by the pan map. Since the second iterate of the OPF exhibits turbulence, there exist $a, b, c \in X$ such that

$$h^2(b) = h^2(a) = a,\ h^2(c) = b, \text{ and either (I) } a < c < b \text{ or (II) } a > c > b.$$

Consider the possibility (I). Either we have (a) $a \leq \hat{x}$, or (b) $a > \hat{x}$. If (a) holds, then $h(a) \geq \hat{x}$, and since h is the pan map, $h^2(a) \geq \hat{x}$. But this means $a = h^2(a) \geq \hat{x}$. Thus $a = \hat{x}$, and $h^2(b) = a = \hat{x}$. Since $b > a = \hat{x}$, we have $h(b) \geq \hat{x}$. Since $h(h(b)) = \hat{x}$, we must have $\hat{x} \leq h(b) \leq k$. Since $\hat{x} = a < c < b$, and h is the pan map, $\hat{x} \leq h(c) \leq h(b)$. Thus $h^2(c) = h(h(c)) = \hat{x}$ also. But this contradicts the fact that $h^2(c) = b > a = \hat{x}$. In case (b), we have $b > c > a > \hat{x}$. Then, since h is the pan map, and $b > c$, we have $a = h^2(b) \geq h^2(c) = b$, a contradiction.

A similar argument establishes a contradiction when possibility (II) occurs, and thereby establishes the claim.

Proposition 8(ii) ensures that (T) is satisfied. Define $F : \mathbb{R} \to \mathbb{R}$ by $F(x) = x^2 - mx - 1$, where $m = [1/(1-d)]$. Clearly $F(0) = -1$, and $F(x) \to \infty$ as $x \to \pm\infty$. Thus, $F(x) = 0$ has a negative root and a positive root. The unique positive root of $F(x) = 0$ is given by $x' = (\sqrt{m^2+4}+m)/2$, and thus by continuity of F, we have $F(x) \geq 0 \Leftrightarrow x \geq x'$. But (T) implies that $F(\xi) \geq 0$, and so, using the fact that $x \geq x'$, we obtain $\xi \geq x' > (\sqrt{5}+1)/2$. Given the claim $\rho\xi \leq 1$, this yields,

$$\rho \leq (1/\xi) < \frac{\sqrt{5}-1}{2} = \mu. \qquad \blacksquare$$

7 Concluding Remarks

We end the paper with two concluding remarks. First, in focusing on the magnitude of the labor-output ratio, a, in the investment good sector (a key technological parameter), our exercise might be seen as neglecting the role of other technological parameters of the two-sector RSS model: the marginal rate of transformation ξ, and the rate of depreciation d. This is certainly the case, and a similar exercise focusing on complementary restrictions in force for the other two parameters, and especially on the depreciation factor, would be extremely valuable. We hope to turn to it in future work. Second, our focus has been exclusively on topological chaos represented by period-three cycles and turbulence, and it would be interesting to consider other representations such as potential necessary and sufficient conditions for optimal period-six cycles, for example, and then to go beyond them to the consideration of ergodic chaos. The point is that these parametric restrictions are important in that they give precise quantitative magnitudes when turnpike theorems and those relating to asymptotic convergence do *not* hold; see [1, 26] for such theorems in both the deterministic and stochastic settings. More generally, as argued in [42, 43], and earlier in [8, 9], these questions have relevance for macroeconomic dynamics, and we hope to turn to them in future work.

Appendix

This appendix collects a medley of results with the principal motivation that they do not interrupt and hinder a substantive reading of the results reported in the text above. The technical difficulty of the results resides principally in what could be referred to as the "necessity theory," which is to say, the proofs of Propositions 7 and 8(ii). The argument for the former can be furnished in a fairly straightforward way if some basic identities, routine but important, are taken out of the way. These identities are gathered here as Lemmas 1–3. The proof of Proposition 8(ii) is long and involved, with a determined verification of a variety of cases. This verification draws on results on the OPF that have not been reported before: (1) a monotone property, and (2) a straight-down-the-turnpike property. These are presented as Lemmas 4–8. The proof also draws on an unpublished result on the OPF that we reproduce for the reader's convenience: this reported as Lemma 9.

Lemmata for the Proof of Proposition 7

We state with their proofs the three lemmata utilized in the proof of Proposition 7.

Lemma 1 *If* $a \geq 1/3$, *then* $[d(1-d)/a] + (1-d)^3 < 1$.

Proof Define $f : [0, 1] \to \mathbb{R}$, $f(z) = [z(1-z)/a] + (1-z)^3$, and note that Taylor's expansion for f around $z \in (0, 1)$, for some $\bar{z} \in [0, z]$, is given by

$$f(z) = f(0) + f'(0)z + (1/2)f''(0)z^2 + (1/6)f'''(\bar{z})z^3.$$

Observe that $f(0) = 1$, $f(1) = 0$, and that $f'(z) = -3(1-z)^2 + (1/a)(1-2z)$, $f''(z) = 6(1-z) - (2/a)$, $f'''(z) = -6$. On factoring the information $f'(0) = -3 + (1/a)$ and $f''(0) = 6 - (2/a)$, into Taylor's expansion yields

$$f(z) = 1 + [-3 + (1/a)]z + (1/2)[6 - (2/a)]z^2 + (1/6)(-6)z^3, \tag{24}$$

and furnishes for all $z \in (0, 1)$,

$$[-3+(1/a)]z+(1/2)[6-(2/a)]z^2 = [-3+(1/a)](z-z^2) = [-3+(1/a)]z(1-z) \leq 0.$$

Using this information in (24), we obtain

$$f(z) \leq 1 - z^3 < 1 \text{ for all } z \in (0, 1),$$

which completes the proof. ■

Next we turn to two useful identities.

Lemma 2 $[d(1-d)/a] + (1-d)^3 = (1-d)[\xi - (\xi - 1)(1-d)]$.

Proof The right hand side equals

$$(1-d)[\xi - (\xi - 1) + d(\xi - 1)] = (1-d)[1 + d(\xi - 1)] = (1-d)[1 + d(\frac{1}{a} - (1-d) - 1]$$
$$= (1-d)[\frac{d}{a} + 1 - d(1-d) - d]$$
$$= (1-d)[\frac{d}{a} + (1-d)^2] = [d(1-d)/a] + (1-d)^3$$ ■

Lemma 3 $(\xi - 1)(1-d)/a = (\xi^2(1-d) - 1) + 1 - ([d(1-d)/a] + (1-d)^3$.

Proof The left hand side equals

$$(\xi - 1)(1-d)(\xi + (1-d)) = (\xi - 1)(1-d)^2 + \xi(\xi - 1)(1-d)$$
$$= (\xi - 1)(1-d)^2 + \xi^2(1-d) - \xi(1-d)$$

$$= [\xi^2(1-d)-1]+1-(1-d)[\xi-(\xi-1)(1-d)]$$
$$= [\xi^2(1-d)-1]+1-([d(1-d)/a]+(1-d)^3),$$

the last equality following from Lemma 2. ■

Lemmata for the Proof of Proposition 8(ii)

A Monotone Property

We present a monotone property of the OPF as a consequence of argumentation in [10], and first appealed to in the context of the two-sector RSS model in [18].

Lemma 4 *For $x \in E$, the optimal policy function is monotone non-decreasing.*

Proof Let $x(0)$ and $x'(0)$ belong to E, with $x'(0) > x(0)$. Let $\{x(t)\}$ be the optimal program from $x(0)$ and let $\{x'(t)\}$ be the optimal program from $x'(0)$. Denote $h(x(0))$ by $x(1)$ and $h(x'(0))$ by $x'(1)$. We want to show that $x'(1) \geq x(1)$. Suppose, on the contrary,

$$x'(1) < x(1) \tag{25}$$

Following [10], we now construct two alternative programs. The first goes from $x(0)$ to $x'(1)$ and then follows the optimal program from $x'(1)$; the second goes from $x'(0)$ to $x(1)$ and then follows the optimal program from $x(1)$. A crucial aspect of this technique in the current context (given the various production constraints) is that one be able to go from $x(0)$ to $x'(1)$, and from $x'(0)$ to $x(1)$. That is, one needs to show that $(x(0), x'(1)) \in \Omega$ and $(x'(0), x(1)) \in \Omega$.

We first check that $(x(0), x'(1)) \in \Omega$. Note that the irreversibility constraint is satisfied, since $x'(1) \geq (1-d)x'(0) > (1-d)x(0)$. Further, using (25), we have

$$a[x'(1)-(1-d)x(0)] < a[x(1)-(1-d)x(0)] \leq 1$$

so that the labor constraint is satisfied if

$$\bar{y} = 1-a[x'(1)-(1-d)x(0)] > 1-a[x(1)-(1-d)x(0)] = y(0) \geq 0$$

is the amount of labor devoted to the production of the consumption good. Finally, the capital constraint is satisfied, since $\bar{y} \leq 1 \leq x(0)$, since $x(0) \in E$, the set E as in Fig. 2.

Next, we check that $(x'(0), x(1)) \in \Omega$. Note that the irreversibility constraint is satisfied, since (by using (25)), we have $x(1) > x'(1) \geq (1-d)x'(0)$. Further, since $x'(0) > x(0)$,

$$a[x(1)-(1-d)x'(0)] < a[x(1)-(1-d)x(0)] \leq 1$$

so that the labor constraint is satisfied if

$$\tilde{y} = 1 - a[x(1) - (1-d)x'(0)] > 1 - a[x(1) - (1-d)x(0)] = y'(0) \geq 0$$

is the amount of labor devoted to the production of the consumption good. Finally, the capital constraint is satisfied, since

$$\tilde{y} = 1 - ax(1) + a(1-d)x'(0) < 1 - ax'(1) + a(1-d)x'(0) = y'(0) \leq x'(0),$$

the strict inequality following from (25).

We can now present a self-contained exposition utilizing the techniques of [10]. First, from the definition of the OPF, we obtain

$$V(x(0)) = y(0) + \rho V(x(1)) \text{ and } V(x'(0)) = y'(0) + \rho V(x'(1)). \tag{26}$$

Second, by the principle of optimality, we have

$$V(x(0)) \geq \bar{y} + \rho V(x'(1)) \text{ and } V(x'(0)) \geq \tilde{y} + \rho V(x(1)).$$

In fact, if $V(x(0)) = \bar{y} + \rho V(x'(1))$, then $(x(0), x'(1), x'(2), \ldots)$ would be an optimal program from $x(0)$. Since (25) holds, and $(x(0), x(1), x(2), \ldots)$ is an optimal program from $x(0)$, this would contradict the fact that an optimal policy function exists. Thus, we must have $V(x(0)) > \bar{y} + \rho V(x'(1))$. For similar reasons, $V(x'(0)) > \tilde{y} + \rho V(x(1))$. This is to say

$$V(x(0)) > \bar{y} + \rho V(x'(1)) \text{ and } V(x'(0)) > \tilde{y} + \rho V(x(1)). \tag{27}$$

Clearly, (26) and (27) yield the inequality $y(0) + y'(0) > \bar{y} + \tilde{y}$. However, note that

$$\bar{y} + \tilde{y} = 1 - a[x'(1) - (1-d)x(0)] + 1 - a[x(1) - (1-d)x'(0)] > y(0) + y'(0),$$

which furnishes a contradiction, and establishes the claim. ■

It will be noted that the only place we make use of the fact that $x(0) \in E$ is in checking that $\bar{y} \leq x(0)$. Thus, if one can verify that this inequality holds, the optimal policy function can be shown to be monotone non-decreasing on an extended domain. The next result exploits this idea, and establishes a "local" monotonicity property of the OPF.

Lemma 5 *Suppose* $x^* \in B$, *and* $h(x^*) > (1/a) - \xi x^*$, *then there is* $\varepsilon > 0$, *such that* $N(\varepsilon) \equiv (x^* - \varepsilon, x^* + \varepsilon) \subset B$, *and the optimal policy function is monotone non-decreasing on* $N(\varepsilon)$.

Proof Denote $\{1 - a[h(x^*) - (1 - d)x^*]\}$ by y^*. Then, we have

$$\begin{aligned} y^* &= 1 - ah(x^*) + a(1 - d)x^* < 1 - a[(1/a) - \xi x^*] + a(1 - d)x^* \\ &= a\xi x^* + a(1 - d)x^* = a[(1/a) - (1 - d)]x^* + a(1 - d)x^* = x^* \end{aligned}$$

Denote $(x^* - y^*)$ by μ. Then, $\mu > 0$, and by continuity of h, we can find $\varepsilon > 0$, such that $N(\varepsilon) \equiv (x^* - \varepsilon, x^* + \varepsilon) \subset B$, and $[1 + a(1 - d)]\varepsilon \leq (\mu/2)$, and $|h(x) - h(x^*)| < (\mu/2a)$ for all $x \in N(\varepsilon)$.

Now, let $x(0)$ and $x'(0)$ belong to $N(\varepsilon)$, with $x'(0) > x(0)$. We have to show that $x'(1) \equiv h(x'(0)) \geq h(x(0)) \equiv x(1)$. Suppose, on the contrary, that $x'(1) < x(1)$. Define $\bar{y}$ and $\tilde{y}$ as in the proof of Lemma 4. Then, one can arrive at a contradiction by following exactly the proof of Lemma 4 if one can show that $\bar{y} \leq x(0)$. Towards this end, note that

$$\begin{aligned} \bar{y} &= 1 - ax'(1) + a(1 - d)x(0) \leq 1 - ah(x^*) + (\mu/2) + a(1 - d)x^* + a(1 - d)\varepsilon \\ &= y^* + (\mu/2) + a(1 - d)\varepsilon = x^* - (\mu/2) + a(1 - d)\varepsilon \\ &= x^* - \varepsilon + [1 + a(1 - d)]\varepsilon - (\mu/2) \leq x^* - \varepsilon < x(0). \end{aligned}$$

This completes the proof of the Lemma. ■

As an application of the monotonicity property, we can say a bit more about the nature of the optimal policy function on the domain $(\hat{x}, 1]$.

Lemma 6 *Suppose there is some $\tilde{x} \in (\hat{x}, 1]$ such that $h(\tilde{x}) = H(\tilde{x})$. Then, $h(x) = H(x)$ for all $x \in [\hat{x}, \tilde{x}]$.*

Proof If not, there is some $x' \in [\hat{x}, \tilde{x}]$ such that $h(x') > H(x')$. Let $x'' = \inf\{x \in [x', \tilde{x}] : h(x) = H(x)\}$. Since $h(\tilde{x}) = H(\tilde{x})$, this is well defined, and by continuity of h and H, we have $x'' > x'$, $h(x'') = H(x'')$ and $h(x) > H(x)$ for all $x \in (x', x'')$. Then by Proposition 2, we have $D_+h(x) \geq 0$ for all $x \in (x', x'')$. Thus, using the continuity of h, we have $h(x'') \geq h(x')$, see [44, Proposition 2, page 99]. But since $H(x'') = h(x'')$ and $h(x') > H(x')$, this implies that $H(x'') > H(x')$, which contradicts the fact that H is decreasing on $[\hat{x}, 1]$. ■

We can collect together the above findings as the following result.

Lemma 7 *Let $\tilde{x} \in (\hat{x}, 1)$. Then, exactly one of the following alternatives holds: (1) $h(\tilde{x}) = H(\tilde{x})$, (2) $h(x) \geq h(\tilde{x})$ for all $x \in [\tilde{x}, k]$*

Proof If (i) does not hold, then $h(\tilde{x}) > H(\tilde{x})$. We claim first that $h(x) > H(x)$ for all $x \in [\tilde{x}, 1]$. If not, there is some $x' \in (\tilde{x}, 1]$ such that $h(x') = H(x')$. But, then by Lemma 6, we must have $h(\tilde{x}) = H(\tilde{x})$, since $\tilde{x} \in (\hat{x}, x')$, a contradiction.

Next, we turn to (ii). Using Lemma 5, we have $D_+h(x) \geq 0$ for all $x \in (\tilde{x}, 1)$. Using the continuity of h, we have $h(x) \geq h(\tilde{x})$ for all $x \in [\tilde{x}, 1]$. For $x \in E = [1, k)$, using Lemma 4, we have $h(x) \geq h(1)$, and since $h(1) \geq h(\tilde{x})$, we must have $h(x) \geq h(\tilde{x})$. This establishes (ii) by continuity of h. ■

A Straight-Down-the-Turnpike Property

Lemma 8 *If $\hat{x} \in h(x)$ for some $x \in (\hat{x}, k)$, then $\hat{x} \in h(x)$ for all $x \in (\hat{x}, k)$.*

Proof Let $x' \in (\hat{x}, k)$ be given such that $\hat{x} \in h(x')$. Then, there is $\varepsilon > 0$, such that for all $z \in I \equiv (\hat{x} - \varepsilon, \hat{x} + \varepsilon)$, we have $(x', z) \in \Omega$ and $\{1 - a[z - (1-d)x']\} < x'$, so that

$$u(x', z) = 1 - a[z - (1-d)x'].$$

Define $F(x') = \{z : (x', z) \in \Omega\}$, and for $z \in F(x')$, define:

$$W(z) = u(x', z) + \rho V(z)$$

For $z \in I$, we have

$$W(z) = 1 - az + a(1-d)x' + \rho V(z).$$

Since $\hat{x} \in I$, we obtain

$$W'_{-}(\hat{x}) = -a + \rho V'_{-}(\hat{x}) \tag{28}$$

For $z \in I$, with $z < \hat{x}$, we must have:

$$W(z) = u(x', z) + \rho V(z) \leq V(x') = W(\hat{x})$$

the second equality following from the fact that $\hat{x} \in h(x')$. Thus, we have the first-order necessary condition $W'_{-}(\hat{x}) \geq 0$. Using this in (28), we obtain

$$V'_{-}(\hat{x}) \geq (a/\rho). \tag{29}$$

Next, let $x \in (\hat{x}, k)$ be given. Then we have $\hat{x} = (1-d)[\hat{x}/(1-d)] > (1-d)x$, and $a[\hat{x} - (1-d)x] < a[\hat{x} - (1-d)\hat{x}] = ad\hat{x} = ad/(1+ad) < 1$. Further, we have

$$\begin{aligned} 1 - a[\hat{x} - (1-d)x] &= 1 - a[\hat{x} - (1-d)\hat{x}] + a(1-d)(x - \hat{x}) \\ &= \hat{x} + a(1-d)(x - \hat{x}) < \hat{x} + (x - \hat{x}) = x \end{aligned}$$

Thus, there is $\varepsilon > 0$, such that for all $z \in I \equiv (\hat{x} - \varepsilon, \hat{x} + \varepsilon)$, we have $(x, z) \in \Omega$ and $\{1 - a[z - (1-d)x]\} < x$, so that

$$u(x, z) = 1 - a[z - (1-d)x]$$

Define $F(x) = \{z : (x, z) \in \Omega\}$, and for $z \in F(x)$, define

$$W(z) = u(x, z) + \rho V(z).$$

Clearly, W is concave on its domain. For $z \in I$, we have

$$W(z) = 1 - az + a(1 - d)x + \rho V(z).$$

Since $\hat{x} \in I$, we obtain from (6) that

$$W'_{+}(\hat{x}) = -a + \rho V'_{+}(\hat{x}) \leq 0. \tag{30}$$

And, we can obtain from (29)

$$W'_{-}(\hat{x}) = -a + \rho V'_{-}(\hat{x}) \geq 0. \tag{31}$$

Now for all $z \in F(x)$ with $z > \hat{x}$, we obtain by (30) and the concavity of W,

$$W(z) - W(\hat{x}) \leq W'_{+}(\hat{x})(z - \hat{x}) \leq 0.$$

Similarly, for $z \in F(x)$ with $z < \hat{x}$, we obtain by (31) and the concavity of W,

$$W(z) - W(\hat{x}) \leq W'_{-}(\hat{x})(z - \hat{x}) \leq 0.$$

Thus, we have $W(z) \leq W(\hat{x})$ for all $z \in F(x)$. This means

$$\max_{(x,z)\in\Omega} [u(x, z) + \rho V(z)] = u(x, \hat{x}) + \rho V(\hat{x}).$$

Since the expression on the left hand side is $V(x)$, we obtain, by the optimality principle, $V(x) = u(x, \hat{x}) + \rho V(\hat{x})$, which means that $\hat{x} \in h(x)$, and completes the proof. ■

An OPF for a Special Case

We now reproduce for the convenience of the reader a result from [20].

Lemma 9 *If* (a, d) *satisfies the restriction for the so-called borderline case,* $(\xi - (1/\xi))(1 - d) = 1$*, then for all values of* $\rho < \sqrt{(a/\xi)}$*, the OPF* h *is given by the check-map,* H*.*

Proof of Proposition 8(ii)

We now turn to a complete proof of Proposition 8(ii).

Since h^2 is turbulent, then there exist $a, b, c \in X$ such that

$$h^2(b) = h^2(a) = a,\ h^2(c) = b, \text{ and either (I) } a < c < b \text{ or (II) } a > c > b.$$

We consider specifically that (I) holds in (1), and analyze this case first. Suppose, contrary to the assertion of Proposition 8(ii), $(\xi - (1/\xi))(1-d) < 1$. Then, by (13) of Proposition 4, we have $H^2(1) < k$.

Let us denote $h(c)$ by c', $h(b)$ by b' and $h(a)$ by a'. Note that all the elements of the set $S = \{a, b, c, a', b', c'\}$ are in the range of h on X. Thus the minimum element of the set must be greater than or equal to $(1-d)$. And the maximum element of the set must be less than or equal to $H(1-d) = H^2(1) < k$. Thus, the set $S \subset [1-d, k)$. We define $A' = [1-d, \hat{x})$, and first claim that

$$a \neq \hat{x} \tag{32}$$

For if $a = \hat{x}$, then since $h(b') = a = \hat{x}$, we cannot have b' in A'. Thus, $b' \in [\hat{x}, k)$. Since $k > b > a = \hat{x}$, and $h(b) = b'$, we must have $b' \leq \hat{x}$. Thus, $b' = \hat{x}$, and so $h(b) = \hat{x}$, with $b \in (\hat{x}, k)$. Since $\hat{x} = a < c < b < k$, we must have $h(c) = \hat{x}$ by Lemma 8. Thus, $h^2(c) = h(\hat{x}) = \hat{x}$. But $h^2(c) = b > \hat{x}$, a contradiction. This establishes the claim (32). It also follows from (32) that $b \neq \hat{x}$, otherwise we get $a = h^2(b) = \hat{x}$, contradicting (4). Further, $c \neq \hat{x}$, otherwise we get $b = h^2(c) = \hat{x}$, a contradiction.

Since $h^2(a) = a$, and (32) holds, we must have $a' = h(a) \neq a$. Thus, we need to consider the two cases: (a) $a' > a$, and (b) $a > a'$.

Consider case (a) $[a' > a]$. Here $h(a) = a' > a$, and so $a \in A'$ and $a' \in B$. Then since $h(b') = h^2(b) = a$ and $a \in A'$, $b' \notin A$ and so $b' \in B$. Since $b' = h(b)$, $b \notin B$, and so $b \in A'$. Finally, since $h(c') = h^2(c) = b$, and $b \in A'$, $c' \notin A$ and so $c' \in B$. Since $c' = h(c)$, $c \notin B$, and so $c \in A'$. To summarize, we have

$$a, b, c \in A' \text{ and } a', b', c' \in B. \tag{33}$$

And since $b > c > a$, (33) implies that

$$b' = h(b) = H(b) < H(c) = h(c) = c' \text{ and } c' = h(c) = H(c) < H(a) = h(a) = a' \tag{34}$$

We show next that $c' < 1$. If not, then since $a' > c'$ by (34), we have $h(a') \geq h(c')$ by Lemma 4. But this yields:

$$a = h^2(a) = h(a') \geq h(c') = h^2(c) = b$$

which contradicts the fact that $b > a$. This establishes the claim. Also, given (34), we have $b' < c' < 1$. Thus $H(b') > H(c')$. Note now that $h(c') = h^2(c) = b > a = h^2(b) = h(b') \geq H(b') > H(c')$, so that $h(c') > H(c')$. Since $k > a' > c'$ by (34), we can use Lemma 7 to obtain $h(a') \geq h(c')$. But this implies:

$$a = h^2(a) = h(a') \geq h(c') = h^2(c) = b$$

which contradicts the fact that $b > a$. Thus, case (a) cannot arise.

Next, we turn to case (b) $[a > a']$. Here $h(a) = a' < a$, and so $a \in B$ and $a' \in A$. If $a' = \hat{x}$, then $a = h^2(a) = h(a') = h(\hat{x}) = \hat{x}$, contradicting (32). Thus $a' \in A'$.Then since $h(b') = h^2(b) = a$ and $a \in B$, $b' \notin B \cup \{\hat{x}\}$, and so $b' \in A'$. Since $b' = h(b)$, $b \notin A$, and so $b \in B$. Finally, since $h(c') = h^2(c) = b$, and $b \in B, c' \notin B \cup \{\hat{x}\}$ and so $c' \in A'$. Since $c' = h(c), c \notin A$, and so $c \in B$. To summarize, we have:

$$a, b, c \in B \text{ and } a', b', c' \in A'$$

And since $b > c > a$, (33) also implies that

$$H(c') = h(c') = h^2(c) = b > a = h^2(a) = h(a') = H(a').$$

Clearly, this implies that $c' < a'$ since $a', c' \in A'$. We now claim that $a < 1$. If not, then since $c > a$, Lemma 4 implies that $c' = h(c) \geq h(a) = a'$, which contradicts $c' < a'$, and establishes the claim.

If $h(a) = H(a)$, then we get $[\hat{x} - a'] = H(\hat{x}) - H(a) = (-\xi)(\hat{x} - a) = (-\xi)(h(\hat{x}) - h(a')) = (-\xi)(H(\hat{x}) - H(a')) = \xi^2(\hat{x} - a')$, so that $\xi = 1$, a contradiction. Thus, we must have $h(a) > H(a)$. Then, using the fact that $a < 1$, and Lemma 7, we obtain $c' = h(c) \geq h(a) = a'$ by virtue of the fact that $k > c > a > \hat{x}$. But this again contradicts $c' < a'$, and establishes that case (b) cannot arise.

Since cases (a) and (b) were the only possible cases, we can conclude that our initial hypothesis was false, and thereby establish Proposition 8(ii) under the possibility (I) in (1).

Next we turn to the consideration of possibility (II) holds in (1). Suppose, contrary to the assertion of part (ii) of the Proposition 8(ii), $(\xi - (1/\xi))(1-d) < 1$. Then, by (13) of Proposition 4, we have $H^2(1) < k$.

Let us denote $h(c)$ by c', $h(b)$ by b' and $h(a)$ by a'. Note that all the elements of the set $S = \{a, b, c, a', b', c'\}$ are in the range of h on X. Thus the minimum element of the set must be greater than or equal to $(1 - d)$. And the maximum element of the set must be less than or equal to $H(1 - d) = H^2(1) < k$. Thus, the set $S \subset [1 - d, k)$. We define $A' = [1 - d, \hat{x})$. Denote, as before, $h(b)$ by b', $h(a)$ by a' and $h(c)$ by c'. We first establish that

$$a \neq \hat{x} \tag{35}$$

Suppose, contrary to (35) that $a = \hat{x}$. Since $h(b') = h^2(b) = a = \hat{x}$, we can infer that $b' \notin A'$. Thus, $b' \in B \cup \{\hat{x}\}$. We consider the two cases: (a) $b' = \hat{x}$, (b) $b' \in B$.

Under case (a), $b' = \hat{x}$, then $h(b) = b' = \hat{x}$. But, $b < a = \hat{x}$, and so $h(b) > \hat{x}$, furnishing a contradiction.

Under case (b), $b' \in B$, then $h(b') = a = \hat{x}$. Thus, $h(x) = \hat{x}$ for all $x \in B$ by Lemma 8. Since $c < a = \hat{x}$, we have $c' = h(c) > \hat{x}$. Thus $c' \in B$, and so $h(c') = \hat{x}$. But $h(c') = h^2(c) = b < a = \hat{x}$, a contradiction. Thus (35) is established. It follows that $b \neq \hat{x}$, otherwise $a = h^2(b) = \hat{x}$, a contradiction. Further, $c \neq \hat{x}$, otherwise $b = h^2(c) = \hat{x}$, a contradiction. Furthermore, since (35) holds, and $h^2(a) = a$, we must have $a' \neq a$, and $h(a) = a'$ and $h(a') = a$. Thus, we need to consider the following two subcases: (A) $a' > a$, (B) $a > a'$.

We begin with the subcase (A) where $a' = h(a) > a$. Thus, $a \in A'$, and $a' \in B$. Since $h(b') = a$ and $a \in A'$, we must have $b' \in B$. Further, since $b' = h(b)$, we must have $b \in A'$. Finally, since $h(c') = b$ and $b \in A'$, we must have $c' \in B$; further since $c' = h(c)$, we must have $c \in A'$. To summarize, we have

$$a, b, c \in A' \text{ and } a', b', c' \in B.$$

Note that $h^2(b') = h(h^2(b)) = h(a) = a'$; $h^2(a') = h(h^2(a)) = h(a) = a'$, and $h^2(c') = h(h^2(c)) = h(b) = b'$. Further, since $a, b, c \in A'$, and $a > c > b$, we have $a' = h(a) < h(c) = c' < h(b) = b'$. Thus, the analysis of Case (I) can be applied to arrive at a contradiction. Consequently subcase (A) cannot occur.

Next, we turn to subcase (B) where $a > a' = h(a)$. Thus, $a \in B$, and $a' \in A$. Since $a' \neq a = h(a')$, we cannot have $a' = \hat{x}$. Thus, $a' \in A'$. Since $h(b') = a$ and $a \in B$, we must have $b' \in A'$. Further, since $b' = h(b)$, we must have $b \in B$. Finally, since $h(c') = b$ and $b \in B$, we must have $c' \in A'$; further since $c' = h(c)$, we must have $c \in B$. To summarize, we have

$$a, b, c \in B \text{ and } a', b', c' \in A'. \tag{36}$$

We now claim that $c < 1$. If not, we must have $a > c \geq 1$. Then by Lemma 4, $a' = h(a) \geq h(c) = c'$. But then by (36), $a = h(a') \leq h(c') = b$, a contradiction to the given condition in (II). This establishes the claim.

If $h(c) = H(c)$, then we get $(\hat{x} - b) = H(\hat{x}) - H(c') = (-\xi)(\hat{x} - c') = (-\xi)(h(\hat{x}) - h(c)) = (-\xi)(H(\hat{x}) - H(c)) = \xi^2(\hat{x} - c)$, so that, using $\xi > 1$, and $c \in (\hat{x}, k)$,

$$(b - \hat{x}) = \xi^2(c - \hat{x}) > (c - \hat{x}) \Longrightarrow b > c,$$

a contradiction. Thus $h(c) > H(c)$. Then, using $a > c$ and Lemma 7, we obtain

$$a' = h(a) \geq h(c) = c'.$$

On using this and (36), we get $a = h(a') \leq h(c') = b$. But this again contradicts the given condition (II). Thus, subcase (B) cannot arise.

Since subcases (A) and (B) were the only possible cases, we can conclude that our initial hypothesis must be false, and thereby establish Proposition 8(ii) under the possibility (II) in (1).

References

1. Arkin V, Evstigneev I (1987) Stochastic models of control and economic dynamics. Academic Press, New York
2. Benhabib J (ed) (1992) Cycles and chaos in economic equilibrium. Princeton University Press, Princeton
3. Benhabib J, Rustichini A (1990) Equilibrium cycling with small discounting. J Econ Theory 52:423–432
4. Block LS, Coppel WA (1992) Dynamics in one dimension. Lecture notes in mathematics, vol 1513. Springer, Berlin
5. Block LS, Guckenheimer J, Misiurewicz M, Young LS (1980) Periodic points and topological entropy of one dimensional maps. In: Global theory of dynamical systems. Lecture notes in mathematics, vol 819. Springer, Berlin, pp 18–34
6. Boldrin M, Deneckere RJ (1990) Sources of complex dynamics in two-sector growth models. J Econ Dyn Control 14:627–653. Also Chapter 10 in [2, pp.228–252]
7. Brooks KM, Bruin H (2004) Topics from one-dimensional dynamics. Cambridge University, Cambridge
8. Day RH, Pianigiani G (1991) Statistical dynamics and economics. J Econ Behav Organ 16:37–83
9. Day RH, Shafer W (1987) Ergodic fluctuations in deterministic economic models. J Econ Behav Organ 8:339–361
10. Dechert WD, Nishimura K (1983) A complete characterization of optimal growth paths in an aggregated model with a non-concave production function. J Econ Theory 31:332–354
11. Deneckere RJ, Judd K Cyclical and chaotic behavior in a dynamic equilibrium model, with implications for fiscal policy. Chapter 14 in [2, pp. 308–329]
12. Gale D (1967) On optimal development in a multi-sector economy. Rev Econ Stud 34:1–18
13. Huntley HE (1970) The divine proportion: a study in mathematical beauty. Dover Publications, New York
14. Khan MA, Mitra T (2005) On choice of technique in the Robinson-Solow-Srinivasan model. Int J Econ Theory 1:83–109
15. Khan MA, Mitra T (2005) On topological chaos in the Robinson-Solow-Srinivasan model. Econ Lett 88:127–133
16. Khan MA, Mitra T (2006) Discounted optimal growth in the two-sector RSS model: a geometric investigation. Adv Math Econ 8:349–381
17. Khan MA, Mitra T. (2005) Discounted optimal growth in a two-sector RSS model: a further geometric investigation. Johns Hopkins University, Mimeo, Baltimore. Now available in (2013) Adv Math Econ 17:1–33
18. Khan MA, Mitra T (2006) Undiscounted optimal growth under irreversible investment: a synthesis of the value-loss approach and dynamic programming. Econ Theory 29:341–362
19. Khan MA, Mitra T (2007) Optimal growth under discounting in the two-sector Robinson-Solow-Srinivasan model: a dynamic programming approach. J Differ Equ Appl 13:151–168
20. Khan MA, Mitra T (2007) Bifurcation Analysis in the two-sector Robinson-Solow-Srinivasan model: a close look at a borderline case. Cornell University, Mimeo, Ithaca

21. Li T, Yorke JA (1975) Period three implies chaos. Am Math Mon 82:985–992
22. Livio M (2002) The golden ratio. Broadway Books, New York
23. Majumdar M, Mitra T, Nishimura K (eds.) (2000) Optimization and chaos. Springer, Berlin
24. May RB (1976) Simple mathematical models with very complicated dynamics. Nature 40:459–467
25. McKenzie LW (1986) Optimal economic growth, turnpike theorems and comparative dynamics. In: Arrow KJ, Intrilligator M (eds) Handbook of mathematical economics III. North-Holland, New York, pp 1281–1355
26. McKenzie LW (2002) Classical general equilibrium theory. MIT, Cambridge
27. Metcalf C (2008) The dynamics of the Stiglitz policy in the RSS model. Chaos Solitons Fractals 37:652–661
28. Mitra T (1996) An exact discount factor restriction for period-three cycles in dynamic optimization models. J Econ Theory 69:281–305
29. Mitra T (1998) On the relation between discounting and complicated behavior in dynamic optimization models. J Econ Behav Organ 33:421–434
30. Mitra T (2001) A sufficient condition for topological chaos with an application to a model of endogenous growth. J Econ Theory 96:133–152
31. Mitra T, Nishimura K, Sorger G (2006) Optimal cycles and chaos. In: Dana R, Le Van C, Mitra T, Nishimura K (eds) Handbook of optimal growth 1, Chapter 6. Springer, Heidelberg, pp. 141–169
32. Nishimura K, Yano M (1992) Business cycles and complex non-linear dynamics. Chaos Solitons Fractals 2:95–102
33. Nishimura K, Yano M (1994) Optimal chaos, nonlinearity and feasibility conditions. Econ Theory 4:689–704. Chapter 4 in [23, pp. 149–172]
34. Nishimura K, Yano M (1995) Nonlinear dynamics and chaos in optimal growth. Econometrica 7:1941–1953
35. Nishimura K, Yano M (1995) Nonlinear dynamics and chaos in optimal growth. J Econ Behav Organ 27:165–181
36. Nishimura K, Yano M. (1996) On the least upper bound on discount factors that are compatible with optimal period-three cycles. J Econ Theory 69:306–333
37. Nishimura K, Yano M (1996) Chaotic solutions in dynamic linear programming. Chaos Solitons Fractals 7:1941–1953
38. Nishimura K, Yano Y (2000) Non-linear dynamics and chaos in optimal growth: a constructive exposition. Optim Chaos 11:258–295. Chapter 8, in [23, pp. 258–295]
39. Nishimura K, Sorger G, Yano M (1994) Ergodic chaos in optimal growth models with low discount rates. Econ Theory 4:705–717. Chapter 9 in [23, pp.296–314]
40. Nishimura K, Shigoka T, Yano M (1998) Interior optimal chaos with arbitrarily low discount rates. Jpn Econ Rev 49:223–233
41. Robinson J. (1969) A model for accumulation proposed by J.E. Stiglitz. Econ J 79:412–413
42. Rosser Jr. JB (1990) Chaos theory and the new Keynesian economics. Manch Sch Econ Soc Stud 58:265–291
43. Rosser Jr. JB (1999) On the complexities of complex economic dynamics. J Econ Perspect 13:169–192
44. Royden HL (1988) Real analysis. Prentice-Hall, New Jersey
45. Saari DG (1995) Mathematical complexity of simple economics. Not Am Math Soc 42:222–230
46. Sharkovsky AN (1964) Coexistence of cycles of a continuous map of a line into itself. Ukrainski Matematicheski Zhurnal 16:61–71 (in Russian). English translation in (1995) Int J Bifur Ch 5:1263–1273
47. Sharkovsky AN, Kolyada SF, Sivak AG, Fedorenko VV (1997) Dynamics of one-dimensional maps. Kluwer Academic, Dordrecht
48. Sorger G (2006) Rationalizability in optimal growth theory. In: Dana R, Le Van C, Mitra T, Nishimura K (eds) Handbook of optimal growth 1. Chapter 4. Springer, Heidlberg, pp. 85–114

49. Sorger G. (2009) Some notes on discount factor restrictions for dynamic optimization problems. J Math Econ 45:435–458
50. Stiglitz JE (1968) A note on technical choice under full employment in a socialist economy. Econ J 78:603–609
51. Stiglitz JE (1970) Reply to Mrs. Robinson on the choice of technique. Econ J 80:420–422

Some Regularity Estimates for Diffusion Semigroups with Dirichlet Boundary Conditions

Shigeo Kusuoka

Abstract In the present paper the author gives regularity estimates for diffusion semigroups with Dirichlet boundary condition whose infinitesimal generator is degenerate elliptic operator satisfying UFG condition. This estimate will enable us to give a new KLNV approximation for expectation of diffusion process with absorbing boundary, which appears in the price of derivatives with knock-out condition.

Keywords Diffusion semi-group · Dirichlet boundary condition · UFG condition

Article type: Research Article
Received: July 18, 2019
Revised: August 1, 2019

1 Introduction

Let $W_0 = \{w \in C([0,\infty); \mathbf{R}^d); \ w(0) = 0\}$, $\mathcal{G}$ be the Borel σ-algebra over W_0 and μ be the Wiener measure on $(W_0, \mathcal{G})$. Let $B^i : [0,\infty) \times W_0 \to \mathbf{R}$, $i = 1, \ldots, d$, be given by $B^i(t, w) = w^i(t)$, $(t, w) \in [0,\infty) \times W_0$. Then $\{(B^1(t), \ldots, B^d(t)); \ t \in [0,\infty)\}$ is a d-dimensional Brownian motion under μ.

JEL Classification: C63, G00
Mathematics Subject Classification (2010): 65C05, 60G40

S. Kusuoka (✉)
Graduate School of Mathematical Sciences, The University of Tokyo, Tokyo, Japan
e-mail: spfu5sh9@car.ocn.ne.jp

T. Maruyama (ed.), *Advances in Mathematical Economics*, Advances in Mathematical Economics 23, https://doi.org/10.1007/978-981-15-0713-7_5

Let $B^0(t) = t$, $t \in [0, \infty)$. Let $\{\mathscr{F}_t\}_{t \geqq 0}$ be the Brownian filtration generated by $\{(B^1(t), \ldots, B^d(t)); t \in [0, \infty)\}$.

Let $V_0, V_1, \ldots, V_d \in C_b^\infty(\mathbf{R}^N; \mathbf{R}^N)$. Here $C_b^\infty(\mathbf{R}^N; \mathbf{R}^n)$ denotes the space of $\mathbf{R}^n$-valued smooth functions defined in $\mathbf{R}^N$ whose derivatives of any order are bounded. We regard elements in $C_b^\infty(\mathbf{R}^N; \mathbf{R}^N)$ as vector fields on $\mathbf{R}^N$.

Now let $X(t, x)$, $t \in [0, \infty)$, $x \in \mathbf{R}^N$, be the solution to the Stratonovich-type stochastic integral equation

$$X(t, x) = x + \sum_{i=0}^{d} \int_0^t V_i(X(s, x)) \circ dB^i(s). \tag{1}$$

Then there is a unique solution to this equation. Moreover we may assume that $X(t, x)$ is continuous in t and smooth in x and $X(t, \cdot) : \mathbf{R}^N \to \mathbf{R}^N$, $t \in [0, \infty)$, is a diffeomorphism with probability one.

Let $A = A_d = \{v_0, v_1, \ldots, v_d\}$, be an alphabet, a set of letters, and A^* be the set of words consisting of A including the empty word which is denoted by 1. For $u = u^1 \cdots u^k \in A^*$, $u^j \in A$, $j = 1, \ldots, k$, $k \geqq 0$, we denote by $n_i(u)$, $i = 0, \ldots, d$, the cardinal of $\{j \in \{1, \ldots, k\}; u^j = v_i\}$. Let $|u| = n_0(u) + \cdots + n_d(u)$, a length of u, and $\| u \| = |u| + n_0(u)$ for $u \in A^*$. Let $\mathbf{R}\langle A \rangle$ be the $\mathbf{R}$-algebra of non-commutative polynomials on A.

Let $r : A^* \setminus \{1\} \to \mathscr{L}(A)$ denote the right bracketing operator inductively given by

$$r(v_i) = v_i, \qquad i = 0, 1, \ldots, d,$$

and

$$r(v_i u) = [v_i, r(u)] = v_i r(u) - r(u) v_i, \qquad i = 0, 1, \ldots, d, \ u \in A^* \setminus \{1\}.$$

Let $A^{**} = \{u \in A^*;\ u \neq 1, v_0\}$, $A^{**}_m = \{u \in A^{**};\ ||u|| = m\}$, and $A^{**}_{\leqq m} = \{u \in A^{**};\ \| u \| \leqq m\}$, $m \geqq 1$.

We can regard vector fields $V_0, V_1, \ldots, V_d$ as first differential operators over $\mathbf{R}^N$. Let $\mathscr{DO}(\mathbf{R}^N)$ denote the set of linear differential operators with smooth coefficients over $\mathbf{R}^N$. Then $\mathscr{DO}(\mathbf{R}^N)$ is a non-commutative algebra over $\mathbf{R}$. Let $\Phi : \mathbf{R}\langle A \rangle \to \mathscr{DO}(\mathbf{R}^N)$ be a homomorphism given by

$$\Phi(1) = Identity, \qquad \Phi(v_{i_1} \cdots v_{i_n}) = V_{i_1} \cdots V_{i_n}, \qquad n \geqq 1,\ i_1, \ldots, i_n = 0, 1, \ldots, d.$$

Then we see that

$$\Phi(r(v_i u)) = [V_i, \Phi(r(u))], \qquad i = 0, 1, \ldots, d,\ u \in A^* \setminus \{1\}.$$

Now we introduce a condition (UFG) for a system of vector field $\{V_0, V_1, \dots, V_d\}$ as follows.
(UFG) There are an integer $\ell_0 \geqq 1$ and $\varphi_{u,u'} \in C_b^\infty(\mathbf{R}^N)$, $u \in A^{**}$, $u' \in A^{**}_{\leqq \ell_0}$, satisfying the following.

$$\Phi(r(u)) = \sum_{u' \in A^{**}_{\leqq \ell_0}} \varphi_{u,u'} \Phi(r(u')), \qquad u \in A^{**}.$$

Let P_t, $t \in [0, \infty)$ be a diffusion semigroup given by

$$P_t f(x) = E[f(X(t,x))], \qquad f \in C_b^\infty(\mathbf{R}^N).$$

Then P_t's are regarded as a linear operators in $C_b^\infty(\mathbf{R}^N)$. We also have the following.

Theorem 1 *Assume that* (UFG) *condition is satisfied. For any* $n, m \geqq 0$, *and* $u_1, \dots, u_{n+m} \in A^{**}$, *there is a* $C \in (0, \infty)$ *such that*

$$||\Phi(r(u_1)\cdots r(u_n)) P_t \Phi(r(u_{n+1})\cdots r(u_{n+m})) f||_\infty \leqq C t^{-(||u_1|| + \cdots + ||u_{n+m}||)/2} ||f||_\infty$$

for any $t \in (0, 1)$, *and* $f \in C_b^\infty(\mathbf{R}^N)$. *Here*

$$||f||_\infty = \sup_{x \in \mathbf{R}^N} |f(x)|.$$

This theorem was shown in [5] under the uniform Hörmander condition and was shown in [3] in general.

In the present paper, we assume (UFG) and the following assumptions (A1) and (A2) throughout.

(A1) $V_1^1(x) = 1$, $V_1^i(x) = 0$, $i = 2, \dots, N$, for any $x \in \mathbf{R}^N$.
(A2) $V_k^1(x) = 0$, $k = 0, 2, \dots, d$, for any $x \in \mathbf{R}^N$.

Then $X^1(t,x) = x^1 + B^1(t)$, $t \geqq 0$. Let $h \in C^\infty(\mathbf{R}^N)$ be given by $h(x) = x^1$, $x \in \mathbf{R}^N$. Then we see that $\Phi(r(v_1))h = 1$, and $\Phi(r(u))h = 0$, $u \in A^* \setminus \{1, v_1\}$. So we see that if (UFG) condition is satisfied, we see that $\varphi_{u,v_1} = 0$, for $u \in A^* \setminus \{1, v_1\}$.

Let $b_k \in C_b^\infty(\mathbf{R}^N)$, $k = 0, \dots, d$, and let

$$P_t^0 f(x) = E[\exp(\sum_{k=0}^d \int_0^t b_k(X(r,x)) \circ dB^k(r)) f(X(t,x)), \min_{r \in [0,t]} X^1(r) > 0],$$

$f \in C_b^\infty(\mathbf{R}^N)$. Then we see that

$$\frac{\partial}{\partial t} P_t^0 f(x) = L^0 P_t f(x), \qquad t > 0,\ x \in (0, \infty) \times \mathbf{R}^{N-1}$$

as generalized functions, and

$$P_t^0 f(x) = 0, \qquad t > 0,\ x \in \{0\} \times \mathbf{R}^{N-1}.$$

Here

$$L^0 = \frac{1}{2}\sum_{k=1}^{d} V_k^2 + V_0 + \sum_{k=1}^{N} b_k V_k + (b_0 + \frac{1}{2}\sum_{k=1}^{d}(b_k^2 + V_k b_k)).$$

Our final purpose is to show the following.

Theorem 2 *Assume that* (UFG) *condition is satisfied. Then for any* $n, m, r \geqq 0$ *and* $u_1, \ldots, u_{n+m} \in A^{**}$, *there is a* $C \in (0, \infty)$ *such that*

$$\sup_{x\in(0,\infty)\times\mathbf{R}^{N-1}} |\Phi(r(u_1)\cdots r(u_n))adj(V_0)^r(P_t^0)\Phi(r(u_{n+1})\cdots r(u_{n+m}))f(x)|$$

$$\leqq Ct^{-(||u_1||+\cdots+||u_{n+m}||/2)-r} \sup_{x\in(0,\infty)\times\mathbf{R}^{N-1}} |f(x)|$$

and

$$\int_{(0,\infty)\times\mathbf{R}^{N-1}} |\Phi(r(u_1)\cdots r(u_n))adj(V_0)^r(P_t^0)\Phi(r(u_{n+1})\cdots r(u_{n+m}))f(x)|dx$$

$$\leqq Ct^{-(||u_1||+\cdots+||u_{n+m}||/2)-r} \int_{(0,\infty)\times\mathbf{R}^{N-1}} |f(x)|dx$$

for any $t \in (0, 1]$ *and* $f \in C_0^\infty((0, \infty) \times \mathbf{R}^{N-1})$.

Here $adj^0(V_0)(P_t^0) = P_t^0$, *and*

$$adj^{n+1}(V_0)(P_t^0) = V_0\, adj(V_0)^n(P_t^0) - adj(V_0)^n(P_t^0)V_0, \qquad n = 0, 1, \ldots.$$

The present paper consists of two parts, because we use two different techniques. We give estimates for horizontal direction in Part I by using Malliavin calculus and give estimates for vertical direction in Part II by using usual bootstrap arguments.

In Part I, we prove the following theorem.

Theorem 3 *Assume that* (UFG) *condition is satisfied. Let* $A^{***} = A^{**} \setminus \{v_1\}$. *Then we have the following.*

(1) *For any* $n, m, r \geqq 0$ *and* $u_1, \ldots, u_{n+m} \in A^{***}$, *there is a* $C \in (0, \infty)$ *such that*

$$\sup_{x\in(0,\infty)\times\mathbf{R}^{N-1}} |\Phi(r(u_1)\cdots r(u_n))adj(V_0)^r P_t^0 \Phi(r(u_{n+1})\cdots r(u_{n+m}))f(x)|$$

$$\leqq Ct^{-(||u_1||+\cdots+||u_{n+m}||/2)-r} \sup_{x\in(0,\infty)\times\mathbf{R}^{N-1}} |f(x)|$$

for any $t \in (0, 1]$ *and* $f \in C_b^\infty(\mathbf{R}^N)$.

(2) *For any* $n, m, r \geqq 0$ *and* $u_1, \ldots, u_{n+m} \in A^{***}$, *there is a* $C \in (0, \infty)$ *such that*

$$\int_{(0,\infty)\times\mathbf{R}^{N-1}} |\Phi(r(u_1)\cdots r(u_n))adj(V_0)^r (P_t^0)\Phi(r(u_{n+1})\cdots r(u_{n+m}))f(x)|dx$$

$$\leqq Ct^{-(||u_1||+\cdots+||u_{n+m}||/2)-r} \int_{(0,\infty)\times\mathbf{R}^{N-1}} |f(x)|dx$$

for any $t \in (0, 1]$ *and* $f \in C_0^\infty(\mathbf{R}^N)$.

2 Part I: Horizontal Direction

2.1 Normed Spaces and Interpolation

From now on, we assume that (UFG) is satisfied. Let $(W_0, \mathscr{G}, \mu)$ be a Wiener space as in Introduction. Let H denote the associated Cameron-Martin space, $\mathscr{L}$ denote the associated Ornstein-Uhlenbeck operator, and $W^{r,p}(E)$, $r \in \mathbf{R}$, $p \in (1, \infty)$, be Watanabe-Sobolev spaces, i.e., $W^{r,p}(E) = (I - \mathscr{L})^{-r/2}(L^p(W_0; E, d\mu))$ for any separable real Hilbert space E. We write $W^{r,p} = W^{r,p}(\mathbf{R})$ for simplicity. Let D denote the gradient operator. Then D is a bounded linear operator from $W^{r,p}(E)$ to $W^{r-1,p}(H \otimes E)$. Let D^* denote the adjoint operator of D. (See Shigekawa [6] for details.)

Let $\tilde{A} = A^{**}_{\leqq \ell_0} \setminus \{v_1\}$. Let $V_u^{(s)} \in C_b^\infty(\mathbf{R}^N; \mathbf{R}^N)$, $u \in \tilde{A}$, $s \in (0, 1]$, be given by

$$V_u^{(s)}(x) = s^{||u||/2}\Phi(r(u))(x), \qquad x \in \mathbf{R}^N.$$

Note that $(V_u^{(u)} h)(x) = 0$, $x \in \mathbf{R}^N$, $u \in \tilde{A}$, $s \in (0, 1]$, where $h(x) = x^1$, $x = (x^1, \ldots, x^N) \in \mathbf{R}^N$.

Proposition 1 *There are* $\tilde{\varphi}_{u_1,u_2,u_3} \in C_b^\infty(\mathbf{R}^N)$, $u_1, u_2, u_3 \in \tilde{A}$, *such that*

$$[V_{u_1}^{(s)}, V_{u_2}^{(s)}] = \sum_{u_3\in\tilde{A}} s^{0\vee(||u_1||+||u_2||-||u_3||)/2} \tilde{\varphi}_{u_1,u_2,u_3} V_{u_3}^{(s)}, \qquad u_1, u_2 \in \tilde{A}.$$

Proof Note that there are $c_{u_1,u_2,u_3} \in \mathbf{R}$, $u_1, u_2 \in \tilde{A}$, $u_3 \in A^{**}$ such that

$$[r(u_1), r(u_2)] = \sum_{u_3 \in A^{**}, ||u_3||=||u_1||+||u_2||} c_{u_1,u_2,u_3} r(u_3).$$

So if $||u_1|| + ||u_2|| \leqq \ell_0$, we have

$$\begin{aligned}
[V_{u_1}^{(s)}, V_{u_2}^{(s)}](x) &= s^{(||u_1||+||u_2||)/2} \Phi([r(u_1), r(u_2)])(x) \\
&= \sum_{u_3 \in \tilde{A}, ||u_3||=||u_1||+||u_2||} c_{u_1,u_2,u_3} s^{||u_3||/2} \Phi(r(u_3))(x) \\
&= \sum_{u_3 \in \tilde{A}, ||u_3||=||u_1||+||u_2||} c_{u_1,u_2,u_3} V_{u_3}^{(s)}(x).
\end{aligned}$$

Also, if $||u_1|| + ||u_2|| > \ell_0$, we have

$$\begin{aligned}
[V_{u_1}^{(s)}, V_{u_2}^{(s)}](x) &= \sum_{u_3 \in \tilde{A}, ||u_3||=||u_1||+||u_2||} c_{u_1,u_2,u_3} s^{(||u_1||+||u_2||)/2} \Phi(r(u_3))(x) \\
&= \sum_{u_4 \in \tilde{A}, ||u_3||=||u_1||+||u_2||} c_{u_1,u_2,u_3} s^{(||u_1||+||u_2||)/2} \varphi_{u_3,u_4}(x) \Phi(r(u_4))(x) \\
&= \sum_{u_4 \in \tilde{A}, ||u_3||=||u_1||+||u_2||} c_{u_1,u_2,u_3} s^{(||u_1||+||u_2||-||u_4||)/2} \varphi_{u_3,u_4}(x) V_{u_4}^{(s)}(x).
\end{aligned}$$

These imply our assertion. ■

Now let $\tilde{B}^u(t)$, $t \in [0, \infty)$, $u \in \tilde{A}$, be independent standard Brownian motions defined on a certain probability space and let $X^{(s)}(t, x)$, $t \in [0, \infty)$, $x \in \mathbf{R}^N$, $s \in (0, 1]$, be a solution to the following stochastic differential equation.

$$dX^{(s)}(t, x) = \sum_{u \in \tilde{A}} V_u^{(s)}(X^{(s)}(t, x)) \circ d\tilde{B}^u(t),$$

$$X^{(s)}(0, x) = x.$$

Note that $h(X^{(s)}(t, x)) = h(x)$, $t \geqq 0$, $x \in \mathbf{R}^N$. Now let $Q_t^{(s)}$, $t \in [0, \infty)$, $s \in (0, 1]$, be linear operators on $C_b^\infty(\mathbf{R}^N)$ given by

$$(Q_t^{(s)} f)(x) = E[f(X^{(s)}(t, x))], \qquad f \in C_b^\infty(\mathbf{R}^N).$$

Let

$$L^{(s)} = \frac{1}{2} \sum_{u \in \tilde{A}} s^{||u||} \Phi(r(u))^2.$$

Then we see that

$$Q_t^{(s)} f = f + \int_0^t L^{(s)} Q_r^{(s)} f dr, \qquad f \in C_b^\infty(\mathbf{R}^N).$$

By Theorem 1 in [4] we have the following.

Proposition 2 *For any $n, m \geqq 0$, and $u_1, \ldots u_{n+m} \in \tilde{A}$, there exists a $C \in (0, \infty)$ such that*

$$s^{(||u_1||+\cdots||u_{n+m}||)/2} ||\Phi(r(u_1)) \cdots \Phi(r(u_n)) Q_t^{(s)} \Phi(r(u_{n+1})) \cdots \Phi(r(u_{n+m})) f||_\infty \leqq C t^{-(n+m)/2} ||f||_\infty$$

for any $f \in C_b^\infty(\mathbf{R}^N)$ and $s, t \in (0, 1]$.

Let $\mathscr{C}$ be the set of bounded measurable functions f defined in $\mathbf{R}^N$ such that $f(x^1, x^2, \ldots, x^N)$ is smooth in $(x^2, \ldots, x^N)$, and that

$$\sup_{x \in \mathbf{R}^N} |\frac{\partial^{\alpha_2 + \ldots + \alpha_N} f}{(\partial x^2)^{\alpha_2} \cdots (\partial x^N)^{\alpha_N}}(x)| < \infty$$

for any $\alpha_2, \ldots, \alpha_N \geqq 0$.

Note that $Q^{(s)} f \in \mathscr{C}$ for any $f \in \mathscr{C}$. Then the following is an easy consequence of Proposition 2.

Corollary 1 *For any $n, m \geqq 0$, and $u_1, \ldots u_{n+m} \in \tilde{A}$, there exists a $C \in (0, \infty)$ such that*

$$s^{(||u_1||+\cdots||u_{n+m}||)/2} ||\Phi(r(u_1)) \cdots \Phi(r(u_n)) Q_t^{(s)} \Phi(r(u_{n+1})) \cdots \Phi(r(u_{n+m})) f||_\infty \leqq C t^{-(n+m)/2} ||f||_\infty$$

for any $f \in \mathscr{C}$ and $s, t \in (0, 1]$.

Let us define normed spaces $\mathscr{D}^1_{(s)}$, $s \in (0, 1]$, and $\mathscr{H}^{-\alpha}_{(s)}$, $s \in (0, 1]$, $\alpha \in [0, 1)$, by the following.

$\mathscr{D}^1_{(s)} = \mathscr{H}^{-\alpha}_{(s)} = \mathscr{C}$ as sets, and their norms are given by

$$||f||_{\mathscr{D}^1_{(s)}} = ||f||_\infty + \sum_{u \in \tilde{A}} s^{||u||/2} ||\Phi(r(u)) f||_\infty$$

and

$$||f||_{\mathscr{H}^{-\alpha}_{(s)}} = \sup_{t \in (0,1]} t^{\alpha/2} ||Q^{(s)}_t f||_\infty$$

for $f \in \mathscr{C}$. Note that

$$||f||_{\mathscr{H}^{0}_{(s)}} = ||f||_\infty, \qquad f \in \mathscr{C}.$$

We have the following as an easy consequence of Corollary 1,

Proposition 3 *There is a* $C_0 \in (0, \infty)$ *such that*

$$||L^{(s)} Q^{(s)}_t f||_\infty \leqq C_0 t^{-1} ||f||_\infty$$

and

$$||Q^{(s)}_t f||_{\mathscr{D}^1_{(s)}} \leqq C_0 t^{-1/2} ||f||_\infty$$

for any $f \in \mathscr{C}$ *and* $s, t \in (0, 1]$.

Then we have the following.

Proposition 4 *Let* $\alpha \in (0, 1)$ *and* $\theta \in (0, 1)$. *If* $\beta = (1-\theta)\alpha - \theta \geqq 0$, *then there is a* $C \in (0, \infty)$ *such that*

$$\sup_{t \in (0,\infty)} t^{-\theta} K(t; f, \mathscr{H}^{-\alpha}_{(s)}, \mathscr{D}^1_{(s)}) \leqq C ||f||_{\mathscr{H}^{-\beta}_{(s)}}$$

for $f \in \mathscr{C}$ *and* $s \in (0, 1]$. *Here*

$$K(t; f, \mathscr{H}^{-\alpha}_{(s)}, \mathscr{D}^1_{(s)}) = \inf\{||g||_{\mathscr{H}^{-\alpha}_{(s)}} + t||f - g||_{\mathscr{D}^1_{(s)}};\ g \in \mathscr{C}\}, \qquad t \in (0, \infty).$$

Remark 1 $K(t; f, \mathscr{H}^{-\alpha}_{(s)}, \mathscr{D}^1_{(s)})$ is a real interpolation (cf. Berph-Löfström [1]).

Proof Let $f \in \mathscr{C}$. Note that

$$||Q^{(s)}_t (Q^{(s)}_r f - f)||_\infty \leqq \int_0^r ||L^{(s)} Q^{(s)}_{t/2} Q^{(s)}_{(t+2z)/2} f||_\infty dz$$

$$\leqq C_0 (t/2)^{-1} \int_0^r ||Q^{(s)}_{(t+2z)/2} f||_\infty dz \leqq C_0 (t/2)^{-1-\beta/2} r ||f||_{\mathscr{H}^{-\beta}_{(s)}}.$$

Here C_0 is as in Corollary 1.

On the other hand,

$$||Q^{(s)}_t (Q^{(s)}_r f - f)||_\infty \leqq 2||Q^{(s)}_t f||_\infty \leqq 2t^{-\beta/2} ||f||_{\mathscr{H}^{-\beta}_{(s)}}.$$

Therefore

$$||Q_t^{(s)}(Q_r^{(s)}f - f)||_\infty \leqq (2 + 4C_0)t^{-\beta/2}(1 \wedge (rt^{-1}))||f||_{\mathscr{H}_{(s)}^{-\beta}}$$

$$\leqq (2 + 4C_0)t^{-\beta/2}(rt^{-1})^{\gamma/2}||f||_{\mathscr{H}_{(s)}^{-\beta}}.$$

Here $\gamma = \theta(1 + \alpha) = \alpha - \beta \in (0, 1)$. Therefore we see that

$$||Q_r^{(s)}f - f||_{\mathscr{H}_{(s)}^{-\alpha}} \leqq (2 + 4C_0)r^{\gamma/2}||f||_{\mathscr{H}_{(s)}^{-\beta}}.$$

Also we have

$$||Q_r^{(s)}f||_{\mathscr{D}_{(s)}^1} \leqq C_0(r/2)^{-1/2}||Q_{r/2}^{(s)}f||_\infty \leqq 4C_0 r^{-(1+\beta)/2}||f||_{\mathscr{H}_{(s)}^{-\beta}}.$$

Since we have

$$f = Q_r^{(s)}f + f - Q_r^{(s)}f, \qquad f \in \mathscr{C},$$

we see that for $t \in (0, 1]$

$$t^{-\theta}K(t; f, \mathscr{H}_{(s)}^{-\alpha}, \mathscr{D}_{(s)}^1) \leqq t^{1-\theta}||Q_r^{(s)}f||_{\mathscr{D}_{(s)}^1} + t^{-\theta}||Q_r^{(s)}f - f||_{\mathscr{H}_{(s)}^{-\alpha}}$$

$$\leqq (2 + 4C_0)(t^{1-\theta}r^{-(1+\beta)/2} + t^{-\theta}r^{\gamma/2})||f||_{\mathscr{H}_{(s)}^{-\beta}}.$$

Let $r = t^{2\theta/\gamma}$. Since $(1 - \theta)(1 + \alpha) = 1 + \beta$, we see that

$$\sup_{t \in (0,1]} t^{-\theta}K(t; f, \mathscr{H}_{(s)}^{-\alpha}, \mathscr{D}_{(s)}^1) \leqq 4(1 + 2C_0)||f||_{\mathscr{H}_{(s)}^{-\beta}}.$$

It is obvious that

$$\sup_{t \in [1,\infty)} t^{-\theta}K(t; f, \mathscr{H}_{(s)}^{-\alpha}, \mathscr{D}_{(s)}^1) \leqq ||f||_{\mathscr{H}_{(s)}^{-\alpha}} \leqq ||f||_{\mathscr{H}_{(s)}^{-\beta}}$$

Therefore we have our assertion. ■

The following has been proved by Watanabe [7], but we give a proof.

Proposition 5 *Let $\theta \in (0, 1)$, $p \in (1, \infty)$ and $r_0, r_1 \in [-1, 0]$. If $r_2 < (1 - \theta)r_0 + \theta r_1$, then there is a $C \in (0, \infty)$ such that*

$$||F||_{W^{r_2,p}} \leqq C \sup_{t \in (0,\infty)} t^{-\theta}K(t; F, W^{r_0,p}, W^{r_1,p})$$

for any $F \in W^{\infty,\infty-} = \bigcap_{r\in\mathbf{R},p\in(1,\infty)} W^{r,p}$. *Here*

$$K(t;\, F,\, W^{r_0,p},\, W^{r_1,p}) = \inf\{||G||_{W^{r_0,p}} + t||F-G||_{W^{r_1,p}};\ G \in W^{\infty,\infty-}\}.$$

Proof Let us take an $F \in W^{\infty,\infty-}$ and fix it. Let T_t, $t \in [0,\infty)$, be the Ornstein-Uhlenbeck semi-group on W_0, and let

$$a = \sup_{t\in(0,\infty)} t^{-\theta} K(t;\, F,\, W^{r_0,p},\, W^{r_1,p}).$$

Then we see that

$$||F||_{W^{r_0\wedge r_1,p}} \leqq a.$$

So we have our assertion if $r_2 \leqq r_1 \wedge r_2$. Therefore we may assume that $r_2 > r_1 \wedge r_2 \geqq -1$.

Note that for any $r \geqq 0$, there is a $C_r > 0$ such that

$$||(I-\mathscr{L})^r T_t g||_{W^{0,p}} \leqq C_r t^{-r} ||g||_{W^{0,p}}$$

for any $t \in (0,1]$ and $g \in W^{\infty,\infty-}$.

For any $t \in (0,1]$ and $\varepsilon > 0$, there is a $G_t \in W^{\infty,\infty-}$ such that

$$(t^{(r_1-r_0)/2})^{-\theta}||G_t||_{W^{r_0,p}} + (t^{(r_1-r_0)/2})^{1-\theta}||F-G_t||_{W^{r_1,p}} \leqq a+\varepsilon.$$

Let $\gamma = ((1-\theta)r_0 + \theta r_1 - r_2)/2 > 0$. Then we have $r_2 - r_1 = -(1-\theta)(r_1 - r_0) - 2\gamma$, and $r_2 - r_0 = \theta(r_1 - r_0) - 2\gamma$. So we see that

$$t^{-(\gamma+(r_2-r_0)/2}||G_t||_{W^{r_0,p}} + t^{-(\gamma+(r_2-r_1)/2)}||F-G_t||_{W^{r_1,p}} \leqq a+\varepsilon.$$

Then we have

$$\begin{aligned}
||(I-\mathscr{L})T_tF||_{W^{r_2,p}} &= ||(I-\mathscr{L})^{1+(r_2/2)}T_tF||_{W^{0,p}} \\
&\leqq ||(I-\mathscr{L})^{1+(r_2/2)}T_tG_t||_{W^{0,p}} + ||(I-\mathscr{L})^{1+(r_2/2)}T_t(F-G_t)||_{W^{0,p}} \\
&\leqq ||(I-\mathscr{L})^{1+((r_2-r_0)/2)}T_t(I-\mathscr{L})^{r_0/2}G_t||_{W^{0,p}} \\
&\quad + ||(I-\mathscr{L})^{1+((r_2-r_1)/2)}T_t(I-\mathscr{L})^{r_1/2}(F-G_t)||_{W^{0,p}} \\
&\leqq C(t^{-(1+(r_2-r_0)/2)}||G_t||_{W^{r_0,p}} + t^{-(1+(r_2-r_1)/2)}||F-G_t||_{W^{r_1,p}}) \\
&\leqq Ct^{-1+\gamma}(a+\varepsilon)
\end{aligned}$$

for any $t \in (0,1]$. Note that

$$F = \int_0^1 e^{-t}(I-\mathscr{L})T_tF\,dt + e^{-1}T_1F.$$

Then we see that

$$||F||_{W^{r_2,p}} \leqq C(a+\varepsilon)\int_0^1 t^{-1+\gamma}dt + ae^{-1}||T_1||_{W^{r_0\wedge r_1,p}\to W^{r_2,p}}.$$

So we have the assertion. ■

Proposition 6 *Let $p \in (1,\infty)$ and $\varepsilon \in (0,1]$. If $p(1-\varepsilon) < 1$, then*

$$\sup_{s\in(0,1],x^1>0} ||1_{(0,\infty)}(\min_{t\in[0,1]}(x^1 + s^{1/2}B^1(t)))||_{W^{1-\varepsilon,p}} < \infty,$$

and so

$$\sup_{s\in(0,1],x^1>0} ||1_{(0,\infty)}(\min_{t\in[0,s]}(x^1 + B^1(t)))||_{W^{1-\varepsilon,p}} < \infty.$$

Proof Let $Y = \min_{t\in[0,1]} B^1(r)$. Then

$$|Y(w+h) - Y(w)| \leqq \max_{t\in[0,1]} |h(t)| \leqq \int_0^1 |\frac{dh^1}{dr}(r)|dr \leqq ||h||_H$$

for any $w \in W_0$ and $h \in H$. Therefore $||DY||_H \leqq 1$ μ-$a.s.$

Let $\varphi \in C_0^\infty(\mathbf{R})$ such that $\varphi \geqq 0$, $\varphi(z) = 0$, $|z| > 1$, and $\int_{\mathbf{R}} \varphi(z)dz = 1$. Also, let

$$\psi_r(z) = \frac{1}{r}\int_{-\infty}^z \varphi(r^{-1}y)dy, \qquad r \in (0,1], z \in \mathbf{R},$$

and

$$G_r(s,x^1) = \psi_r(s^{-1/2}x^1 + Y), \qquad r, s \in (0,1],\ x^1 > 0.$$

Then we see that $0 \leqq \psi_r \leqq 1$, $\psi_r(z) = 0$, $z \in (-\infty,-r]$, and $\psi_r(z) = 1$, $z \in [r,\infty)$. Also, we see that

$$DG_r(s,x^1) = \frac{1}{r}\varphi(r^{-1}(s^{-1/2}x^1 + Y))DY,$$

and so

$$E^\mu[||DG_r(s,x^1)||_H^p] \leqq r^{-p}E^\mu[\varphi(r^{-1}(s^{-1/2}x^1 + Y))^p]$$

$$\leqq r^{-p}||\varphi||_\infty^p \mu(|s^{-1/2}x^1 + Y| \leqq r).$$

Note that

$$\mu(|s^{-1/2}x_1 + Y| \leqq r) = \mu(Y \in [-s^{-1/2}x_1 - r, -s^{-1/2}x_1 + r])$$
$$\leqq 4(2\pi)^{-1/2} r \leqq 2r.$$

So we have

$$E^{\mu}[||DG_r(s, x^1)||_H^p]^{1/p} \leqq 2r^{-(1-1/p)}||\varphi||_\infty.$$

Also, note that

$$|1_{(0,\infty)}(\min_{t\in[0,1]}(x^1 + s^{1/2}B^1(t))) - G_r(s, x^1)|$$
$$= |1_{(0,\infty)}(s^{-1/2}x^1 + Y) - \psi_r(s^{-1/2}x^1 + Y)| \leqq 1_{(-r,r)}(s^{-1/2}x^1 + Y).$$

Then we have

$$||1_{(0,\infty)}(x^1 + s^{1/2}Y) - G_r(s, x^1)||^p_{L^p(d\mu)} \leqq 2r.$$

So we see that

$$\sup_{r\in(0,1]}(r^{-1/p}||1_{(0,\infty)}(\min_{t\in[0,1]}(x^1 + s^{1/2}B^1(t))) - G_r(s, x^1)||_{W^{0,p}} + r^{1-1/p}||G_r(s, x^1)||_{W^{1,p}})$$
$$\leqq 2 + (2 + 2||\varphi||_\infty).$$

Also, it is obvious that

$$\sup_{r\in[1,\infty)} r^{-1/p}||1_{(0,\infty)}(\max_{t\in[0,t]}(x^1 + s^{1/2}B^1(t))||_{W^{0,p}} \leqq 1.$$

Since $1 - \varepsilon < 1/p$, we have our assertion by Proposition 5. ∎

2.2 Basic Results

Let $V_{s,0}(x) = sV_0(x)$, $V_{s,i}(x) = s^{1/2}V_i(x)$, $i = 1, \ldots, d$, $s \in (0, 1]$. Let us think of the following SDE with a parameter $s \in (0, 1]$.

$$dX_s(t, x) = \sum_{i=0}^{d} V_{s,i}(X_s(t, x)) \circ dB^i(t),$$
$$X_s(0, x) = x \in \mathbf{R}^N.$$

Let us define a homomorphism $\Phi_s : \mathbf{R}\langle A \rangle \to \mathscr{DO}(\mathbf{R}^N)$, $s \in (0, 1]$, by

$$\Phi_s(1) = Identity$$

and

$$\Phi_s(v_{i_1} \cdots v_{i_n}) = V_{s,i_1} \cdots V_{s,i_n}, \qquad n \geqq 1,\ i_1, \ldots, i_n = 0, 1, \ldots, d.$$

Then we see the following.

$$\Phi_s(r(u))(x) = \sum_{u' \in A^{**}_{\leqq \ell_0}} s^{(||u||-||u'||)/2} \varphi_{u,u'}(x) \Phi_s(r(u'))(x), \qquad s \in (0, 1],\ x \in \mathbf{R}^N$$

for any $u \in A^{**} \setminus A^{**}_{\leqq \ell_0}$. Here $\varphi_{v_k u, u'}$'s are as in the assumption (UFG).

From now on, we follow results in [4] basically . For any C_b^∞ vector field W on $\mathbf{R}^N$, we define $(X_s(t)_* W)(X(t, x)) = \sum_{i,j=1}^N \frac{\partial}{\partial x^j} X_s^i(t, x) W^j(x) \frac{\partial}{\partial x^i}$. Then $X_s(t)_*$ is a push-forward operator with respect to the diffeomorphism $X_s(t, \cdot) : \mathbf{R}^N \to \mathbf{R}^N$ for any $s \in (0, 1]$. Also we see that

$$\begin{aligned} &d(X_s(t)_*^{-1} \Phi_s(r(u)))(x) \\ &= \sum_{i=0}^d (X_s(t)_*^{-1} \Phi_s(r(v_i u)))(x) \circ dB^i(t) \end{aligned}$$

for any $u \in A^* \setminus \{1\}$.

Let $c_k^{(s)}(\cdot, u, u') \in C_b^\infty(\mathbf{R}^N, \mathbf{R})$, $k = 0, 1, \ldots, d$, $u, u' \in A^{**}_{\leqq \ell_0}$, be given by

$$c_k^{(s)}(x; u, u') = \begin{cases} 1, & \text{if } ||v_k u|| \leqq \ell_0 \text{ and } u' = v_k u, \\ s^{(||v_k u||-||u'||)/2} \varphi_{v_k u, u'}(x), & \text{if } ||v_k u|| > \ell_0 \text{ and } ||u'|| \leqq \ell_0, \\ 0, & \text{otherwise.} \end{cases}$$

Then we have

$$\begin{aligned} &d(X_s(t)_*^{-1} \Phi_s(r(u)))(x) \\ &= \sum_{k=0}^d \sum_{u' \in A^{**}_{\leqq \ell_0}} c_k^{(s)}(X(t, x); u, u')(X_s(t)_*^{-1} \Phi_s(r(u')))(x) \circ dB^k(t), \\ &\qquad\qquad\qquad\qquad u \in A^{**}_{\leqq \ell_0}. \end{aligned}$$

There exists a unique solution $a_s(t,x;u,u')$, $u,u' \in A^{**}_{\leqq \ell_0}$, $s \in (0,1]$, to the following SDE

$$da_s(t,x;u,u') = \sum_{k=0}^{d} \sum_{u'' \in A^{**}_{\leqq \ell_0}} (c_k^{(s)}(X_s(t,x);u,u'')a_s(t,x;u'',u')) \circ dB^k(t) \tag{2}$$

$$a_s(0,x;u,u') = \delta_{u,u'}.$$

Then the uniqueness of SDE implies

$$(X_s(t)_*^{-1}\Phi_s(r(u)))(x) = \sum_{u' \in A^{**}_{\leqq \ell_0}} a_s(t,x;u,u')\Phi_s(r(u'))(x) \tag{3}$$

for $u \in A^{**}_{\leqq \ell_0}$, $s \in (0,1]$.

Similarly we see that there exists a unique solution $b_s(t,x;u,u')$, $u,u' \in A^{**}_{\leqq \ell_0}$, to the SDE

$$b_s(t,x;u,u')$$
$$= \delta_{u,u'} - \sum_{k=0}^{d} \sum_{u'' \in A^{**}_{\leqq \ell_0}} \int_0^t (b_s(r,x;u,u'')c_k^{(s)}(X_s(r,x);u'',u')) \circ dB^k(r). \tag{4}$$

Then we see that

$$\sum_{u'' \in A^{**}_{\leqq \ell_0}} a_s(t,x,u,u'')b_s(t,x,u'',u') = \delta_{u,u'}, \qquad u,u' \in A^{**}_{\leqq \ell_0},$$

and that

$$\Phi_s(r(u))(x) = \sum_{u' \in A^{**}_{\leqq \ell_0}} b_s(t,x;u,u')(X_s(t)_*^{-1}\Phi_s(r(u')))(x), \qquad u \in A^{**}_{\leqq \ell_0}. \tag{5}$$

Furthermore we see by Proposition 1 (1) that

$$a_s(t,x,u,v_1) = b_s(t,x,u,v_1) = 0, \ a.s. \qquad u \in \tilde{A}.$$

Also, we see that

$$(X_s(t)_*^{-1}\Phi_s(v_0))(x)$$
$$= \Phi_s(v_0) + \sum_{k=1}^{d} \int_0^t (X_s(r)_*^{-1}\Phi_s(r(v_k v_0))) \circ dB^k(r).$$

So we have

$$(X_s(t)_*^{-1}\Phi_s(v_0))(x) = \Phi_s(v_0)(x) + \sum_{u\in\tilde{A}} \hat{a}_s(t, x; u)\Phi_s(r(u))(x), \tag{6}$$

and

$$\Phi_s(v_0)(x) = (X_s(t)_*^{-1}\Phi_s(v_0))(x) + \sum_{u\in\tilde{A}} \hat{b}_s(t, x; u)(X_s(t)_*^{-1}\Phi_s(r(u))(x), \tag{7}$$

where

$$\hat{a}_s(t, x; u) = \sum_{k=1}^{d} \int_0^t a_s(r, x; v_k v_0, u) \circ dB^k(r)$$

and

$$\hat{b}_s(t, x; u) = -\sum_{u'\in\tilde{A}} b_s(t, x; u, u')\hat{a}(t, x; u').$$

Note that

$$\Phi_s(r(u))(f(X_s(t, x)) = \langle X_s(t)^* df, \Phi_s(r(u))\rangle_x.$$

So we have

$$\Phi_s(r(u))(f(X_s(t, x))) = \sum_{u'\in A^{**}_{\leqq \ell_0}} b_s(t, x; u, u')(\Phi_s(r(u'))f)(X_s(t, x)) \tag{8}$$

and

$$(\Phi_s(r(u))f)(X_s(t, x)) = \sum_{u'\in A^{**}_{\leqq \ell_0}} a_s(t, x; u, u')\Phi_s(r(u'))(f(X_s(t, x))) \tag{9}$$

for $u \in A^{**}_{\leqq \ell_0}$. Also we see that

$$\begin{aligned}&\Phi_s(v_0)(f(X_s(t, x)) - (\Phi_s(v_0)f)(X_s(t, x)) \\ &\quad = \sum_{u'\in \mathscr{A}^{**}_{\leqq \ell_0}} \hat{b}_s(t, x; u')(\Phi_s(r(u'))f)(X_s(t, x)).\end{aligned} \tag{10}$$

Let us define $k_s : [0,\infty) \times \mathbf{R}^N \times A^{**}_{\leqq \ell_0} \times W_0 \to H$ by

$$k_s(t,x;u) = (\int_0^{t\wedge\cdot} a_s(r,x;v_k,u)dr)_{k=1,\ldots,d}.$$

Let $M_s(t,x) = \{M_s(t,x;u,u')\}_{u,u' \in A^{**}_{\leqq \ell_0}}$ be a matrix-valued random variable given by

$$M_s(t,x;u,u') = t^{-(||u||+||u'||)/2}(k_s(t,x;u), k_s(t,x;u'))_H.$$

Then we have

$$\sup_{s\in(0,1]}\sup_{t\in(0,T]}\sup_{x\in\mathbf{R}^N} E^\mu[|\det M_s(t,x)|^{-p}] < \infty \text{ for any } p \in (1,\infty) \text{ and } T > 0.$$

Let $M_s^{-1}(t,x) = \{M_s^{-1}(t,x;u,u')\}_{u,u' \in A^{**}_{\leqq \ell_0}}$ be the inverse matrix of $M_s(t,x)$.

For any separable real Hilbert space E, let $\hat{\mathscr{K}}_0(E)$ be the set of $\{F_s\}_{s\in(0,1]}$ such that

(1) $F_s : (0,\infty) \times \mathbf{R}^N \times W_0 \to E$ is measurable map for all $s \in (0,1]$,
(2) $F_s(t,\cdot,w) : \mathbf{R}^N \to E$ is smooth for any $s \in (0,1]$, $t \in (0,\infty)$ and $w \in W_0$,
(3) $(\partial^\alpha F_s/\partial x^\alpha)(\cdot,*,w) : (0,\infty) \times \mathbf{R}^N \to E$ is continuous for any $s \in (0,1]$, $w \in W_0$ and $\alpha \in \mathbf{Z}^N_{\geqq 0}$,
(4) $(\partial^\alpha F_s/\partial x^\alpha)(t,x,\cdot) \in W^{r,p}$ for any $s \in (0,1]$, $r,p \in (1,\infty)$, $\alpha \in \mathbf{Z}^N_{\geqq 0}$, $t \in (0,\infty)$ and $x \in \mathbf{R}^N$,
and
(5) for any $r,p \in (1,\infty)$, $\alpha \in \mathbf{Z}^N_{\geqq 0}$, and $T > 0$

$$\sup_{s\in(0,1],t\in(0,T]}\sup_{x\in\mathbf{R}^N} ||\frac{\partial^\alpha}{\partial x^\alpha}F_s(t,x)||_{W^{r,p}} < \infty.$$

Then we have the following.

Proposition 7

(1) $\{t^{-(||u'||-||u||)/2}a_s(t,x;u,u')\}_{s\in(0,1]}$ *and* $\{t^{-(||u'||-||u||)/2}b_s(t,x;u,u')\}_{s\in(0,1]}$ *belong to* $\hat{\mathscr{K}}_0(\mathbf{R})$ *for any* $u, u' \in A^{**}_{\leqq \ell_0}$.
(2) $\{t^{-||u||/2}k_s(t,x;u)\}_{s\in(0,1]}$ *belongs to* $\hat{\mathscr{K}}_0(H)$ *for any* $u \in A^{**}_{\leqq \ell_0}$.
(3) $\{M_s(t,x;u,u')\}_{s\in(0,1]}$, *and* $\{M_s^{-1}(t,x;u,u')\}_{s\in(0,1]}$ *belong to* $\hat{\mathscr{K}}_0(\mathbf{R})$ *for any* $u, u' \in A^{**}_{\leqq \ell_0}$.
(4) $\{\hat{a}_s(t,x;u)\}_{s\in(0,1]}$ *and* $\{\hat{b}_s(t,x;u)\}_{s\in(0,1]}$ *belong to* $\hat{\mathscr{K}}_0(\mathbf{R})$ *for any* $u \in \tilde{A}$.

Finally we have the following basic equation.

$$t^{-\|u\|/2}(\Phi_s(u)f)(X_s(t,x))$$
$$= \sum_{u_1,u_2\in A^{**}_{\leqq \ell_0}} a_s(t,x;u,u_1)M_s^{-1}(t,x;u_1,u_2)$$
$$(D(f(X_s(t,x)), t^{-\|u_2\|/2}k_s(t,x;u_2))_H \qquad (11)$$

for any $f \in \mathscr{C}$ and $u \in \tilde{A}$.

By Proposition 7 and Eq. (11), we easily see the following.

Proposition 8 *For any $p \in (1,\infty)$, there is a constant $C \in (0,\infty)$ such that*

$$\|(\Phi_s(u)f)(X_s(t,x))\|_{W^{0,p}} \leqq C\|f\|_{\mathscr{D}^1_{(s)}},$$

and

$$\|t^{\|u\|/2}(\Phi_s(u)f)(X_s(t,x))\|_{W^{-1,p}} \leqq C\|f\|_\infty,$$

for any $u \in \tilde{A}$, $f \in \mathscr{C}$ and $s,t \in (0,1]$.

Proposition 9 *For any $\alpha \in [0,1)$ and $p \in (1,\infty)$, there is a constant $C \in (0,\infty)$ such that*

$$\|f(X_s(t,x))\|_{W^{-1,p}} \leqq Ct^{-\ell_0/2}\|f\|_{\mathscr{H}^{-\alpha}_{(s)}}$$

for any $f \in \mathscr{C}$, $s,t \in (0,1]$ and $x \in \mathbf{R}^N$.

Proof Note that

$$f = Q^{(s)}_1 f - \int_0^1 L^{(s)}Q^{(s)}_r f dr = Q^{(s)}_1 f - \frac{1}{2}\sum_{u\in\tilde{A}} \Phi_s(r(u))f_u,$$

where

$$f_u = \int_0^1 \Phi_s(r(u))Q^{(s)}_t f dt.$$

By definition, we have

$$\|Q^{(s)}_1 f\|_\infty \leqq \|f\|_{\mathscr{H}^{-\alpha}_{(s)}},$$

and

$$||f_u||_\infty \leqq \int_0^1 ||\Phi_s(r(u))Q^{(s)}_{t/2}Q^{(s)}_{t/2}f||_\infty dt \leqq C_0 \int_0^1 (\frac{t}{2})^{-1/2}||Q^{(s)}_{t/2}f||_\infty dt$$
$$\leqq C_0(\int_0^1 (\frac{t}{2})^{-(1+\alpha)/2}dt)||f||_{\mathscr{H}^{-\alpha}_{(s)}}.$$

Since

$$f(X_s(t,x)) = (Q^{(s)}_1 f)(X_s(t,x)) - \frac{1}{2}\sum_{u\in\tilde{A}}(\Phi_s(r(u))f_u)(X_s(t,x)).$$

we have our assertion from Proposition 8. ■

2.3 Main Lemma

For any $K = \{K_s\}_{s\in(0,1]} \in \hat{\mathscr{K}}_0(\mathbf{R})$, let $P^{s,K}_{(t)}$, $t > 0$, be linear operators defined in $\mathscr{C}$ given by

$$(P^{s,K}_{(t)} f)(x) = E[K_s(t,x)f_s(X_s(t,x)), \min_{r\in[0,t]} X^1_s(t,x) > 0], \qquad f \in \mathscr{C}.$$

Since $\min_{r\in[0,t]}(X^1_s(t,x)) = \min_{r\in[0,t]}(s^{1/2}B^1(t) + x^1)$ and it does not depend on $x^2,\ldots,x^N$, we see that $P^{s,K}_{(t)} f \in \mathscr{C}$ for any $f \in \mathscr{C}$ and $t \geqq 0$.

In this section, we prove the following.

Lemma 1 *For any $K_1, K_2 \in \hat{\mathscr{K}}_0(\mathbf{R})$, there is a $C \in (0,\infty)$ such that*

$$||P^{s,K_1}_{(t)} P^{s,K_2}_{(t)} \Phi_s(r(u))f||_\infty \leqq Ct^{-\ell_0}||f||_\infty$$

for any $s,t \in (0,1]$, $f \in \mathscr{C}$ and $u \in \tilde{A}$.

We need some preparations to prove this lemma.

Proposition 10 *For any $K \in \hat{\mathscr{K}}_0(\mathbf{R})$, $\varepsilon \in (0,1)$ and $p \in (1/\varepsilon,\infty)$, there is a $C \in (0,\infty)$ such that*

$$||P^{s,K}_{(t)} f||_\infty \leqq C||f(X_s(t,x))||_{W^{-1+\varepsilon,p}}, \qquad s,t \in (0,1],\ f \in \mathscr{C}.$$

Proof There is a $q \in (1, (1-\varepsilon)^{-1})$ and $r \in (1, \infty)$ such that $q^{-1}+r^{-1}+p^{-1} = 1$. Then there is a $C_1 \in (0, \infty)$ such that

$$|P^{s,K}_{(t)} f(x)| \leqq C_1 ||1_{(0,\infty)}(\min_{r\in[0,t]} (s^{-1/2}x^1 + B^1(r)))||_{W^{1-\varepsilon,q}} \times ||K_s(t, x)||_{W^{1,r}} ||f(X_s(t, x))||_{W^{-1+\varepsilon,p}}$$

for any $s, t \in (0, 1]$, $x \in \mathbf{R}^N$ and $f \in \mathscr{C}$. So we have our assertion from Proposition 6. ∎

Proposition 11 *Let $K \in \hat{\mathscr{K}}_0(\mathbf{R})$. Then for any $\alpha \in (0, 1)$ there is a $C \in (0, \infty)$ such that*

$$||P^{s,K}_{(t)} f||_\infty \leqq Ct^{-\ell_0/2}||f||_{\mathscr{H}^{-\alpha}_{(s)}}$$

for any $s, t \in (0, 1]$ and $f \in \mathscr{C}$.

Proof Let $\alpha \in (0, 1)$. Then if we take a sufficiently small $\theta \in (0, 1)$, there is a $\beta \in (0, 1)$ such that $\alpha = (1-\theta)\beta-\theta$. Let us take an $\varepsilon \in (0, \theta)$, and a $p \in (1/\varepsilon, \infty)$. Then $-1+\varepsilon < -(1-\theta)$.

First note that

$$||f(X_s(t, x))||_{W^{0,p}} \leqq ||f||_\infty \leqq ||f||_{\mathscr{D}^1_{(s)}}.$$

for any $s \in (0, 1]$, $p \in (1, \infty)$ and $f \in \mathscr{C}$.

Also, by Proposition 9 there is a constant $C_1 \in (0, \infty)$ such that

$$||f(X_s(t, x))||_{W^{-1,p}} \leqq C_1 t^{-\ell_0/2}||f||_{\mathscr{H}^{-\beta}_{(s)}}$$

for any $f \in \mathscr{C}$, $s, t \in (0, 1]$ and $x \in \mathbf{R}^N$. Then by Propositions 3, 5, 7, and 8, we see that there are constants $C_2, C_3 \in (0, \infty)$ such that

$$||f(X_s(t, x))||_{W^{-1+\varepsilon,p}} \leqq C_2 \sup_{r\in(0,\infty)} r^{-\theta} K(r; f(X_s(t, x)); W^{-1,p}, W^{-0,p})$$

$$\leqq C_1C_2t^{-\ell_0/2} \sup_{r\in(0,\infty)} r^{-\theta} K(r; f; \mathscr{H}^{-\beta}_{(s)}, \mathscr{D}^1_{(s)}) \leqq C_3t^{-\ell_0/2}||f||_{\mathscr{H}^{-\alpha}_{(s)}}$$

for any $f \in \mathscr{C}$, $s, t \in (0, 1]$ and $x \in \mathbf{R}^N$. Then by Proposition 10 we have our assertion. ∎

Now let us prove Lemma 1.

Let $K_1, K_2 \in \hat{\mathscr{K}_0}(\mathbf{R})$. By Propositions 8, 11, and Eq. (13) we see that for any $p \in (1, \infty)$ and $\alpha \in [0, 1)$, there is a constant $C_1 \in (0, \infty)$ such that

$$||(P_{(t)}^{s,K_2}\Phi_s(r(u))f)(X_s(t,x))||_{W^{-1,p}} \leqq C_1 t^{-\ell_0/2}||f||_{\mathscr{H}_{(s)}^{-1/2}},$$

for any $u \in \tilde{A}$, $f \in \mathscr{C}$ and $s \in (0, 1]$. It is obvious that for any $p \in (1, \infty)$, there is a constant $C > 0$ such that

$$||(P_{(t)}^{s,K_2}\Phi_s(r(u))f)(X_s(t,x))||_{W^{0,p}} \leqq ||f||_{\mathscr{D}_{(s)}^1},$$

for any $u \in \tilde{A}$, $f \in \mathscr{C}$, and $s \in (0, 1]$.

Take an $\varepsilon \in (0, 1/3)$. Then $-1 + \varepsilon < -(1 - 1/3)$. Let us take a $p \in (1/\varepsilon, \infty)$. By Propositions 4 and 5, we see that there are constants $C_2, C_3 \in (0, \infty)$ such that

$$\begin{aligned}
&||(P_{(t)}^{s,K_2}\Phi_s(r(u))f)(X_s(t,x))||_{W^{-1+\varepsilon,p}} \\
&\leqq C_2 t^{-\ell_0/2} \sup_{r\in(0,\infty)} r^{-1/3} K(r; (P_{(t)}^{s,K_2}\Phi_s(r(u))f)(X_s(t,x)); W^{-1,p}, W^{0,p}) \\
&\leqq C_2 t^{-\ell_0} \sup_{r\in(0,\infty)} r^{-1/3} K(r; f; \mathscr{H}_{(s)}^{-1/2}, \mathscr{D}_{(s)}^1) \leqq C_3 t^{-\ell_0}||f||_\infty
\end{aligned}$$

for any $f \in \mathscr{C}$, $s, t \in (0, 1]$ and $x \in \mathbf{R}^N$. Then by Proposition 10 we have Lemma 1.

This completes the proof of Lemma 1.

2.4 Proof of Theorem 3(1)

The following is an easy consequence of Lemma 1, Eqs. (12) and (13).

Corollary 2 *Let $K_1, K_2 \in \hat{\mathscr{K}_0}(\mathbf{R})$. Then for any $n \geqq 0$ there is a $C > 0$ such that*

$$\sum_{k=0}^{n+1} \sum_{u_1,\ldots,u_k\in\tilde{A}} ||\Phi_s(r(u_1))\ldots\Phi_s(r(u_k))P_{(t)}^{s,K_1}P_{(t)}^{s,K_2}f||_\infty$$

$$\leqq Ct^{-\ell_0}\sum_{k=0}^{n} \sum_{u_1,\ldots,u_k\in\tilde{A}} ||\Phi_s(r(u_1))\ldots\Phi_s(r(u_k))f)||_\infty$$

for any $s, t \in (0, 1]$ and $f \in \mathscr{C}$.

For linear operators A and B in $\mathscr{C}$ we define $adj(A)^n(B)$, $n = 0, 1, \dots,$ inductively by $adj(A)^0(B) = B$, and

$$adj(A)^n(B) = A(adj(A)^{n-1}(B)) - (adj(A)^{n-1}(B))A.$$

Then we see that for linear operators A, B, C in $\mathscr{C}$

$$adj(A)^n(BC) = \sum_{k=0}^{n} \binom{n}{k} adj(A)^k(B) adj(A)^{n-k}(C).$$

Now by Eqs. (8) and (9) we have

$$(\Phi_s(r(u))P_{(t)}^{s,K} f)(x) = (P_{(t)}^{s,K_{00}(u)} f)(x) + \sum_{u' \in \tilde{A}} (P_{(t)}^{s,K_0(u;u')} \Phi_s(r(u')) f)(x) \tag{12}$$

and

$$(P_{(t)}^{s,K} \Phi_s(r(u)) f)(x) = (P_{(t)}^{s,K_{10}(u)} f)(x) + \sum_{u' \in \tilde{A}} (\Phi_s(r(u') P_{(t)}^{s,K_1(u;u')} f)(x), \tag{13}$$

for any $u \in \tilde{A}$, $f \in \mathscr{C}$, $s, t \in (0, 1]$ and $x \in \mathbf{R}^N$. Here

$$K_{00}(u)_s(t, x) = (\Phi_s(r(u))K_s(t, \cdot))|_{\cdot = x},$$

$$K_0(u; u')_s(t, x) = b_s(t, x; u, u')K_s(t, x), \qquad u' \in \tilde{A}.$$

$$K_{10}(u)_s(t, x) = -\sum_{u' \in \tilde{A}} (\Phi_s(r(u))(a_s(t, \cdot; u, u')K(t, \cdot))|_{\cdot = x},$$

and

$$K_1(u; u')_s(t, x) = a_s(t, x; u, u')K(t, x), \qquad u' \in \tilde{A}.$$

Also, note that by Eq. (10)

$$(adj(\Phi_s(v_0))(P_{(t)}^{s,K}) f)(x) = (\Phi_s(v_0) P_{(t)}^{s,K} f)(x) - (P_{(t)}^{s,K} \Phi_s(v_0) f)(x)$$

$$= (P_{(t)}^{s,\hat{K}_0} f)(x) + \sum_{u \in \tilde{A}} (P_{(t)}^{s,\hat{K}(u)} \Phi_s(r(u)) f)(x) \tag{14}$$

for any $f \in \mathscr{C}$, $s, t \in (0, 1]$ and $x \in \mathbf{R}^N$. Here

$$\hat{K}_{0s}(t, x) = (\Phi_s(v_0)K_s(t, \cdot))|_{\cdot=x},$$

$$\hat{K}(u)_s(t, x) = \hat{b}_s(t, x; u)K_s(t, x), \qquad u' \in \tilde{A}.$$

By Proposition 7, we see that $K_{00}(u)$, $K_0(u; u')$, $K_{10}(u)$, $K_1(u; u')$, $\hat{K}_0$, $\hat{K}(u)$ $\in \hat{\mathscr{K}_0}(\mathbf{R})$ for any $u, u' \in \tilde{A}$.

So by using Eqs. (12), (13) and (14) we have the following.

Lemma 2 *Let $n \geqq 0$ and $K_1, \ldots, K_{6n} \in \hat{\mathscr{K}_0}(\mathbf{R})$. Then there is a $C \in (0, \infty)$ such that*

$$\sum_{k,j,\ell=0}^{n} \sum_{u_1,\ldots,u_k \in \tilde{A}} \sum_{u'_1,\ldots,u'_\ell \in \tilde{A}} ||\Phi_s(r(u_1)\ldots r(u_k))adj(\Phi_s(v_0))^j$$

$$(P_{(t)}^{s,K_1} \cdots P_{(t)}^{s,K_{6n}})\Phi_s(r(u'_1)\ldots r(u'_\ell))f||_\infty \leqq Ct^{-3n\ell_0}||f||_\infty$$

for any $s, t \in (0, 1]$ and $f \in \mathscr{C}$.

Now we introduce the following notion.

Definition 1 We say that $\{K_s\}_{s\in(0,1]} \in \hat{\mathscr{K}_0}(\mathbf{R})$ is multiplicative, if for any $m \geqq 1$ there are $n \geqq 1$ and $\{K_s^{i,j}\} \in \hat{\mathscr{K}_0}(\mathbf{R})$, $i = 1, \ldots, n$, $j = 1, \ldots, m$, such that

$$\begin{aligned} &K_s(t_m, x, w) \\ &= \sum_{i=1}^{n} K_s^{i,1}(t_1, x, w)K_s^{i,2}(t_2 - t_1, X_s(t_1, x), \theta_{t_1}w)\cdots \\ &\quad \times K_s^{i,m}(t_m - t_{m-1}, X_s(t_{m-1}, x), \theta_{t_{m-1}}w) \end{aligned}$$

for any $s \in (0, 1]$, $0 < t_1 < \ldots < t_m$ and $x \in \mathbf{R}^N$.

Here $\theta_r : W_0 \to W_0$, $r \in [0, \infty)$, is given by $(\theta_r w)(t) = w(t + r) - w(r)$, $w \in W_0$, $t \in [0, \infty)$.

Proposition 12 *Let $\{K_s\}_{s\in(0,1]}, \{L_s\}_{s\in(0,1]} \in \hat{\mathscr{K}_0}(\mathbf{R})$ be multiplicative. Then $\{K_s + L_s\}_{s\in(0,1]}$ and $\{K_sL_s\}_{s\in(0,1]}$ are multiplicative.*

Proof Let $m \geqq 2$. Since K_s and L_s are multiplicative, there are $n_1, n_2 \geqq 1$, $\{K_s^{i,j}\} \in \hat{\mathscr{K}_0}(\mathbf{R})$, $i = 1, \ldots n_1$, $j = 1, \ldots, m$, and $\{L_s^{ij}\} \in \hat{\mathscr{K}_0}(\mathbf{R})$, $i = 1, \ldots n_2$, $j = 1, \ldots, m$, such that

$$\begin{aligned} &K_s(t_n, x, w) \\ &= \sum_{i=1}^{n_1} K_s^{i,1}(t_1, x, w)K_s^{i,2}(t_2 - t_1, X_s(t_1, x), \theta_{t_1}w)\cdots \\ &\quad \times K_s^{i,m}(t_m - t_{m-1}, X_s(t_{m-1}, x), \theta_{t_{m-1}}w), \end{aligned}$$

and

$$\begin{aligned}
&L_s(t_n, x, w) \\
&= \sum_{i=1}^{n_2} L_s^{i,1}(t_1, x, w) L_s^{i,2}(t_2 - t_1, X_s(t_1, x), \theta_{t_1} w) \cdots \\
&\times L_s^{i,m}(t_m - t_{m-1}, X_s(t_{m-1}, x), \theta_{t_{m-1}} w),
\end{aligned}$$

for any $s \in (0, 1]$ $0 < t_1 < \ldots < t_m$ and $x \in \mathbf{R}^N$.

Then we have

$$\begin{aligned}
&K_s(t_n, x, w) + L_s(t_n, x, w) \\
&= \sum_{i=1}^{n_1} K_s^{i,1}(t_1, x, w) K_s^{i,2}(t_2 - t_1, X_s(t_1, x), \theta_{t_1} w) \cdots \\
&\times K_s^{i,m}(t_m - t_{m-1}, X_s(t_{m-1}, x), \theta_{t_{m-1}} w) \\
&+ \sum_{i=1}^{n_2} L_s^{i,1}(t_1, x, w) K_s^{i2}(t_2 - t_1, X_s(t_1, x), \theta_{t_1} w) \cdots \\
&\times L_s^{i,m}(t_m - t_{m-1}, X_s(t_{m-1}, x), \theta_{t_{m-1}} w),
\end{aligned}$$

and

$$\begin{aligned}
&K_s(t_n, x, w) L_s(t_n, x, w) \\
&= \sum_{i=1}^{n_1} \sum_{j=1}^{n_2} (K_s^{i,1}(t_1, x, w) L_s^{j,1}(t_1, x, w)) \\
&\times (K_s^{i,2}(t_2 - t_1, X_s(t_1, x), \theta_{t_1} w) L_s^{j,2}(t_2 - t_1, X_s(t_1, x), \theta_{t_1} w))) \cdots \\
&\times (K_s^{i,m}(t_m - t_{m-1}, X_s(t_{m-1}, x), \theta_{t_{m-1}} w) L_s^{j,m}(t_m - t_{m-1}, X_s(t_{m-1}, x), \theta_{t_{m-1}} w))).
\end{aligned}$$

So we have our assertion. ■

Proposition 13 *Let* $M \geqq 1$ *and* $d_s^{ijk} \in C_b^\infty(\mathbf{R}^N)$, $i, j = 1, \ldots, M$, $k = 0, 1, \ldots, d$, $s \in (0, 1]$. *and assume that*

$$\sup_{s \in (0,1]} \sup_{x \in \mathbf{R}^N} |\frac{\partial^{|\alpha|}}{\partial x^\alpha} d_s^{ijk}(x)| < \infty$$

for any $\alpha \in \mathscr{Z}_{\geqq 0}^N$.

Let $y^i \in \mathbf{R}$, and $Y_s^i(t,x)$, $i = 1, \ldots, M$, $s \in (0,1]$, $t \geqq 0$, $x \in \mathbf{R}^N$, be the solution to the following SDE.

$$Y_s^i(t,x) = y^i + \sum_{k=0}^{d}\sum_{\ell=1}^{M}\int_0^t d_s^{i\ell k}(X_s(r,x))Y_s^{\ell}(r,x)\circ dB^k(r), \qquad i = 1, \ldots, M.$$

Then we see that $\{Y_s^i\}_{s\in(0,1]}$ belongs to $\hat{\mathscr{K}}_0$, and is multiplicative for $i, j = 1, \ldots, M$.

Also, $\{\int_0^t Y_s^i(r,x)dr\}$ belongs to $\hat{\mathscr{K}}_0$, and is multiplicative.

Proof Let $E_s^{i,j}(t,x)$, $i, j = 1, \ldots, M$, $s \in (0,1]$, $t \geqq 0$, $x \in \mathbf{R}^N$, be the solution to the following SDE.

$$E_s^{i,j}(t,x) = \delta_{ij} + \sum_{k=0}^{d}\sum_{\ell=1}^{M}\int_0^t d_s^{i\ell k}(X_s(r,x))E_s^{\ell,j}(r,x)\circ dB^k(r) \qquad i, j = 1, \ldots, M.$$

Then it is easy to see that $\{E_s^{i,j}\}_{s\in(0,1]} \in \hat{\mathscr{K}}_0$, and

$$Y_s^i(t,x) = \sum_{j=1}^{M} E_s^{i,j}(t,x)y_j.$$

Note that for $t_2 > t_1 \geqq 0$,

$$E_s^{i,j}(t_2,x,w) = \sum_{\ell=1}^{M} E_s^{i,\ell}(t_2-t_1, X(t_1,x,w), \theta_{t_1}w)E_s^{\ell,j}(t_1,x,w), \ i, j = 1, \ldots, M.$$

So we see that $\{E_s^{i,j}\}_{s\in(0,1]}$, $i, j = 1, \ldots, M$, are multiplicative.

Also, we see that

$$\int_0^{t_2} E_s^{i,j}(r,x,w)dr = \int_0^{t_1} E_s^{i,j}(r,x,w)dr + \sum_{\ell=1}^{M}\left(\int_0^{t_2-t_1} E_s^{i,\ell}(r, X(t_2-t_1,x,w), \theta_{t_1}w)dr\right)E_s^{\ell,j}(t_1,x,w),$$

$i, j = 1, \ldots, M$. So we see that $\{\int_0^t E_s^{i,j}(r,x)dr\}_{s\in(0,1]}$, $i, j = 1, \ldots, M$, are multiplicative. These imply our assertion. ■

Proposition 14 *Let $\{K_s\}_{s\in(0,1]} \in \hat{\mathscr{K}}_0(\mathbf{R})$. be multiplicative. Then $\{\frac{\partial}{\partial x^i}K_s\}_{s\in(0,1]}$ is multiplicative for any $i = 1, 2, \ldots, N$.*

Proof Let $m \geqq 1$, and $0 < t_1 < \ldots < t_m$ and $x \in \mathbf{R}^N$. Then from the assumption there are $n \geqq 1$ and $\{K_s^{i,j}\} \in \hat{\mathscr{K}_0}(\mathbf{R})$, $i = 1, \ldots m$, $j = 1, \ldots, n$, such that

$$\begin{aligned} &K_s(t_m, x, w) \\ &\quad = \sum_{i=1}^{n} K_s^{i,1}(t_1, x, w) K_s^{i,2}(t_2 - t_1, X_s(t_1, x), \theta_{t_1} w) \cdots \\ &\quad \times K_s^{i,m}(t_m - t_{m-1}, X_s(t_{m-1}, x), \theta_{t_{m-1}} w). \end{aligned}$$

Note that

$$X(t_{k+1}, x) = X(t_{k+1} - t_k, X(t_k, x), \theta_{t_k} w), \qquad k = 0, 1, \ldots, m-1.$$

Here $t_0 = 0$. Then we have

$$\begin{aligned} &\frac{\partial}{\partial x^j} X^i(t_{k+1}, x) \\ &\quad = \sum_{\ell=1}^{N} \frac{\partial X^i}{\partial x^\ell}(t_{k+1} - t_k, X(t_k, x), \theta_{t_k} w) \frac{\partial X^\ell}{\partial x^j}(t_{k+1}, x). \end{aligned}$$

This implies that

$$\begin{aligned} &\frac{\partial}{\partial x^j} X^i(t_{k+1}, x) \\ &= \sum_{\ell_k, \ell_{k-1}, \ldots, \ell_1 = 1}^{N} \frac{\partial X^{\ell_1}}{\partial x^j}(t_1, x) (\prod_{r=1}^{k-1} \frac{\partial X^{\ell_r}}{\partial x^{\ell_{r-1}}}(t_{r+1} - t_r, X(t_r, x), \theta_{t_r} w)) \\ &\quad \times \frac{\partial X^i}{\partial x^{\ell_k}}(t_{k+1} - t_k, X(t_k, x), \theta_{t_k} w). \end{aligned}$$

Also, we see that

$$\begin{aligned} &\frac{\partial}{\partial x^j} K_s(t_n, x, w) \\ &= \sum_{k=1}^{n} \sum_{\ell=1}^{N} \sum_{i=1}^{m_1} K_s^{i,1}(t_1, x, w) K_s^{i,2}(t_2 - t_1, X_s(t_1, x), \theta_{t_1} w) \cdots \\ &\quad \times K_s^{i,k-1}(t_n - t_{n-1}, X_s(t_{n-1}, x), \theta_{t_{n-1}} w) \\ &\quad \times \frac{\partial K_s^{i,k}}{\partial x^\ell}(t_k - t_{k-1}, X_s(t_{k-1}, x), \theta_{t_1} w) \frac{\partial X_s^\ell}{\partial x^j}(t_k - t_{k-1}, x) \\ &\quad \times K_s^{i,k+1}(t_2 - t_1, X_s(t_1, x), \theta_{t_1} w) \cdots K_s^{i,n}(t_n - t_{n-1}, X_s(t_{n-1}, x), \theta_{t_{n-1}} w). \end{aligned}$$

These observation imply our assertion. ∎

We see that if $\{K_s\}_{s\in(0,1]} \in \hat{\mathscr{K}}_0(\mathbf{R})$ is multiplicative, then

$$P^{s,K}_{(nt)} = \sum_{i=1}^{m} P^{s,K_s^{i,1}}_{(t)} P^{s,K_s^{i,2}}_{(t)} \cdots P^{s,K_s^{i,n}}_{(t)},$$

where $\{K_s^{i,j}\} \in \mathscr{K}_0(\mathbf{R})$, $i = 1, \dots m$, $j = 1, \dots, n$, are as in Definition 1.

So by Lemma 2 we have the following.

Theorem 4 *Suppose that* $\{K_s\}_{s\in(0,1]} \in \hat{\mathscr{K}}_0(\mathbf{R})$ *is multiplicative. Then for any* $n, m, r \geqq 0$, *and* $u_1, \dots, u_{n+m} \in \tilde{A}$, *there is a* $C \in (0, \infty)$ *such that*

$$||\Phi_s(u_1)\dots\Phi_s(u_n)(adj(\Phi_s(v_0))^r(P^{s,K}_{(t)}))\Phi_s(u_{n+1})\dots\Phi_s(u_{n+m})f)||_\infty$$

$$\leqq Ct^{-(n+m+r)\ell_0}||f||_\infty$$

for any $s, t \in (0, 1]$ *and* $f \in \mathscr{C}$.

Now let us prove Theorem 3(1). Let $\rho_s(t, x)$ be the solution to the following SDE.

$$\begin{aligned}&\rho_s(t, x)\\ &= \exp(s^{1/2}\sum_{k=1}^{d}\int_0^t b_k(X_s(r, x))dB^k(r)) + s\int_0^t b_0(X_s(r, x))dB^0(r)),\\ &x \in \mathbf{R}^N,\ t \geqq 0.\end{aligned}$$

Then we see that

$$\begin{aligned}\rho_s(t, x) = 1 + s^{1/2}&\sum_{k=1}^{d}\int_0^t b_k(X_s(r, x))\rho_s(r, x)\circ dB^k(r)\\ &+ s\int_0^t (b_0(X_s(r, x)) + \frac{1}{2}\sum_{k=1}^{d} b_k(X_s(r, x))^2)\rho_s(r, x)dB^0(r).\end{aligned}$$

So we see that $\{\rho_s\}_{s\in(0,1]} \in \hat{\mathscr{K}}_0$ and is multiplicative. Moreover, by using scale invariance of Wiener process, we can easily see that

$$P^0_s f(x) = E[\rho_s(1, x)f(X_s(1, x)),\ \min_{r\in[0,1]} X^1_s(r, x) > 0] = (P^{s,\rho}_{(1)} f)(x)$$

for any $s \in (0, 1]$, and $f \in C^\infty_b(\mathbf{R}^N)$.

This observation and Theorem 4 imply that for any $n, m, r \geqq 0$, $u_1, \ldots, u_{n+m} \in \tilde{A}$ there is a $C \in (0, \infty)$ such that

$$s^{(||u_1||+\cdot+||u_{n+m}||)/2+r} ||\Phi(r(u_1)) \cdots \Phi(r(u_n))\, adj(V_0)^r (P_s^0)$$
$$\Phi(r(u_{n+1})) \cdots \Phi(r(u_{n+m})) f||_\infty \leqq C||f||_\infty$$

for any $s \in (0, 1]$ and $f \in C_b^\infty$.
This proves Theorem 3 (1).

2.5 Dual Operators

Let $T \in (0, 1]$, and $\hat{B}^k(w)(t) = -B^k(T-t)$, $t \in [0, T]$, $k = 0, 1, \ldots, d$. Also, let $\hat{X} : [0, T] \times \mathbf{R}^N \times W_0 \to \mathbf{R}^N$ be the solution of the following SDE.

$$\hat{X}(t, x) = x + \sum_{k=0}^{d} \int_0^t V_k(\hat{X}(t, x)) \circ d\hat{B}^k(t) - \int_0^t V_0(\hat{X}(t, x)) \circ d\hat{B}^0(t),$$

$t \in [0, T]$, $x \in \mathbf{R}^N$. We may assume that $\hat{X}(\cdot, *, w) : [0, T] \times \mathbf{R}^N \to \mathbf{R}^N$ is continuous for μ-$a.s$. Then we see that with probability one

$$X(t, x) = \hat{X}(T-t, X(T, x)), \quad t \in [0, T], \ x \in \mathbf{R}^N$$

(c.f. Kunita [2]). So we see that for any $f, g \in C_0^\infty(\mathbf{R}^N)$

$$\int_{(0,\infty)\times\mathbf{R}^{N-1}} g(x)(P_T^0 f)(x)dx$$
$$= E^\mu[\int_{(0,\infty)\times\mathbf{R}^{N-1}} dx\ g(x) \exp(\sum_{k=0}^{d} \int_0^T b_k(X(r, x)) \circ dB^k(r))$$
$$\times f(X(T, x)) 1_{(0,\infty)}(\min_{r\in[0,T]} X^1(r, x))]$$
$$= E^\mu[\int_{(0,\infty)\times\mathbf{R}^{N-1}} dy\ g(\hat{X}(T, y))$$
$$\times \exp(\sum_{k=1}^{d} \int_0^T b_k(\hat{X}(r, y)) \circ d\hat{B}^k(r) - \int_0^T b_0(\hat{X}(r, y)) \circ d\hat{B}^0(r)) f(y)$$
$$\times \det(\{\frac{\partial \hat{X}^i}{\partial y^j}(T, y)\}_{i,j=1,\ldots,N} 1_{(0,\infty)}(\min_{r\in[0,T]} \hat{X}^1(y, r))].$$

Let $\bar{X} : [0, \infty) \times \mathbf{R}^N \times W^d \to \mathbf{R}^N$ be the solution of the following SDE.

$$\bar{X}(t, x) = x + \sum_{k=1}^{d} \int_0^t V_k(\bar{X}(t, x)) \circ dB^k(t) - \int_0^t V_0(\bar{X}(t, x)) \circ dB^k(0), \qquad t \in [0, \infty),\ x \in \mathbf{R}^N.$$

Then we have

$$\int_{(0,\infty)\times\mathbf{R}^{N-1}} g(x)(P_T^0 f)(x)dx$$
$$= \int_{(0,\infty)\times\mathbf{R}^{N-1}} f(x)E^{\mu}[\exp(\sum_{k=1}^{d} \int_0^T b_k(\bar{X}(r, x)) \circ dB^k(r)$$
$$- \int_0^T b_0(\bar{X}(r, x)) \circ dB^0(r)) \det(\bar{J}(T, x))g(\bar{X}(T, x)),\ \min_{r\in[0,T]} (x^1 + B^1(r)) > 0].$$

Here

$$\bar{J}(t, x) = \{\bar{J}_j^i(t, x)\}_{i,j=1,\ldots,N}) = \{\frac{\partial \bar{X}_k^i}{\partial x^j}(t, x)\}_{i,j=1,\ldots,N}.$$

Since we have

$$d\bar{J}_i^j(t, x)$$
$$= \sum_{\ell=1}^{N} \sum_{k=1}^{d} \frac{\partial V_k^i}{\partial x^\ell}(\bar{X}(t, x))\bar{J}_j^\ell(t, x) \circ dB^k(t)$$
$$- \sum_{\ell=1}^{N} \frac{\partial V_0^i}{\partial x^\ell}(\bar{X}(t, x))\bar{J}_j^\ell(t, x) \circ dB^0(t),$$

we see that

$$d \det \bar{J}(t, x) = \sum_{k=1}^{d} (div\ V_k)(\bar{X}(t, x)) \det \bar{J}(t, x) \circ dB^k(t) - (div\ V_0)(\bar{X}(t, x)) \det \bar{J}(t, x) \circ dB^0(t),$$

where

$$div\ V_k(x) = \sum_{i=1}^{N} \frac{\partial V_k^i}{\partial x^i}(x), \qquad x \in \mathbf{R}^N.$$

So we have

$$\det \bar{J}(t,x)$$
$$= \exp(\sum_{k=1}^{d} \int_0^t (div\ V_k)(\bar{X}(r, \bar{X}(t,x)) \circ dB^k(r)$$
$$- \int_0^t (div\ V_0)(\bar{X}(r, \bar{X}(t,x)) \circ dB^0(r)).$$

Let $\bar{b}_k \in C_b^\infty(\mathbf{R}^N)$, $k = 0, 1, \ldots, d$, be given by

$$\bar{b}_0(x) = -b_0(x) - div\ V_0(x),$$

and

$$\bar{b}_k(x) = b_k(x) + div\ V_k(x), \qquad k = 1, \ldots, d,$$

and let $\bar{P}_t^0$, $t \in [0, \infty)$ be a linear operator given by

$$(\bar{P}_t^0 f)(x)$$
$$= E^\mu[\exp(\sum_{k=0}^{d} \int_0^t \bar{b}_k(\bar{X}(r,x)) \circ dB^k(r)) f(\bar{X}(t,x)),$$
$$\min_{r \in [0,t]} (x^1 - B^1(r)) > 0],$$

$f \in C_b^\infty(\mathbf{R}^N)$. Then we have

$$\int_{(0,\infty)\times\mathbf{R}^{N-1}} g(x)(P_t^0 f)(x)dx = \int_{(0,\infty)\times\mathbf{R}^{N-1}} f(x)(\bar{P}_t^0 g)(x)dx \tag{15}$$

for any $t > 0$ and $f, g \in C_0^\infty(\mathbf{R}^N)$.

Since a system $\{-V_0, V_1, \ldots, V_d\}$ of vector fields satisfies the assumptions (UFG), (A1) and (A2), we see by Theorem 4, that for any $n, m, r \geqq 0$, $u_1, \ldots, u_{n+m} \in \tilde{A}$, there is a $C \in (0, \infty)$ such that

$$t^{(||u_1||+\cdot+||u_{n+m}||)/2+r} \sup_{x\in(0,\infty)\times\mathbf{R}^{N-1}} |(\Phi(r(u_1)) \cdots \Phi(r(u_n))\ adj(V_0)^r(\bar{P}_t^0)$$
$$\Phi(r(u_{n+1})) \cdots \Phi(r(u_{n+m}))f)(x)|$$
$$\leqq C \sup_{x\in(0,\infty)\times\mathbf{R}^{N-1}} |f(x)|, \qquad t \in (0,1],\ f \in C_b^\infty(\mathbf{R}^N)$$

for any $t \in (0, 1]$ and $f \in C_b^\infty$.

Let us denote by $\mathscr{D}_n$, $n \geqq 0$, the space of linear differential operators A in $\mathbf{R}^N$ such that there are $c_0 \in C_b^\infty(\mathbf{R}^N)$, $a_{u_1,\ldots,u_k} \in C_b^\infty(\mathbf{R}^N)$, $k \leqq n$, $u_1, \ldots, u_k \in A^{***}$, with $||u_1|| + \cdots + ||u_k|| \leqq n$, such that

$$(Af)(x) = c_0(x) f(x) + \sum_{k=1}^{n} \sum_{u_1,\ldots,u_k \in A^{***}, ||u_1||+\cdots+||u_k|| \leqq n} a_{u_1,\ldots,u_k}(x) (\Phi(r(u_1) \cdots r(u_k)) f)(x),$$

for $x \in \mathbf{R}^N$ and $f \in C_b^\infty(\mathbf{R}^N)$.

It is easy to see the following.

Proposition 15

(1) *If $A \in \mathscr{D}_n$, and $B \in \mathscr{D}_m$, $n, m \geqq 0$, then $AB \in \mathscr{D}_{n+m}$.*
(2) *If $A \in \mathscr{D}_n$, $n \geqq 0$, then $[V_1, A] \in \mathscr{D}_{n+1}$, and $[V_0, A] \in \mathscr{D}_{n+2}$.*
(3) *If $A \in \mathscr{D}_n$, $n \geqq 0$, then a formal dual operator $A^* \in \mathscr{D}_n$.*

Also, we have the following by Theorem 4.

Proposition 16 *Let $n_i \geqq 0$, $i = 1, 2$, $m \geqq 0$, and $A_i \in \mathscr{D}_{n_i}$, $i = 1, 2$. Then there is a $C \in (0, \infty)$ such that*

$$\sup_{x \in (0,\infty) \times \mathbf{R}^{N-1}} |(A_1\, adj^m(V_0)(\bar{P}_t^0) A_2 f)(x)|$$
$$\leqq C t^{-m-(n_1+n_2)/2} \sup_{x \in (0,\infty) \times \mathbf{R}^{N-1}} |f(x)|.$$

for any $t \in (0, 1]$ and $f \in C_0^\infty(\mathbf{R}^N)$.

Note that if $W \in C_b^\infty(\mathbf{R}^N; \mathbf{R}^N)$ and if we regard W as a vector field over $\mathbf{R}^N$, then the formal adjoint operator W^* is given by

$$W^* = -W - \sum_{i=1}^{N} \frac{\partial W^i}{\partial x^i}.$$

Let $h \in C^\infty(\mathbf{R}^N)$ be given by $h(x) = x^1$, $x \in \mathbf{R}^N$. Note that if $Wh = 0$, we see that

$$\int_{(0,\infty)\times\mathbf{R}^{N-1}} g(x)(Wf)(x)dx = \int_{(0,\infty)\times\mathbf{R}^{N-1}} (W^* g)(x) f(x) dx$$

for any $f, g \in C_0^\infty(\mathbf{R}^N)$.

Then we have the following.

Proposition 17 *Let $m \geqq 0$. Then there are for any linear operator B in $\mathscr{C}$, there are $n_{m,k,i}, n'_{m,k,i} \geqq 0$, $k = 0, \ldots, m-1$, $i = 1, \ldots, 5^m$, and $A_{m,k,i} \in \mathscr{D}_{n_{m,k,i}}$, $A'_{m,k,i} \in$*

$\mathscr{D}_{n'_{m,k,i}}$, $i = 1, \ldots, 5^m$, *such that* $n_{m,k,i} + n'_{m,k,i} + 2k \leqq 2m$, $k = 0, \ldots, m-1$, $i = 1, \ldots, 5^m$, *and that*

$$\begin{aligned} &adj(V_0^*)^m(B) \\ &\quad = (-1)^m adj(V_0)^m(B) + \sum_{k=0}^{m-1}\sum_{i=1}^{5^m} A_{m,k,i} adj(V_0)^k(B)A'_{m,k,i}. \end{aligned}$$

Proof It is obvious that our assertion is valid for $m = 0$. Note that

$$\begin{aligned} &adj(V_0^*)^{m+1}(B) \\ &\quad = -adj(V_0)(adj(V_0^*)^m(B)) - (div\ V_0)(adj(V_0^*)^m(B)) \\ &\quad\quad + adj(V_0^*)^m(B)(div\ V_0). \end{aligned}$$

So if our assertion is valid for m, we have

$$\begin{aligned} &adj(V_0)(adj(V_0^*)^m(B)) \\ &\quad = (-1)^m adj(V_0)^{m+1}(B) \\ &\quad\quad + \sum_{k=0}^{m-1}\sum_{i=1}^{5^m}(adj(V_0)(A_{m,k,i})adj(V_0)^k(B)A'_{m,k,i} \\ &\quad\quad + \sum_{k=0}^{m-1}\sum_{i=1}^{5^m} A_{m,k,i} adj(V_0)^{k+1}(B)A'_{m,k,i} \\ &\quad\quad + \sum_{k=0}^{m-1}\sum_{i=1}^{5^m} A_{m,k,i} adj(V_0)^k(B) adj(V_0)(A'_{m,k,i}). \end{aligned}$$

So we see that our assertion is valid for $m+1$. This completes the proof. ∎

Now let us prove Theorem 3 (2).

Let $n_i \geqq 0$, $i = 1, 2$, and $B_i \in \mathscr{D}_{n_i}$, $i = 1, 2$. Then we see that for $f, g \in C_0^\infty(\mathbf{R}^N)$

$$\begin{aligned} &\int_{(0,\infty)\times\mathbf{R}^{N-1}} g(x)(B_1\ adj(V_0)^m(P_t^0)B_2 f)(x)dx \\ &\quad = \sum_{k=0}^{m}(-1)^k\binom{n}{k}\int_{(0,\infty)\times\mathbf{R}^{N-1}} g(x)(B_1 V_0^k P_t^0 V_0^{m-k} B_2 f)(x)dx \\ &\quad = \sum_{k=0}^{m}(-1)^k\binom{n}{k}\int_{(0,\infty)\times\mathbf{R}^{N-1}}((V_0^*)^k B_1^* g)(x)(P_t^0 V_0^{m-k} B_2 f)(x)dx \end{aligned}$$

$$= \sum_{k=0}^{m} (-1)^k \binom{n}{k} \int_{(0,\infty)\times\mathbf{R}^{N-1}} (B_2^*(V_0^*)^{n-k} \bar{P}_t^0 (V_0^*)^{m-k} B_1^* g)(x) f(x) dx$$

$$= (-1)^m \int_{(0,\infty)\times\mathbf{R}^{N-1}} (B_2^* adj(V_0^*) \bar{P}_t^0) B_1^* g)(x) f(x) dx.$$

So by Theorem 4 and Proposition 15 we see that there is a $C \in (0, \infty)$ such that

$$|\int_{(0,\infty)\times\mathbf{R}^{N-1}} g(x)(B_1\, adj(V_0)^m (P_t^0) B_2 f)(x) dx|$$

$$\leqq Ct^{-m-(n_1+n_2)/2} (\sup_{x\in(0,\infty)\times\mathbf{R}^{N-1}} |g(x)|)(\int_{(0,\infty)\times\mathbf{R}^{N-1}} |f(x)| dx)$$

for any $t \in (0, 1)$, and $f, g \in C_0^\infty(\mathbf{R}^N)$. This implies that

$$\int_{(0,\infty)\times\mathbf{R}^{N-1}} |(B_1\, adj(V_0)^m (P_t^0) B_2 f)(x)| dx$$

$$\leqq Ct^{-m-(n_1+n_2)/2} (\int_{(0,\infty)\times\mathbf{R}^{N-1}} |f(x)| dx)$$

for any $t \in (0, 1)$, and $f \in C_0^\infty(\mathbf{R}^N)$.

This proves Theorem 3 (2).

3 Part II: Vertical Direction

3.1 Banach Spaces

Let E be a separable Banach space, and let $C_c^\infty((0,\infty); E)$ be a space of smooth functions $f : (0,\infty) \to E$ such that $f(x) = 0$, $x \geqq R$, for some $R \in (0,\infty)$ and

$$\sup_{x\in(0,\infty)} \|\frac{d^n}{dx^n} f(x)\|_E < \infty \text{ for any } n = 0, 1, \ldots$$

Let $B_b((0,\infty); E)$ be a space of measurable functions $f : (0,\infty) \to E$ such that there are measurable functions $f_i : (0,\infty) \to E$, $i = 1, 2$, for which $f = f_1 + f_2$, $\sup_{x\in(0,\infty)} \|f_1(x)\|_E < \infty$ and $\int_0^\infty \|f_2(x)\|_E dx < \infty$. It is obvious that $C_c^\infty((0,\infty); E) \subset B_b((0,\infty); E)$.

Let us define norms $||\cdot||_{1,n}$, $||\cdot||_{\infty,n}$, $n = 0, 1, \ldots$, on $C_c^\infty((0,\infty); E)$ by

$$||f||_{1,n} = \sum_{k=0}^{n} \int_0^\infty ||\frac{d^k}{dx^k} f(x)||_E dx,$$

and

$$||f||_{\infty,n} = \sum_{k=0}^{n} \sup_{x\in(0,\infty)} ||\frac{d^k}{dx^k} f(x)||_E$$

for $f \in C_c^\infty((0,\infty); E)$. We denote by $W_1^n(E)$, (resp. $W_\infty^n(E)$) the completion of $C_c^\infty((0,\infty); E)$ by the norm $||\cdot||_{1,n}$ (resp. $||\cdot||_{\infty,n}$) for $n = 0, 1, \ldots$.

It is easy to see that the linear operator $\frac{d}{dx} : C_c^\infty((0,\infty); E) \to C_c^\infty((0,\infty); E)$ is extendable to a bounded linear operator $D_x : W_p^{n+1}(E) \to W_p^n(E)$ for $n \geqq 0$, and $p = 1, \infty$.

Also we have the following.

Proposition 18

(1) *There is a continuous linear map* $I_0 : W_\infty^0(E) \to C([0,\infty); E)$ *such that* $(I_0 f)(x) = f(x)$, $f \in C_c^\infty((0,\infty); E)$, $x \in (0,\infty)$, *and*

$$\sup_{x\in[0,\infty)} ||(I_0 f)(x)||_E = ||f||_{\infty,0}, \qquad f \in W_\infty^0(E).$$

(2) *There is a continuous linear map* $J_0 : W_1^1(E) \to W_\infty^0(E)$ *such that* $J_0(f) = f$, $f \in C_c^\infty((0,\infty); E)$, *and*

$$||J_0 f||_{\infty,0} \leqq ||f||_{1,1}, \qquad f \in W_0^1(E).$$

Proof Let $f \in C_c^\infty((0,\infty); E)$. Since $||\frac{df}{dx}(x)||_E$ is a bounded function in x, there is a $\tilde{f} \in C([0,\infty); E)$ such that $\tilde{f}|_{(0,\infty)} = f$ and $||f||_{\infty,0} = \sup_{x\in[0,\infty)} ||\tilde{f}(x)||_E$. So we have Assertion (1).

Note that

$$\begin{aligned} ||f(x)||_E &\leqq ||f(x) - \int_0^1 f(y)dy||_E + \int_0^\infty ||f(y)||_E dy \\ &\leqq \int_0^1 dy|| \int_y^x \frac{df}{dz}(z)dz||_E + \int_0^\infty ||f(y)||_E dy \\ &\leqq \int_0^\infty ||\frac{df}{dz}(z)||_E dz + \int_0^\infty ||f(y)||_E dy = ||f||_{1,1}. \end{aligned}$$

This implies Assertion (2). ■

We will use the following notions throughout for the simplicity. We denote $(I_0 f)(0)$ by $f(0)$ for any $f \in W^0_\infty(E)$, and denote $(I_0 J_0 f)(0)$ by $f(0)$ for any $f \in W^1_\infty(E)$.

Let

$$g(t,x) = \frac{1}{\sqrt{2\pi t}} \exp(-\frac{x^2}{2t}), \qquad t > 0,\ x \in \mathbf{R}.$$

Also, let

$$g_i(t,x,y) = g(t,y-x) - (-1)^i g(t,x+y), \qquad t > 0,\ x,y \in [0,\infty),\ i = 0,1.$$

Then we have for $i = 0, 1$,

$$\frac{\partial}{\partial t} g_i(t,x,y) = \frac{1}{2}\frac{\partial^2}{\partial x^2} g_i(t,x,y), \qquad t > 0,\ x,y \in (0,\infty),$$

$$\frac{\partial}{\partial x} g_i(t,x,y) = -\frac{\partial}{\partial y} g_{1-i}(t,x,y), \qquad t > 0,\ x,y \in (0,\infty),$$

and

$$\frac{\partial^{n+m}}{\partial t^n \partial x^m} g(t,x) = t^{-(2n+m+1)/2} \frac{\partial^{n+m} g}{\partial t^n \partial x^m}(1, t^{-1/2}x) \qquad t > 0,\ x \in \mathbf{R}.$$

So we have

$$\left|\frac{\partial^{n+m}}{\partial t^n \partial x^m} g(t,x)\right| \leqq C_{0,n,m} t^{-(2n+m+1)/2} \exp(-\frac{x^2}{4t}), \qquad t > 0,\ x \in \mathbf{R},$$

$$\int_0^\infty \left|\frac{\partial^{n+m}}{\partial t^n \partial x^m} g(t,x)\right| dx \leqq C_{1,n,m} t^{-(2n+m)/2} \qquad t > 0,$$

$$\int_0^\infty \left|\frac{\partial^{n+m+k}}{\partial t^n \partial x^m \partial y^k} g_i(t,x,y)\right| dy \leqq C_{1,n,m+k} t^{-(2n+m+k)/2}, \qquad t > 0,\ x > 0,\ i =, 0.1,$$

and

$$\int_0^\infty \left|\frac{\partial^{n+m+k}}{\partial t^n \partial x^m \partial y^k} g_i(t,x,y)\right| dx \leqq C_{1,n,m+k} t^{-(2n+m+k)/2}, \qquad t > 0,\ y > 0,\ i = 0.1.$$

Here

$$C_{0,n,m} = \sup_{x \in [0,\infty)} \exp(\frac{x^2}{4}) \left|\frac{\partial^{n+m}}{\partial t^n \partial x^m} g(1,x)\right|$$

and $C_{1,n,m} = 4\sqrt{\pi} C_{0,n,m}$.

Then we have the following.

Proposition 19

(1) *For $t > s > 0$ and $x > 0$,*

$$\int_0^\infty |g_i(t,x,y) - g_i(s,x,y)|dy \leqq 2 \wedge (C_{1,1,0}s^{-1}(t-s)).$$

Also, for $n \geqq 1$, $t > s > 0$ and $x > 0$,

$$\int_0^\infty |\frac{\partial^n}{\partial x^n} g_i(t,x,y) - \frac{\partial^n}{\partial x^n} g_i(s,x,y)|dy \leqq (2C_{1,0,n}s^{-n/2}) \wedge (C_{1,1,n}s^{-(n+2)/2}(t-s)).$$

(2) *For $n \geqq 0$, $t > s > 0$ and $x > 0$,*

$$\int_0^\infty |\frac{\partial^n}{\partial x^n} g_i(t,x,y) - \frac{\partial^n}{\partial x^n} g_i(s,x,y)|dx \leqq (2C_{1,0,n}s^{-n/2}) \wedge (C_{1,1,n}s^{-(n+2)/2}(t-s)).$$

(3) *For $n \geqq 0$, $t > s > 0$ and $x > 0$,*

$$|\frac{\partial^n}{\partial x^n} g(t,x) - \frac{\partial^n}{\partial x^n} g(s,x)| \leqq (2C_{0,0,n}s^{-(n+1)/2}) \wedge (C_{0,1,n}s^{-(n+3)/2}(t-s)).$$

Proof Note that

$$\int_0^\infty |g_i(t,x,y) - g_i(s,x,y)|dy \leqq 2 \wedge (C_{1,1,0}\int_s^t r^{-1}dr)$$
$$= 2 \wedge (C_{1,1,0}\log(t/s)) \leqq 2 \wedge (C_{1,1,0}s^{-1}(t-s)),$$

and

$$\int_0^\infty |\frac{\partial^n}{\partial x^n} g_i(t,x,y) - \frac{\partial^n}{\partial x^n} g_i(s,x,y)|dy$$
$$\leqq (2C_{1,0,n}s^{-n/2}) \wedge (C_{1,1,n}\int_s^t r^{-(n+2)/2}dr)$$
$$\leqq (2C_{1,0,n}s^{-n/2}) \wedge (C_{1,1,n}s^{-(n+2)/2}(t-s)).$$

So we have Assertion (1).

The proofs of Assertions (2) and (3) are similar. ■

Let $Q_{i,t}$, $t > 0$, $i = 0, 1$, be a operator in $B_b((0,\infty); E)$ given by

$$(Q_{i,t}f)(x) = \int_0^\infty g_i(t,x,y)f(y)dy, \qquad f \in B_b((0,\infty); E).$$

Then we have the following.

Proposition 20

(1) *For any* $t > 0$, $x \in (0, \infty)$, $f \in C_c^\infty((0, \infty); E)$, *and* $i = 0, 1$,

$$\frac{d}{dx}(Q_{i,t} f)(x) = (Q_{1-i,t}\frac{df}{dx})(x) + 2\delta_{i0} g(t, x) f(0).$$

Here δ_{ij} *is Kronecker's delta.*

(2) *For any* $t > 0$, $x \in (0, \infty)$, $f \in C_c^\infty((0, \infty); E)$, $n \geqq 0$, *and* $i = 0, 1$,

$$\sup_{x\in(0,\infty)} ||(\frac{d}{dx})^n (Q_{i,t} f)(x)||_E \leqq C_{1,0,n} t^{-n/2} \sup_{x\in(0,\infty)} ||f(x)||_E,$$

and

$$\int_0^\infty ||(\frac{d}{dx})^n (Q_{i,t} f)(x)||_E dx \leqq C_{1,0,n} t^{-n/2} \int_0^\infty ||f(x)||_E dx.$$

(3) *For any* $t > s > 0$, $x \in (0, \infty)$, $f \in C_c^\infty((0, \infty); E)$, $n \geqq 0$, *and* $i = 0, 1$,

$$\sup_{x\in(0,\infty)} ||(\frac{d}{dx})^n (Q_{i,t} f)(x) - (\frac{d}{dx})^n (Q_{i,s} f)(x)||_E$$
$$\leqq s^{-n/2}((C_{1,1,n} s^{-1}(t - s)) \wedge (2C_{1,0,n})) \sup_{x\in(0,\infty)} ||f(x)||_E,$$

and

$$\int_0^\infty ||(\frac{d}{dx})^n (Q_{i,t} f)(x) - (\frac{d}{dx})^n (Q_{i,s} f)(x)||_E dx$$
$$\leqq s^{-n/2}((C_{1,1,n} s^{-1}(t - s)) \wedge (2C_{1,0,n})) \int_0^\infty ||f(x)||_E dx.$$

Therefore we have the following.

Proposition 21

(1) *Let* $t \in (0, 1)$, $n \geqq 0$, $i = 0, 1$, *and* $p = 1, \infty$. *Then* $Q_{i,t} f \in W_p^n(E)$ *for any* $f \in W_p^0(E)$, *and,*

$$||D_x^n Q_{i,t} f||_{p,0} \leqq C_{1,0,n} t^{-n/2} ||f||_{p,0}, \qquad f \in W_p^0(E).$$

(2) *Let* $0 < s < t < 1$, $n \geqq 0$, $i = 0, 1$, *and* $p = 1, \infty$. *Then we have*

$$||D_x^n Q_{i,t} f - D_x^n Q_{i,s} f||_{p,0} \leqq s^{-n/2}(C_{1,1,n} + 2C_{1,0,n}))((s^{-1}(t-s)) \wedge 1)||f||_{p,0}$$

for $f \in W_p^0(E)$.

3.2 Some Normed Spaces

Let $\alpha \in [0, \infty)$ and $p = 1, \infty$. Let $\mathscr{C}_{p,\alpha}^0$ be the vector space consisting of continuous functions $\psi : (0,1) \to W_p^0(E)$ such that

$$\sup_{t \in (0,1)} t^{\alpha} ||\psi(t)||_{p,0} < \infty.$$

We define a norm $||\cdot||_{p,\alpha,0}$ on $\mathscr{C}_{p,\alpha}^{0,0}$ by

$$||\psi||_{p,\alpha,0} = \sup_{t \in (0,1)} t^{\alpha} ||\psi(t)||_{p,0}.$$

Let $\mathscr{C}_{p,\alpha}^{1/4}$ be the vector space consisting of $\psi \in \mathscr{C}_{p,\alpha}^0$ such that

$$\sup_{t,s \in (0,1), t>s} s^{\alpha+1/4}(t-s)^{-1/4}||\psi(t) - \psi(s)||_{p,0} < \infty.$$

We define a norm $||\cdot||_{p,\alpha,1/4}$ on $\mathscr{C}_{p,\alpha}^{1/4}$ by

$$||\psi||_{p,\alpha,1/4} = ||\psi||_{p,\alpha,0} + \sup_{t,s \in (0,1), t>s} s^{\alpha+1/4}(t-s)^{-1/4}||\psi(t) - \psi(s)||_{p,0}.$$

Let $\mathscr{C}_{p,\alpha}^{1/2}$ be the vector space consisting of continuous functions $\psi : (0,1) \to W_p^1(E)$ satisfying the following two conditions.

(1) $\psi \in \mathscr{C}_{p,\alpha}^0$ and $D_x \psi : (0,1) \to W_p^0(E)$ belongs to $\mathscr{C}_{p,\alpha+1/2}^0$.
(2) $\sup_{t,s \in (0,1), t>s} s^{\alpha+1/2}(t-s)^{-1/2}||\psi(t) - \psi(s)||_{p,0} < \infty.$

We define a norm $||\cdot||_{\alpha,1/2}$ on $\mathscr{C}^{1/2}_{p,\alpha}$ by

$$||\psi||_{p,\alpha,1/2} = ||\psi||_{p,\alpha,0} + ||D_x\psi||_{p,\alpha+1/2,0} + \sup_{t,s\in(0,1),t>s} s^{\alpha+1/2}(t-s)^{-1/2}||\psi(t)-\psi(s)||_{p,0}.$$

Let $\mathscr{C}^{3/4}_{p,\alpha}$ be the vector space consisting of $\psi \in \mathscr{C}^{1/2}_{p,\alpha}$ satisfying the following two conditions.

(1) $D_x\psi : (0,1) \to W^0_p(E)$ belongs to $\mathscr{C}^{1/4}_{p,\alpha+1/2}$.
(2) $\sup_{t,s\in(0,1),t>s} s^{\alpha+3/4}(t-s)^{-3/4}||\psi(t)-\psi(s)||_{p,0} < \infty.$

We define a norm $||\cdot||_{p,\alpha,0}$ on $\mathscr{C}^{3/4}_{p,\alpha}$ by

$$||\psi||_{p,\alpha,3/4} = ||\psi||_{p,\alpha,1/2} + ||D_x\psi||_{p,\alpha+1/2,1/4} + \sup_{t,s\in(0,1),t>s} s^{\alpha+3/4}(t-s)^{-3/4}||\psi(t)-\psi(s)||_{p,0}.$$

Let $\mathscr{C}^{n/4}_{p,\alpha}$ $n \geqq 4$ be the vector space consisting of continuous functions ψ : $(0,1) \to W^2_p(E)$ satisfying the following two conditions.

(1) $\psi : (0,1) \to W^0_p(E)$ is continuously differentiable in t, and $\psi \in \mathscr{C}^{(n-1)/4}_{p,\alpha}$.
(2) $\dfrac{\partial\psi}{\partial t}$ belongs to $\mathscr{C}^{(n-4)/4}_{p,\alpha+1}$, and $D_x\psi$ belongs to $\mathscr{C}^{(n-2)/4}_{p,\alpha+1/2}$.

We define a norm $||\cdot||_{p,\alpha,n/4}$ on $\mathscr{C}^{n/4}_{p,\alpha}$ by

$$||\psi||_{p,\alpha,n/4} = ||\psi||_{p,\alpha,(n-1)/4} + ||\frac{\partial}{\partial t}\psi||_{p,\alpha+1,(n-4)/4} + ||D_x\psi||_{p,\alpha+1/2,(n-2)/4}.$$

First we observe the following.

Proposition 22 *Let $\beta \in [0,1)$ and $\gamma > 0$.*

(1) *If $\beta+\gamma \geqq 1$, there is a $C_{2,\beta,\gamma} \in (0,\infty)$ such that for any $s,t \in (0,1)$ with $s<t$,*

$$\int_s^t (t-r)^{-\beta}r^{-\gamma}dr \leqq C_{2,\beta,\gamma}s^{-\gamma}(((t-s)^{1-\beta}) \wedge s^{1-\beta}).$$

(2) *If* $\beta + \gamma < 1$, *there is a* $C_{2,\beta,\gamma} \in (0, \infty)$ *such that for any* $s, t \in (0, 1)$ *with* $s < t$,

$$\int_s^t (t-r)^{-\beta} r^{-\gamma} dr \leqq C_{2,\beta,\gamma}(t-s)^{1-\beta}(s^{-\gamma} \wedge (t-s)^{-\gamma}).$$

(3) *Let* $\delta \in [0, 1-\beta]$. *Then for any* $s, t \in (0, 1)$ *with* $s < t$,

$$\int_s^t (t-r)^{-\beta} r^{-\gamma} dr \leqq C_{2,\beta,\gamma}(t-s)^{1-\beta-\delta} s^{-((\gamma-\delta)\vee 0)}.$$

Proof (1) Suppose that $\beta + \gamma \geqq 1$. Note that

$$\int_s^t (t-r)^{-\beta} r^{-\gamma} dr = t^{-\beta-\gamma+1} \int_{s/t}^1 (1-u)^{-\beta} u^{-\gamma} du$$

$$= t^{-\beta-\gamma+1} \int_{(s/t)\vee(1/2)}^1 (1-u)^{-\beta} u^{-\gamma} du$$

$$+ t^{-\beta-\gamma+1} \int_{(s/t)}^{(s/t)\vee(1/2)} (1-u)^{-\beta} u^{-\gamma} du.$$

We see that

$$t^{-\beta-\gamma+1} \int_{(s/t)\vee(1/2)}^1 (1-u)^{-\beta} u^{-\gamma} du \leqq t^{-\beta-\gamma+1} 2^{\gamma} \int_{(s/t)\vee(1/2)}^1 (1-u)^{-\beta} du$$

$$\leqq 2^{\gamma} t^{-\beta-\gamma+1} (1-\beta)^{-1} (1-(s/t))^{1-\beta} \leqq 2^{\gamma} (1-\beta)^{-1} (s^{-\gamma-\beta+1} \wedge s^{-\gamma}(t-s)^{1-\beta}).$$

Also, we have

$$t^{-\beta-\gamma+1} \int_{(s/t)}^{(s/t)\vee(1/2)} (1-u)^{-\beta} u^{-\gamma} du \leqq t^{-\beta-\gamma+1} 2^{\beta} (\int_{(s/t)}^{1/2} u^{-\gamma} du) 1_{(0,1/2)}(s/t).$$

If $\gamma > 1$, then we have

$$t^{-\beta-\gamma+1} 2^{\beta} (\int_{(s/t)}^{1/2} u^{-\gamma} du) 1_{(0,1/2)}(s/t)$$

$$\leqq 2^{\beta} t^{-\beta-\gamma+1} (\gamma-1)^{-1} (s/t)^{-\gamma+1} 1_{(0,1/2)}(s/t)$$

$$= 2^{\beta} (\gamma-1)^{-1} s^{-\gamma} s t^{-\beta} 1_{(0,1/2)}(s/t)$$

$$\leqq 2^{\beta} (\gamma-1)^{-1} s^{-\gamma} (s^{1-\beta} \wedge ((2(1-(s/t)))^{1-\beta} t^{1-\beta}).$$

If $\gamma < 1$, then

$$t^{-\beta-\gamma+1}2^{\beta}(\int_{(s/t)}^{1/2} u^{-\gamma}du)1_{(0,1/2)}(s/t) \leqq 2^{\beta}t^{-\beta-\gamma+1}(1-\gamma)^{-1}1_{(0,1/2)}(s/t)$$

$$\leqq (1-\gamma)^{-1}2^{\beta}(s^{-\beta-\gamma+1} \wedge ((2(1-(s/t)))^{1-\beta}t^{1-\beta}s^{-\gamma})).$$

If $\gamma = 1$, we have

$$t^{-\beta-\gamma+1}2^{\beta}(\int_{(s/t)}^{1/2} u^{-\gamma}du)1_{(0,1/2)}(s/t) \leqq 2^{\beta}t^{-\beta}\log(t/s)1_{(0,1/2)}(s/t)$$

$$\leqq (2^{\beta}s^{-\beta}(t/s)^{-\beta}\log(t/s)1_{(0,1/2)}(s/t))$$

$$\wedge(2(1-(s/t))^{1-\beta}t^{1-\beta}s^{-1}(t/s)^{-1}\log(t/s)1_{(0,1/2)}(s/t)))$$

$$\leqq 2(\sup_{u>2} u^{-\beta}\log u)s^{-1}(s^{1-\beta} \wedge (t-s)^{1-\beta}).$$

So we have Assertion (1).

(2) Suppose that $\beta + \gamma < 1$. Note that

$$\int_s^t (t-r)^{-\beta}r^{-\gamma}dr \leqq s^{-\gamma}\int_s^t (t-r)^{-\beta}dr = s^{-\gamma}(1-\beta)^{-1}(t-s)^{1-\beta}.$$

If $s/t \geqq 1/2$, then we have $s \geqq t-s$, and so

$$\int_s^t (t-r)^{-\beta}r^{-\gamma}dr \leqq (1-\beta)(t-s)^{1-\beta-\gamma}.$$

If $s/t < 1/2$, then we have $t < 2(t-s)$, and so

$$\int_s^t (t-r)^{-\beta}r^{-\gamma}dr = t^{1-\beta-\gamma}\int_{s/t}^1 (1-r)^{-\beta}r^{-\gamma}dr$$

$$\leqq (\int_0^1 (1-r)^{-\beta}r^{-\gamma}dr)2^{1-\beta-\gamma}(t-s)^{1-\beta-\gamma}$$

These imply Assertion (2).

Now let us prove Assertion (3).

Let us think of the case that $\beta + \gamma \geqq 1$. Let $\xi = (1-\beta)^{-1}\delta \in [0,1]$. Note that $\gamma \geqq 1-\beta \geqq \delta$. So we see that

$$s^{-\gamma}((t-s)^{1-\beta} \wedge s^{1-\beta}) \leqq s^{-\gamma}(t-s)^{(1-\beta)(1-\xi)}s^{(1-\beta)\xi}$$

$$= s^{-\gamma+\delta}(t-s)^{1-\beta-\delta} = (t-s)^{1-\beta-\delta}s^{-((\gamma-\delta)\vee 0)}$$

Let us think of the case that $\beta + \gamma < 1$. If $\gamma \leqq \delta$, then we see that

$$(t-s)^{1-\beta}(s^{-\gamma} \wedge (t-s)^{-\gamma}) \leqq (t-s)^{1-\beta-\delta}((s^{-\gamma}(t-s)^{\delta}) \wedge (t-s)^{\delta-\gamma})$$
$$\leqq (t-s)^{1-\beta-\delta} = (t-s)^{1-\beta-\delta} s^{-((\gamma-\delta)\vee 0)}.$$

Suppose that $\gamma > \delta$. Then letting $\xi = \gamma^{-1}\delta \in [0, 1)$, we see that

$$(t-s)^{1-\beta}(s^{-\gamma} \wedge (t-s)^{-\gamma})$$
$$\leqq (t-s)^{1-\beta} s^{-\gamma(1-\xi)}(t-s)^{-\gamma\xi} = (t-s)^{1-\beta-\delta} s^{-((\gamma-\delta)\vee 0)}.$$

These imply Assertion (3). ■

For $\psi \in \mathscr{C}^0_{p,\alpha}$, $i = 0, 1$, $p = 1, \infty$, and $q \in (0, 1/2)$, let $G_{i,q}\psi : (2q, 1) \to W^0_p(E)$ be given by

$$(G_{i,q}\psi)(t) = \int_q^t (Q_{i,t-s}\psi(s))ds \qquad t \in (2q, 1).$$

Proposition 23 *Let* $\alpha \geqq 0$. *If* $\psi \in \mathscr{C}^0_{p,\alpha}$, $p = 1, \infty$, $i = 0, 1$, $q \in (0, 1/2)$, *then* $(G_{i,q}\psi)(t) \in W^1_p(E)$, $t \in (2q, 1)$. *Moreover,*

$$||(G_{i,q}\psi)(t)||_{p,0} \leqq C_{1,0,0}C_{2,0,\alpha}q^{-((\alpha-1)\vee 0)}||\psi||_{p,\alpha,0},$$

and

$$||D_x(G_{i,q}\psi)(t)||_{p,0} \leqq C_{1,0,1}C_{2,1/2,\alpha}q^{-((\alpha-1/2)\vee 0)}||\psi||_{p,\alpha,0}$$

for any $t \in (2q, 1)$.

Furthermore,

$$||(G_{i,q}\psi)(t) - (G_{i,q}\psi)(s)||_{p,0}$$
$$\leqq (C_{1,1,0} + 2C_{1,0,0})(C_{2,3/4,\alpha}$$
$$+ C_{2,0,\alpha})(t-s)^{3/4}q^{-((\alpha-1/4)\vee 0)}||\psi||_{p,\alpha,0},$$

and

$$||D_x(G_{i,q}\psi)(t) - D_x(G_{i,q}\psi)(s)||_{p,0}$$
$$\leqq (C_{1,1,1} + 2C_{1,0,1})(C_{2,3/4,\alpha} + C_{2,1/2,\alpha})(t-s)^{1/4}q^{-((\alpha-1/4)\vee 0)}||\psi||_{p,\alpha,0}$$

for any $s, t \in (2q, 1)$, *with* $s < t$. *In particular, the mapping* $t \in (2q, 1) \to (G_{i,q}\psi)(t) \in W^1_p(E)$ *is continuous.*

Proof By Proposition 21, we see that

$$||Q_{i,t-r}\psi(r)||_{p,0} \leqq C_{1,0,0}||\psi(r)||_{p,0}$$

and

$$||D_x Q_{i,t-r}\psi(r)||_{p,0} \leqq C_{1,0,1}(t-r)^{-1/2}||\psi(r)||_{p,0}$$

for $r \in (0,t)$. So by Proposition 21 we have

$$||(G_{i,q}\psi)(t)||_{p,0} \leqq C_{1,0,0}(\int_q^t r^{-\alpha} dr)||\psi||_{p,\alpha,0} \leqq C_{1,0,0}C_{2,0,\alpha}q^{-((\alpha-1)\vee 0)}||\psi||_{p,\alpha,0}$$

and

$$||D_x(G_{i,q}\psi)(t)||_{p,0} \leqq C_{1,0,1}(\int_q^t (t-r)^{-1/2} r^{-\alpha} dr)||\psi||_{p,\alpha,0}$$
$$\leqq C_{1,0,1}C_{2,1/2,\alpha}q^{-((\alpha-1/2)\vee 0)}||\psi||_{p,\alpha,0}.$$

Note that

$$(G_{i,q}\psi)(t) - (G_{i,q}\psi)(s)$$
$$= \int_q^s (Q_{i,t-r}\psi(r)) - (Q_{i,s-r}\psi(r))dr + \int_s^t (Q_{i,t-r}\psi(r))dr.$$

So we have by Propositions 21 and 22

$$||(G_{i,q}\psi)(t) - (G_{i,q}\psi)(s)||_{p,0}$$
$$\leqq (\int_q^s (C_{1,1,0} + 2C_{1,0,0})((t-s)(s-r)^{-1})^{3/4} r^{-\alpha} dr)$$
$$||\psi||_{p,\alpha,0} + C_{1,0,0}(\int_s^t r^{-\alpha} dr)||\psi||_{p,\alpha,0}.$$
$$\leqq (C_{1,1,0} + 2C_{1,0,0})(t-s)^{3/4} C_{2,3/4,\alpha} q^{-((\alpha-1/4)\vee 0)}||\psi||_{p,\alpha,0}$$
$$+ C_{1,0,0}C_{2,0,\alpha}(t-s)^{3/4} C_{2,0,\alpha} q^{-((\alpha-1/4)\vee 0)}||\psi||_{p,\alpha,0},$$

and

$$||D_x(G_{i,q}\psi)(t) - D_x(G_{i,q}\psi)(s)||_{p,0}$$
$$\leqq (C_{1,1,1} + 2C_{1,0,1})(\int_q^s (t-s)^{1/4}(s-r)^{-3/4}) r^{-\alpha} dr)||\psi||_{p,\alpha,0}$$

$$+ C_{1,0,1}(\int_s^t (t-r)^{-1/2} r^{-\alpha} dr)||\psi||_{p,\alpha,0}$$
$$\leqq (C_{1,1,1} + 2C_{1,0,1})(t-s)^{1/4} C_{2,3/4,\alpha} q^{-((\alpha-1/4)\vee 0)}||\psi||_{p,\alpha,0}$$
$$+ C_{1,0,1} C_{2,1/2,\alpha}(t-s)^{1/4} q^{-((\alpha-1/4)\vee 0)}||\psi||_{p,\alpha,0}.$$

These imply our assertion. ■

Proposition 24

(1) *Let* $\psi \in \mathscr{C}^0_{\infty,\alpha}$, $\alpha \geqq 0$, *and* $q \in (0, 1/2)$. *Let* $(B_q\psi) : (2q, 1) \times (0, \infty) \to E$ *be given by*

$$(B_q\psi)(t, x) = \int_q^t g(t-s, x)\psi(s)(0)ds, \qquad t \in (2q, 1),\ x \in (0, \infty).$$

Then $(B_q\psi)(t, \cdot) \in W^0_\infty(E)$, $t \in (2q, 1)$, *and the map* $t \in (2q, 1) \to (B_q\psi)(t, \cdot) \in W^0_\infty(E)$ *is continuous. Moreover, we have*

$$||(B_q\psi)(t)||_{\infty,0} \leqq C_{0,0,0} C_{2,1/2,\alpha} q^{-((\alpha-1/2)\vee 0)}||\psi||_{\infty,\alpha,0}, \qquad t \in (2q, 1).$$

(2) *Let* $\psi \in \mathscr{C}^{1/2}_{1,\alpha}$, $\alpha \geqq 0$. *and let* $(B_q\psi) : (2q, 1) \times (0, \infty) \to E$ *be given by*

$$(B_q\psi)(t, x) = \int_q^t g(t-s, x)\psi(s)(0)ds, \qquad t \in (2q, 1),\ x \in (0, \infty).$$

Then $(B_q\psi)(t, \cdot) \in W^1_1(E)$, $t \in (2q, 1)$,

$$||(B_q\psi)(t)||_{1,0} \leqq C_{2,0,\alpha+1/2} q^{-((\alpha-1/2)\vee 0)}||\psi||_{1,\alpha,1/2},$$

and

$$||D_x(B_q\psi)(t)||_{1,0} \leqq C_{1,0,1} C_{2,1/2,\alpha} q^{-\alpha}||\psi||_{1,\alpha,1/2}$$

for $t \in (2q, 1)$. *Moreover, we have*

$$||(B_q\psi)(t, \cdot) - (B_q\psi)(s, \cdot)||_{1,0}$$
$$\leqq (C_{1,1,0} + 2C_{1,0,0})(C_{2,3/4,\alpha+1/2}) + C_{2,0,\alpha+1/2}$$
$$(t-s)^{3/4} q^{-((\alpha+1/4)\vee 0)}||\psi||_{1,\alpha,1/2},$$

and

$$\begin{aligned}&||D_x(B_q\psi)(t,\cdot)-D_x(B_q\psi)(s,\cdot)||_{1,0}\\&\leqq (C_{1,1,1}+2C_{1,0,1})(C_{2,1/4,\alpha+1/2}\\&\quad + C_{2,0,\alpha+1/2})(t-s)^{1/4}q^{-(\alpha+1/4)}||\psi||_{1,\alpha,1/2}\end{aligned}$$

for $s,t\in(2q,1)$ *with* $s<t$. *In particular, the map* $t\in(2q,1)$ *to* $(B_q\psi)(t,\cdot)\in W_1^1(E)$ *is continuous.*

Proof (1) We see that

$$||(B_q\psi)(t)||_{\infty,0}\leqq (\int_q^t C_{0,0,0}(t-r)^{-1/2}r^{-\alpha}dr)||\psi||_{\infty,\alpha,0}.$$

So by Proposition 22 we have the inequality in Assertion (1). Note that

$$\begin{aligned}&(B_q\psi)(t,x)-(B_q\psi)(s,x)\\&=\int_q^s(g(t-r,x)-g(s-r,x))\psi(r)(0)dr\\&\quad+\int_s^t g(t-r,x)\psi(r)(0)dr\end{aligned}$$

for $s,t\in(2q,1)$ with $s<t$. Then by Proposition19 we have

$$\begin{aligned}&||(B_q\psi)(t)-(B_q\psi)(s)||_{\infty,0}\\&\leqq (\int_q^s C_{0,1,0}^{1/4}C_{0,0,0}^{3/4}(s-r)^{-3/4}(t-s)^{1/4}r^{-\alpha}dr)||\psi||_{\infty,\alpha,0}\\&\quad+(\int_s^t C_{0,0,0}(t-r)^{-1/2}r^{-\alpha}dr)||\psi||_{\infty,\alpha,0}.\end{aligned}$$

So we see the continuity. These imply Assertion (1).

(2) Let $\psi\in\mathscr{C}_{1,\alpha}^{1/2}$. Then we have

$$\begin{aligned}||(B_q\psi)(t)||_{1,0}&\leqq\int_q^t(\int_0^\infty g(t-r,x)dx)||\psi(r,0)||_E dr\\&\leqq(\int_q^t r^{-(\alpha+1/2)}dr)||\psi||_{1,\alpha,1/2},\end{aligned}$$

$$||D_x(B_q\psi)(t)||_{1,0}\leqq\int_q^t C_{1,0,1}(t-r)^{-1/2}r^{-(\alpha+1/2)}dr)||\psi||_{1,\alpha,1/2},$$

$$||(B_q\psi)(t,\cdot)-(B_q\psi)(s,\cdot)||_{1,0}$$

$$\leqq (C_{1,1,0}+2C_{1,0,0})\int_q^s (t-s)^{3/4}r^{-3/4}r^{-(\alpha+1/2)}dr||\psi||_{1,\alpha,1/2}$$
$$+\int_s^t C_{1,0,0}r^{-(\alpha+1/2)}dr||\psi||_{1,\alpha,1/2},$$

and

$$||D_x(B_q\psi)(t,\cdot)-D_x(B_q\psi)(s,\cdot)||_{1,0}$$
$$\leqq (C_{1,1,1}+2C_{1,0,1})\int_q^s (t-s)^{1/4}r^{-3/4}r^{-(\alpha+1/2)}dr||\psi||_{1,\alpha,1/2}$$
$$+\int_s^t C_{1,0,0}r^{-(\alpha+1/2)}dr||\psi||_{1,\alpha,1/2}.$$

These and Proposition 22 imply Assertion (2). ■

The following is an easy consequence of Proposition 20 (1).

Proposition 25 *For any* $\psi \in \mathscr{C}^{1/2}_{p,\alpha}$, $\alpha \geqq 0$, $p=1,\infty$, $i=0,1$, *and* $q\in(0,1/2)$,

$$D_x(G_{i,q}\psi)(t)=(G_{1-i,q}D_x\psi)(t)+2\delta_{i0}(B_q\psi)(t), \qquad t\in(2q,1).$$

Let $\Phi:(0,\infty)\times\mathbf{R}\to\mathbf{R}$ be given by

$$\Phi(t,x)=\int_x^\infty g(t,y)dy, \qquad t>0,\ x>0.$$

Then we have

$$0\leqq\Phi(t,x)\leqq 1/2, \qquad t>0,\ x\in\mathbf{R},$$

$$\int_0^\infty \Phi(t,x)dx=\int_0^\infty yg(t,y)dy=-t\int_0^\infty \frac{\partial g}{\partial y}(t,y)dy=(2\pi)^{-1/2}t^{1/2} \qquad t>0,$$

and

$$\lim_{\varepsilon\downarrow 0}\Phi(\varepsilon,x)=0, \qquad x>0.$$

Also, we see that

$$\frac{\partial\Phi}{\partial t}(t,x)=\int_x^\infty 2\frac{\partial^2}{\partial y^2}g(t,y)dy=-2\frac{\partial g}{\partial x}(t,x), \qquad t,x>0.$$

Proposition 26 *Let $q \in (0, 1/2)$ and $\alpha \geqq 0$. Then we have the following.*

(1) *For any $\psi \in \mathscr{C}^{1/2}_{\infty,\alpha}$, and $s, t \in (2q, 1)$, $s < t$,*

$$\begin{aligned}
&\|(B_q\psi)(t) - (B_q\psi)(s)\|_{\infty,0} \\
&\leqq (C_{0,1,0} + 2C_{0,0,0})(C_{2,3/4,\alpha+1/2} + 8) \\
&\times (t-s)^{3/4} q^{-(\alpha+1/4)} \|\psi\|_{\infty,\alpha,1/2}.
\end{aligned}$$

(2) *Let $\psi \in \mathscr{C}^{1/4}_{\infty,\alpha}$. Then $(B_q\psi)(t) \in W^1_\infty(E)$, $t \in (2q, 1)$, and*

$$\begin{aligned}
&D_x(B_q\psi)(t) \\
&= \int_q^t \frac{\partial g}{\partial x}(t-r, \cdot)(\psi(r)(0) - \psi(t)(0))dr - 2\Phi(t-q, \cdot)\psi(t)(0), \\
&t \in (2q, 1).
\end{aligned}$$

Moreover,

$$\|D_x(B_q\psi)(t)\|_{\infty,0} \leqq (C_{0,0,1}C_{2,3/4,\alpha+1/4} + 1)q^{-\alpha}\|\psi\|_{\infty,\alpha,1/4}.$$

(3) *Let $\gamma = 1/2, 3/4$, or 1. Then for any $\psi \in \mathscr{C}^{\gamma}_{\infty,\alpha}$, and $s, t \in (2q, 1)$, $s < t$,*

$$\begin{aligned}
&\|D_x(B_q\psi)(t) - D_x(B_q\psi)(s)\|_{\infty,0} \\
&\leqq (11C_{0,0,1} + 1)(C_{2,3/4,0} + 1)(t-s)^{\gamma-1/4} \\
&\times q^{-(\alpha+\gamma-1/4)}\|\psi\|_{\infty,\alpha,\gamma}.
\end{aligned}$$

In particular, $t \in (2q, 1) \to D_x(B_q\psi)(t) \in W^0_\infty(E)$ is continuous.

(4) *Let $\gamma = 3/4$ or 1. For any $\psi \in \mathscr{C}^{\gamma}_{1,\alpha}$, and $s, t \in (2q, 1)$, $s < t$,*

$$\begin{aligned}
&\|D_x(B_q\psi)(t) - D_x(B_q\psi)(s)\|_{1,0} \\
&\leqq (C_{1,1,1} + 2C_{1,0,1})(C_{2,3/4,\alpha+\gamma} + 6 \\
&+ C_{2,1/2,0})(t-s)^{\gamma-1/4} q^{-(\alpha+\gamma-1/4)}\|\psi\|_{1,\alpha,\gamma}.
\end{aligned}$$

In particular, $t \in (2q, 1) \to D_x(B_q\psi)(t) \in W^0_1(E)$ is continuous.

Proof By Proposition 24, we have

$$\begin{aligned}
&(B_q\psi)(t, x) - (B_q\psi)(s, x) \\
&= \int_q^s (g(t-r, x) - g(s-r, x))(\psi(r)(0) - \psi(s)(0))dr
\end{aligned}$$

$$+\int_s^t g(t-r,x)(\psi(r)(0)-\psi(s)(0))dr + (\int_q^t g(t-r,x)dr)(\psi(t)(0)$$
$$-\psi(s)(0)) - (\int_q^s g(s-r,x)dr)\psi(s)(0).$$

So we see that

$$\begin{aligned}
&||(B_q\psi)(t,x)-(B_q\psi)(s,x)||_E \\
&\leqq \int_q^s |g(t-r,x)-g(s-r,x)|||\psi(r)(0)-\psi(s)(0)||_E dr \\
&\quad + \int_q^t g(r,x)||\psi(r)(0)-\psi(t)(0)||_E dr \\
&\quad + (\int_q^t g(t-r,x)dr)||\psi(t)(0)-\psi(s)(0)||_E dr \\
&\quad + (\int_0^{t-s} g(s-q+r,x)dr)||\psi(s)(0)||_E \\
&\leqq (C_{0,1,0}+2C_{0,0,0})\int_q^s (t-s)^{3/4}(s-r)^{-5/4}(s-r)^{1/2} \\
&\quad \times r^{-(\alpha+1/2)}||\psi||_{\infty,\alpha,1/2} \\
&\quad + (\int_0^{t-s} C_{0,0,0}r^{-1/2}r^{1/4}q^{-(\alpha+1/4)})dr||\psi||_{\infty,\alpha,1/4} \\
&\quad + (\int_0^{t-s} C_{0,0,0}r^{-1/2}dr)(t-s)^{1/4}q^{-(\alpha+1/4)}||\psi||_{\infty,\alpha,1/4} \\
&\quad + (\int_0^{t-s} C_{0,0,0}r^{-1/4}q^{-1/4}dr)q^{-\alpha}||\psi||_{\infty,\alpha,0} \\
&\leqq (C_{0,1,0}+2C_{0,0,0})(C_{2,3/4,\alpha+1/2}+8)(t-s)^{3/4}q^{-(\alpha+1/4)}||\psi||_{\infty,\alpha,1/2}.
\end{aligned}$$

This implies Assertion (1).

Let $\varepsilon \in (0, q/2)$, and

$$u_\varepsilon(t)(x) = \int_q^{t-\varepsilon} g(t-r,x)\psi(r)(0)dr, \qquad t \in (2q,1),\ x \in (0,\infty).$$

Then we have

$$
\begin{aligned}
\frac{\partial}{\partial x}u_\varepsilon(t)(x) &= \int_q^{t-\varepsilon} \frac{\partial g}{\partial x}(t-r,x)\psi(r)(0)dr \\
&= \int_q^{t-\varepsilon} \frac{\partial g}{\partial x}(t-r,x)(\psi(r)(0)-\psi(t)(0))dr \\
&\quad - 2(\int_q^{t-\varepsilon} \frac{\partial \Phi}{\partial t}(t-r,x)dr)\psi(t)(0) \\
&= \int_q^{t-\varepsilon} \frac{\partial g}{\partial x}(t-r,x)(\psi(r)(0)-\psi(t)(0))dr \\
&\quad - 2(\Phi(t-q,x)-\Phi(\varepsilon,x))\psi(t)(0).
\end{aligned}
$$

It is easy to see that

$$
\begin{aligned}
&\int_q^t ||\frac{\partial g}{\partial x}(t-r,\cdot)(\psi(r)(0)-\psi(t)(0))||_{\infty,0}dr \\
&\leqq \int_q^t C_{0,0,1}(t-r)^{-1}(t-r)^{1/4}r^{-(\alpha+1/2)}dr||\psi||_{\infty,\alpha,1/4} \\
&\leqq C_{0,0,1}C_{2,3/4,\alpha+1/4}q^{-\alpha}||\psi||_{\infty,\alpha,1/4}.
\end{aligned}
$$

Note that $u_\varepsilon(t) \to (B_q\psi)(t)$ in $W^0_\infty(E)$ as $\varepsilon \downarrow 0$. So we have

$$
\begin{aligned}
&(B_q\psi)(t)(x) - (B_q\psi)(t)(y) \\
&= \int_y^x (\int_q^t \frac{\partial g}{\partial x}(t-r,z)(\psi(r)(0)-\psi(t)(0))dr)dz \\
&\quad + 2(\int_y^x \Phi(t-q,x)dz)\ \psi(t)(0)
\end{aligned}
$$

for $t \in (2q, 1)$, and $x, y \in (0, \infty)$. This proves Assertion (2).

Let $\gamma = 1/2, 3/4$, or 1, and $\psi \in \mathscr{C}^\gamma_{1,\alpha}$. Let $s, t \in (2q, 1)$ with $s < t$. We have by Assertion (2),

$$
\begin{aligned}
&||D_x(B_q\psi)(t)(x) - D_x(B_q\psi)(s)(x)||_E \\
&\leqq \int_0^{s-q} |\frac{\partial g}{\partial x}(r,x)|||\psi(t-r)(0)-\psi(t)(0) \\
&\quad - (\psi(s-r)(0)-\psi(s)(0))||_E dr
\end{aligned}
$$

$$+\int_{s-q}^{t-q} |\frac{\partial g}{\partial x}(r,x)| ||\psi(t-r)(0)-\psi(t)(0)||_E dr$$

$$+2||\Phi(t-q,x)\psi(t)(0)-\Phi(s-q,x)\psi(s)(0)||_E.$$

It is easy to see that

$$\begin{aligned}&||\psi(t-r)(0)-\psi(t)(0)-\psi(s-r)(0)-\psi(s)(0))||_E\\ &\leqq 2(r^{\gamma}\wedge(t-s)^{\gamma})q^{-(\alpha+\gamma)}||\psi||_{\infty,\alpha,\gamma}\\ &\leqq 2r^{1/4}(t-s)^{\gamma-1/4}q^{-(\alpha+\gamma)}||\psi||_{\infty,\alpha,\gamma},\end{aligned}$$

and

$$\begin{aligned}&||\Phi(t-q,x)\psi(t)(0)-\Phi(s-q,x)\psi(s)(0)||_E\\ &\leqq (\int_{s-q}^{t-q}|\frac{\partial g}{\partial x}(r,x)|dr)||\psi(t)(0)||_E+||\psi(t)(0)-\psi(s)(0)||_E\end{aligned}$$

Note that

$$\begin{aligned}\int_{s-q}^{t-q}|\frac{\partial g}{\partial x}(r,x)|dr &\leqq C_{0,0,1}(\int_{s-q}^{t-q}(q^{-1}\wedge(r^{-3/4}q^{-1/4}))dr)\\ &\leqq C_{0,0,1}(C_{2,3/4,0}+1)(t-s)^{\gamma-1/4}q^{-(\gamma-1/4)}.\end{aligned}$$

So we have

$$\begin{aligned}&||D_x(B_q\psi)(t)-D_x(B_q\psi)(s)||_{\infty,0}\\ &\leqq C_{0,1,0}(\int_0^{s-q}r^{-1/2}r^{\gamma-3/4}dr)q^{-(\alpha+\gamma)}||\psi||_{\infty,\alpha,\gamma}\\ &\quad+3C_{0,0,1}(C_{2,3/4,0}+1)(t-s)^{\gamma-1/4}q^{-(\gamma-1/4)}q^{-\alpha}||\psi||_{\infty,\alpha,0}\\ &\quad+(t-s)^{\gamma-1/4}q^{-(\alpha+\gamma-1/4)}||\psi||_{\infty,\alpha,\gamma-1/4}\\ &\leqq 8C_{0,1,0}(t-s)^{\gamma-1/4}q^{-(\alpha+\gamma-1/4)}||\psi||_{\infty,\alpha,\gamma}\\ &\quad+(3C_{0,0,1}+1)(C_{2,3/4,0}+1)(t-s)^{\gamma-1/4}q^{-(\alpha+\gamma-1/4)}||\psi||_{\infty,\alpha,\gamma}.\end{aligned}$$

This shows Assertion (3).

Let $\gamma=3/4$ or 1, and $\psi\in\mathscr{C}^{\gamma}_{1,\alpha}$. Let $s,t\in(2q,1)$ with $s<t$. Then we have

$$\begin{aligned}&||D_x(B_q\psi)(t)-D_x(B_q\psi)(s)||_{1,0}\\ &\leqq \int_q^s(\int_0^{\infty}|\frac{\partial g}{\partial x}(t-r,x)-\frac{\partial g}{\partial x}(s-r,x)|dx)||\psi(r)(0)-\psi(s)(0)||_E dr\end{aligned}$$

$$+\int_0^{t-s}(\int_0^{\infty}|\frac{\partial g}{\partial x}(r,x)|dx)||\psi(t-r)(0)-\psi(t)(0)||_E dr$$
$$+\int_0^{t-s}(\int_0^{\infty}|\frac{\partial g}{\partial x}g(r,x)|dx)||\psi(t)(0)-\psi(s)(0)||_E dr$$
$$+\int_0^{t-s}|\frac{\partial g}{\partial x}(t-r+q,x)|||\psi(r)(0)||_E dr$$
$$\leqq (C_{1,1,1}+2C_{1,0,1})(\int_q^s((t-s)^{\gamma-1/4}(s-r)^{-(\gamma+1/4)})$$
$$\times (s-r)^{\gamma-1/4}r^{-(\alpha+\gamma-1/4)}dr)||\psi||_{1,\alpha,\gamma}$$
$$+C_{1,0,1}(\int_0^{t-s}r^{-1/2}r^{\gamma-3/4}dr)q^{-(\alpha+\gamma-1/4)}||\psi||_{1,\alpha,\gamma}$$
$$+C_{1,0,1}(\int_0^{t-s}r^{-1/2}dr)(t-s)^{\gamma-3/4}q^{-(\alpha+\gamma-1/4)}||\psi||_{1,\alpha,\gamma}$$
$$+C_{1,0,1}(\int_{s-q}^{t-q}r^{-1/2}dr)q^{-(\alpha-1/2)}||\psi||_{1,\alpha,\gamma}.$$
$$\leqq (C_{1,1,1}+2C_{1,0,1})(C_{2,1/2,\alpha+\gamma-1/4}+(\gamma-1/4)^{-1}+2+C_{2,1/2,0})$$
$$\times (t-s)^{\gamma-1/4}q^{-(\alpha+\gamma-1/4)}||\psi||_{1,\alpha,\gamma}.$$

This implies Assertion (4). ■

Proposition 27 *Let* $q \in (0,1/2)$, $\alpha \geqq 0$, *and* $p = 1, \infty$.

(1) $||D_x(G_{i,q}\psi)(t) - D_x(G_{i,q}\psi)(s)||_{p,0}$

$$\leqq (2(C_{1,1,1}+2C_{1,0,1})C_{2,3/4,\alpha}$$
$$+16C_{1,0,1}+C_{1,0,1}C_{2,0,1/2})(t-s)^{1/4}q^{-((\alpha-1/4)\vee 0))}||\psi||_{p,\alpha,0}$$

for any $\psi \in \mathscr{C}^0_{p,\alpha}$, *and any* $s,t \in (2q,1)$ *with* $s<t$. *Also, we have*

$$||D_x(G_{i,q}\psi)(t) - D_x(G_{i,q}\psi)(s)||_{p,0}$$
$$\leqq (C_{1,1,1}+2C_{1,0,1})(C_{2,3/4,\alpha+\gamma}+6+C_{2,0,1/2})$$
$$\times (t-s)^{\gamma+1/4}q^{-((\alpha+\gamma-1/4)\vee 0)}||\psi||_{p,\alpha,\gamma}$$

for any $\psi \in \mathscr{C}^{\gamma}_{p,\alpha}$, $\gamma = 1/4, 1/2$, *and any* $s,t \in (2q,1)$ *with* $s<t$.

(2) *Let* $\psi \in \mathscr{C}^{1/4}_{p,\alpha}$. *Then* $D_x(G_{i,q}\psi)(t) \in W^1_p(E)$, $t \in (2q, 1)$, *and*

$$D_x^2(G_{i,q}\psi)(t) = \int_q^t (D_x^2 Q_{i,t-r}(\psi(r) - \psi(t))dr + 2Q_{i,t-q}\psi(t) - 2\psi(t) \tag{16}$$

for $t \in (2q, 1)$. *Moreover, the mapping* $t \in (2q, 1) \to D_x^2(G_{i,q}\psi)(t) \in W^0_p(E)$ *is continuous, and*

$$||D_x^2(G_{i,q}\psi)(t)||_{p,0} \leqq (C_{1,0,2}C_{2,1/4,\alpha+1/4} + 4)q^{-\alpha}||\psi||_{p,\alpha,1/4}.$$

(3) *Let* $\gamma = 1/2, 3/4$, *or* 1. *Then we see that*

$$\begin{aligned}&||D_x^2(G_{i,q}\psi)(t) - D_x^2(G_{i,q}\psi)(s)||_{p,0}\\&\quad \leqq 8(C_{1,0,0} + C_{1,1,0} + C_{1,0,2} + 1)q^{-(\alpha+\gamma)}(t-s)^{\gamma-1/4}||\psi||_{p,\alpha,\gamma}\end{aligned}$$

for any $\psi \in \mathscr{C}^{\gamma}_{p,\alpha}$, $s, t \in (q, 1)$ *with* $s < t$.

(4) *Let* $\psi \in \mathscr{C}^{1/4}_{p,\alpha}$. *Then the mapping* $t \in (2q, 1) \to (G_{i,q}\psi)(t) \in W^0_p(E)$ *is continuously differentiable, and*

$$\frac{\partial}{\partial t}(G_{iq}\psi)(t) = \frac{1}{2}D_x^2(G_{i,q}\psi)(t) + \psi(t) \tag{17}$$

for $t \in (2q, 1)$. *Moreover, we have*

$$||\frac{\partial}{\partial t}(G_{iq}\psi)(t)||_{p,0} \leqq (C_{1,0,2}C_{2,1/4,\alpha+1/4} + 5)q^{-\alpha}||\psi||_{p,\alpha,1/4}.$$

Proof Let $\psi \in \mathscr{C}^0_{p,\alpha}$. Then by Proposition 21, we have

$$\begin{aligned}&||D_x(G_{i,q}\psi)(t) - D_x(G_{i,q}\psi)(s)||_{p,0}\\&\leqq \int_q^s ||D_x Q_{i,t-r}(\psi(r) - \psi(s)) - D_x Q_{i,s-r}(\psi(r) - \psi(s))||_{p,0}dr\\&\quad + \int_0^{t-s} ||D_x Q_{i,r}(\psi(t-r) - \psi(t))||_{p,0}dr\\&\quad + \int_0^{t-s} ||D_x Q_{i,r}(\psi(t) - \psi(s))||_{p,0}dr\\&\quad + \int_{s-q}^{t-q} ||D_x Q_{i,r}\psi(s)||_{p,0}dr.\end{aligned}$$

If $\gamma = 0$, then

$$
\begin{aligned}
&||D_x(G_{i,q}\psi)(t) - D_x(G_{i,q}\psi)(s)||_{p,0} \\
&\quad \leqq (\int_q^s (C_{1,1,1} + 2C_{1,0,1})(t-s)^{1/4} \\
&\quad \times (s-r)^{-1/4}(s-r)^{-1/2} 2r^{-\alpha})dr)||\psi||_{p,\alpha,0} \\
&\quad + 4(\int_0^{t-s} C_{1,0,1} r^{-1/2} r^{-1/4} q^{-((\alpha-1/4)\vee 0)} dr)||\psi||_{p,\alpha,0} \\
&\quad + C_{1,0,1}(\int_{s-q}^{t-q} r^{-1/2} dr) q^{-\alpha} ||\psi||_{p,\alpha,0} \\
&\quad \leqq 2(C_{1,1,1} + 2C_{1,0,1}) C_{2,3/4,\alpha}(t-s)^{1/4} q^{-((\alpha-1/4)\vee 0))}||\psi||_{p,\alpha,0} \\
&\quad + 16 C_{1,0,1}(t-s)^{1/4} q^{-((\alpha-1/4)\vee 0)}||\psi||_{p,\alpha,0} \\
&\quad + C_{1,0,1} C_{2,0,1/2}(t-s)^{1/4} q^{-((\alpha-1/4)\vee 0)}||\psi||_{p,\alpha,0}.
\end{aligned}
$$

If $\gamma = 1/4$, or $1/2$, then

$$
\begin{aligned}
&||D_x(G_{i,q}\psi)(t) - D_x(G_{i,q}\psi)(s)||_{p,0} \\
&\quad \leqq (\int_q^s (C_{1,1,1} + 2C_{1,0,1})(t-s)^{\gamma+1/4} \\
&\quad \times (s-r)^{-(\gamma+3/4)}(s-r)^{\gamma} r^{-(\alpha+\gamma)} dr)||\psi||_{p,\alpha,\gamma} \\
&\quad + (\int_0^{t-s} C_{1,0,1} r^{-1/2} r^{\gamma-1/4} q^{-(\alpha+\gamma-1/4)} dr)||\psi||_{p,\alpha,\gamma-1/4} \\
&\quad + (\int_0^{t-s} C_{1,0,1} r^{-1/2} dr)(t-s)^{\gamma-1/4} q^{-(\alpha+\gamma-1/4)}||\psi||_{p,\alpha,\gamma-1/4} \\
&\quad + C_{1,0,1}(\int_{s-q}^{t-q} r^{-1/2} dr) q^{-\alpha}||\psi||_{p,\alpha,0} \\
&\quad \leqq (C_{1,1,1} + 2C_{1,0,1}) C_{2,3/4,\alpha+\gamma}(t-s)^{\gamma+1/4} q^{-(\alpha+\gamma-1/4)}||\psi||_{p,\alpha,\gamma} \\
&\quad + C_{1,0,1}((3/4-\gamma)^{-1} + 2)(t-s)^{\gamma+1/4} q^{-(\alpha+\gamma-1/4)}||\psi||_{p,\alpha,\gamma-1/4} \\
&\quad + C_{1,0,1} C_{2,0,1/2}(t-s)^{\gamma+1/4} q^{-(\alpha+\gamma-1/4)}||\psi||_{p,\alpha,\gamma-1/4}.
\end{aligned}
$$

This implies Assertion (1).

Let $\varepsilon \in (0, q/2)$ and let

$$
u_\varepsilon(t) = \int_q^{t-\varepsilon} (D_x Q_{i,t-r}\psi(r)) dr, \qquad t \in (2q, 1).
$$

Then we have

$$
\begin{aligned}
&(D_x u_\varepsilon(t))(x) \\
&= \int_q^{t-\varepsilon} D_x^2 Q_{i,t-r}(\psi(r) - \psi(t))(x)dr \\
&\quad + 2\int_q^{t-\varepsilon} dr(\int_0^\infty \frac{\partial g_i}{\partial t}(t-r,x,y)\psi(t)(y)dy) \\
&= \int_q^{t-\varepsilon} (D_x^2 Q_{i,t-r}(\psi(r) - \psi(t))(x)dr \\
&\quad + 2\int_0^\infty (g_i(t-q,x,y) - g_i(\varepsilon,x,y))\psi(t)(y)dy.
\end{aligned}
$$

So we have

$$
D_x u_\varepsilon(t) = \int_q^{t-\varepsilon} (D_x^2 Q_{i,t-r}(\psi(r) - \psi(t))dr + 2Q_{i,t-q}\psi(t) - 2Q_{i,\varepsilon}\psi(t)
$$

By Proposition 21, we have

$$
\begin{aligned}
&\int_q^t ||D_x^2 Q_{i,t-r}(\psi(r) - \psi(t))||_{p,0}dr \\
&\leqq \int_q^t C_{1,0,2}(t-r)^{-1}(t-r)^{3/4}r^{-(\alpha+1/4)}||\psi||_{p,\alpha,1/4}dr \\
&\leqq C_{1,0,2}C_{2,1/4,\alpha+1/4}q^{-\alpha}||\psi||_{p,\alpha,1/4}.
\end{aligned} \tag{18}
$$

In the case that $p = 1$,

$$
||Q_{i,\varepsilon}\psi(t) - \psi(t)||_{1,0} \to 0, \qquad \varepsilon \downarrow 0.
$$

In the case that $p = \infty$, since $u_\varepsilon(t) \to D_x(G_{iq}\psi)(t)$ in $W^0_\infty(E)$ as $\varepsilon \downarrow 0$, we have

$$
\begin{aligned}
&D_x(G_{iq}\psi)(t)(x') - D_x(G_{iq}\psi)(t)(x) \\
&= \int_x^{x'} dy(\int_q^t (D_x^2 Q_{i,t-r}(\psi(r) - \psi(t))(y)dr \\
&\quad + 2((Q_{i,t-q}\psi(t))(y) - \psi(t)(y)))
\end{aligned}
$$

for $x, x' \in (0,\infty)$. So in both cases we have Eq. (16).

Also, we see that for $s, t \in (2q, 1)$ with $s < t$,

$$\begin{aligned}
&D_x^2(G_{i,q}\psi)(t) - D_x^2(G_{i,q}\psi)(s)\\
&\quad = \int_0^{t-q} D_x^2 Q_{i,r}(\psi(t-r) - \psi(t))dr\\
&\qquad - \int_0^{s-q} D_x^2 Q_{i,r}(\psi(s-r) + \psi(s))dr\\
&\qquad + 2(Q_{i,t-q}\psi(t) - Q_{i,s-q}\psi(s)) - 2(\psi(t) - \psi(s))\\
&\quad = \int_0^{s-q} D_x^2 Q_{i,r}(\psi(t-r) - \psi(t) - \psi(s-r) + \psi(s))dr\\
&\qquad + \int_{s-q}^{t-q} D_x^2 Q_{i,r}(\psi(t-r) - \psi(t))dr\\
&\qquad + 2(Q_{i,t-q}\psi(t) - Q_{i,s-q}\psi(s)) - 2(\psi(t) - \psi(s)). \qquad (19)
\end{aligned}$$

Note that

$$||\psi(t-r) - \psi(t) - \psi(s-r) + \psi(s))||_{p,0} \leqq 2r^{1/8}(t-s)^{1/8}q^{-(\alpha+1/4)}||\psi||_{p,\alpha,1/4}.$$

So we see that

$$\begin{aligned}
&\int_0^{s-q} ||(D_x^2 Q_{i,s-r}(\psi(t-r) - \psi(t) - \psi(s-r) + \psi(s))||_{p,0}dr\\
&\leqq 2C_{1,0,2}\int_0^{s-q} r^{-1}r^{1/8}dr\ (t-s)^{1/8}q^{-(\alpha+1/4)}||\psi||_{p,\alpha,1/4}.
\end{aligned}$$

Also, we see that

$$\int_{s-q}^{t-q} ||D_x^2 Q_{i,r}(\psi(t-r) - \psi(t))||_{p,0}dr \leqq 2C_{1,0,2}\int_{s-q}^{t-q} q^{-3/4}q^{-\alpha}||\psi||_{p,\alpha,0}dr.$$

So we see that $D_x^2(G_{i,q}\psi)(t)$ is continuous in t. The last inequality of Assertion (2) follows from Eqs. (16) and (18).

Let $\gamma = 1/2, 3/4$, or 1, and let $\psi \in \mathscr{C}^{\gamma}_{p,\alpha}$. Then we see that

$$||\psi(t-r) - \psi(t) - \psi(s-r) + \psi(s))||_{p,0} \leqq 2(r^{\gamma} \wedge (t-s)^{\gamma})q^{-(\alpha+\gamma)}||\psi||_{p,\alpha,\gamma},$$

and so we see that

$$\int_0^{s-q} ||(D_x^2 Q_{i,s-r}(\psi(t-r) - \psi(t) - \psi(s-r) + \psi(s))||_{p,0} dr$$
$$\leqq 2C_{1,0,2} \int_0^{s-q} r^{-1} r^{1/4} dr\ (t-s)^{\gamma-1/4} q^{-(\alpha+\gamma)} ||\psi||_{p,\alpha,\gamma}$$
$$\leqq 8C_{1,0,2}(t-s)^{\gamma-1/4} q^{-(\alpha+\gamma-1/4)} ||\psi||_{p,\alpha,\gamma}.$$

Also, we see that

$$|| \int_{s-q}^{t-q} D_x^2 Q_{i,r}(\psi(t-r) - \psi(t)) dr ||_{p,0}$$
$$\leqq C_{1,0,2}(\int_{s-q}^{t-q} r^{\gamma-1-1/4} dr) q^{-(\alpha+\gamma-1/4)} ||\psi||_{p,\alpha,\gamma}.$$

Also, by Proposition 21 we see that

$$||Q_{i,t-q}\psi(t) - Q_{i,s-q}\psi(s)||_{0,p}$$
$$\leqq ||Q_{i,t-q}\psi(t) - Q_{i,s-q}\psi(t)||_{0,p} + ||Q_{i,s-q}(\psi(t) - \psi(s))||_{0,p}$$
$$\leqq (C_{1,1,0} + 2C_{1,0,0})(t-s)^{\gamma-1/4} q^{-(\alpha+\gamma-1/4)} ||\psi||_{p,\alpha,\gamma}.$$

These imply Assertion (3).

Let $\varepsilon \in (0, q/2)$ and let

$$v_\varepsilon(t) = \int_q^{t-\varepsilon} (Q_{i,t-r}\psi(r)) dr, \qquad t \in (2q, 1).$$

Then we have

$$\frac{\partial}{\partial t} v_\varepsilon(t)(x)$$
$$= \int_q^{t-\varepsilon} (\int_0^\infty \frac{\partial g_i}{\partial t}(t-r, x, y)\psi(r)(y) dy) dr + Q_{i,\varepsilon}\psi(t-\varepsilon)(x)$$
$$= \frac{1}{2} D_x u_\varepsilon + Q_{i,t-q}\psi(q)(x).$$

So by a similar argument to the proof of Assertion (2), we have for $s, t \in (2q, 1)$, with $s < t$,

$$(G_{i,q}\psi)(t) - (G_{i,q}\psi)(s) = \frac{1}{2} \int_s^t D_x^2 (G_{i,q}\psi)(r) dr + \int_s^t \psi(r) dr.$$

So we have Assertion (4) similarly. ∎

Proposition 28 *Let $q \in (0, 1/2)$ and $\alpha \geqq 0$.*

(1) *Let $\psi \in \mathscr{C}^{3/4}_{\infty,\alpha}$. Then $(B_q\psi)(t) \in W^2_\infty(E)$, $t \in (2q, 1)$, and*

$$D_x^2(B_q\psi)(t) = \int_q^t \frac{\partial^2 g}{\partial x^2}(t-r,\cdot)(\psi(r)(0) - \psi(t)(0)) + 2g(t-q,\cdot)\psi(t)(0), \qquad t \in (2q, 1).$$

Moreover, the mapping $t \in (2q, 1) \to D_x^2(B_q\psi)(t) \in W^0_\infty(E)$ is continuous, and

$$||D_x^2(B_q\psi)(t)||_{\infty,0} \leqq (C_{0,0,2}C_{2,3/4,\alpha+3/4} + 2C_{0,0,0})q^{-(\alpha+1/2)}||\psi||_{\infty,\alpha,3/4}.$$

Also, the mapping $t \in (2q, 1) \to (B_q\psi)(t) \in W^0_\infty(E)$ is continuously differentiable in t, and

$$\frac{d}{dt}(B_q\psi)(t) = \frac{1}{2}D_x^2(B_q\psi)(t), \qquad t \in (2q, 1).$$

(2) *Let $\psi \in \mathscr{C}^1_{\infty,\alpha}$. Then Also, the mapping $t \in (2q, 1) \to (B_q\psi)(t) \in W^0_\infty(E)$ is continuously differentiable in t, and*

$$\frac{\partial}{\partial t}(B_q\psi)(t) = \frac{1}{2}D_x^2(B_q\psi)(t) = (B_q\frac{\partial\psi}{\partial t})(t) + g(t-q,\cdot)\psi(q)(0), \qquad t \in (2q, 1).$$

(3) *Let $\alpha \geqq 0$, $\gamma = 1, 5/4$, or $3/2$. Then there is a constant $\hat{C}_{\alpha,\gamma} \in (0, \infty)$ such that*

$$||D_x^2(B_q\psi)(t) - D_x^2(B_q\psi)(s)||_{\infty,0} \leqq \hat{C}_{\alpha,\gamma}(t-a)^{\gamma-3/4}q^{-(\alpha+\gamma-1/4)}||\psi||_{\infty,\alpha,\gamma}$$

for any $q \in (0, 1/2)$, $\psi \in \mathscr{C}^\gamma_{\infty,\alpha}$, and $s, t \in (2q, 1)$ with $s < t$. (4) *Let $\psi \in \mathscr{C}^{3/4}_{1,\alpha}$. Then $(B_q\psi)(t) \in W^2_1(E)$, $t \in (2q, 1)$, and*

$$D_x^2(B_q\psi)(t) = \int_q^t \frac{\partial^2 g}{\partial x^2}(t-r,\cdot)(\psi(r)(0) - \psi(t)(0)) + 2g(t-q,\cdot)\psi(t)(0), \qquad t \in (2q, 1).$$

Moreover, the mapping $t \in (2q, 1) \to D_x^2(B_q\psi)(t) \in W^0_1(E)$ is continuous, and

$$||D_x^2((B_q\psi)(t))||_{1,0} \leqq (C_{1,0,2}C_{2,3/4,\alpha+3/4} + 2)q^{-(\alpha+1/2)}||\psi||_{1,\alpha,3/4}.$$

Also, the mapping $t \in (2q,1) \to (B_q\psi)(t) \in W_1^0(E)$ *is continuously differentiable in* t, *and*

$$\frac{\partial}{\partial t}(B_q\psi)(t) = \frac{1}{2}D_x^2(B_q\psi)(t), \qquad t \in (2q,1).$$

(5) *Let* $\psi \in \mathscr{C}^{\gamma}_{1,\alpha}$, $\gamma = 1, 5/4$, *or* $3/2$. *Then for any* $s, t \in (2q,1)$, *with* $s < t$,

$$\begin{aligned}&||D_x^2((B_q\psi)(t)) - D_x^2((B_q\psi)(s))||_{1,0}\\ &\leqq 12(C_{1,0,2} + C_{1,0,0})(t-s)^{\gamma-3/4}q^{-(\alpha+\gamma-1/4)}||\psi||_{1,\alpha,\gamma}.\end{aligned}$$

(6) *Let* $\psi \in \mathscr{C}^{3/2}_{1,\alpha}$. *Then*

$$D_x^2(B_q\psi)(t) = 2(B_q\frac{d\psi}{dt}(t) + 2g(t-q,\cdot)\psi(q)(0), \qquad t \in (2q,1).$$

Proof Let $\psi \in \mathscr{C}^{3/4}_{\infty,\alpha}$. Let $\varepsilon \in (0, q/2)$, and

$$u_\varepsilon(t)(x) = \int_q^{t-\varepsilon} g(t-r,x)\psi(r)(0)dr, \qquad t \in (2q,1),\ x \in (0,\infty).$$

Then we have

$$\begin{aligned}\frac{\partial^2}{\partial x^2}u_\varepsilon(t)(x) &= \int_q^{t-\varepsilon}\frac{\partial^2 g}{\partial x^2}(t-r,x)\psi(r)(0)dr,\\ &= \int_q^{t-\varepsilon}\frac{\partial^2 g}{\partial x^2}(t-r,x)(\psi(r)(0)-\psi(t)(0))dr\\ &\quad - 2(\int_q^{t-\varepsilon}\frac{\partial g}{\partial t}(t-r,x)dr)\psi(t)(0)\\ &= \int_q^{t-\varepsilon}\frac{\partial^2 g}{\partial x^2}(t-r,x)(\psi(r)(0)-\psi(t)(0))dr\\ &\quad - 2(g(t-q,x)-g(\varepsilon,x))\psi(t)(0).\end{aligned}$$

It is easy to see that

$$\begin{aligned}&\int_q^t ||\frac{\partial^2 g}{\partial x^2}(t-r,\cdot)(\psi(r)(0)-\psi(t)(0))||_{\infty,0}dr\\ &\leqq \int_q^t C_{0,0,2}(t-r)^{-3/2}(t-r)^{3/4}r^{-(\alpha+3/4)}||\psi||_{\infty,\alpha,3/4}\\ &\leqq C_{0,0,2}C_{2,3/4,\alpha+3/4}q^{-(\alpha+1/2)}||\psi||_{\infty,\alpha,3/4}.\end{aligned}$$

Thus we have the first part of (1) similarly to the proof of Proposition 27 (2).

Note that

$$\frac{\partial}{\partial t} u_\varepsilon(t)(x) = \int_q^{t-\varepsilon} \frac{\partial g}{\partial t}(t-r,x)\psi(r)(0)dr + g(\varepsilon,x)\psi(t-\varepsilon)(0)$$

$$= \frac{1}{2}(D_x^2 u_\varepsilon(t))(x) + g(\varepsilon,x)(I_0\psi(t-\varepsilon))(0).$$

Taking $\varepsilon \downarrow 0$, we have

$$(B_q\psi)(t,x) - (B_q\psi)(s,x) = \frac{1}{2}\int_s^t D_x^2(B_q\psi)(r)(x)dr.$$

So this implies the second part of Assertion (1).

Let $\psi \in \mathscr{C}^1_{\infty,\alpha}$. Then by Assertion (1), we have

$$D_x^2(B_q\psi)(t)(x)$$
$$= -2\int_q^t \frac{\partial g}{\partial r}(t-r,x)(\psi(r)(0)-\psi(t)(0))dr + 2g(t-q,x)\psi(t)(0)$$
$$= 2\int_q^t g(t-r,x)\frac{d\psi}{dr}(r)(0)dr + 2g(t-q,x)\psi(q)(0).$$

So we have Assertion (2).

Let $\psi \in \mathscr{C}^\gamma_{\infty,\alpha}$, $\gamma = 1, 5/4$, or $3/2$. Then $\frac{d\psi}{dt} \in \mathscr{C}^{\gamma-1}_{\infty,\alpha+1}$. So we have Assertion (3) by Assertion (2) and Proposition 24.

Let $\psi \in \mathscr{C}^{3/4}_{1,\alpha}$. Note that

$$\int_q^t ||\frac{\partial^2 g}{\partial x^2}(t-r,\cdot)(\psi(r)(0)-\psi(t)(0))||_{1,0}dr$$
$$\leqq \int_q^t C_{1,0,2}(t-r)^{-1}(t-r)^{1/4}r^{-(\alpha+3/4)}||\psi||_{1,\alpha,3/4}$$
$$\leqq C_{1,0,2}C_{2,3/4,\alpha+3/4}q^{-(\alpha+1/2)}||\psi||_{1,\alpha,3/4}.$$

So we have Assertion (4) similarly to the proof of Assertion (1).

Let $\psi \in \mathscr{C}^\gamma_{1,\alpha}$, $\gamma = 1, 5/4$, or $3/2$. Then for $s, t \in (2q, 1)$ with $s < t$,

$$||D_x^2(B_q\psi)(t) - D_x^2(B_q\psi)(s)||_{1,0}$$
$$\leqq \int_0^{s-q}\left(\int_0^\infty |\frac{\partial^2 g}{\partial x^2}(r,x)|dx\right)||\psi(t-r)(0)-\psi(t)(0)$$
$$+\psi(s)(0)-\psi(s-r)(0)||_E dr$$

$$+\int_{s-q}^{t-q}(\int_0^\infty |\frac{\partial^2 g}{\partial x^2}(r,x)|dx)||\psi(t-r)(0)-\psi(t)(0)||_E dr$$

$$+2(\int_0^\infty |g(t-q,x)-g(s-q,x)|dx)||\psi(t)(0)||_E$$

$$\leqq C_{1,0,2}(\int_0^{s-q} r^{-1}r^{1/4}(t-s)^{\gamma-3/4}dr)q^{-(\alpha+\gamma)}||\psi||_{1,\alpha,\gamma}$$

$$+C_{1,0,2}(\int_0^{t-s} r^{\gamma-7/4}q^{-\gamma+3/4}dr)2q^{-(\alpha+1/2)}||\psi||_{1,\alpha,1/2}$$

$$+2C_{1,0,0}q^{-(\alpha+1/2)}||\psi||_{1,\alpha,1/2}$$

$$\leqq 12(C_{1,0,2}+C_{1,0,0})(t-s)^{\gamma-3/4}q^{-(\alpha+\gamma-1/4)}||\psi||_{1,\alpha,\gamma}.$$

This shows Assertion (5). The proof of Assertion (6) is similar to that of Assertion (2).

This completes the proof. ■

We summarize results in the following.

Proposition 29

(1) *There is a* $C_{3,\alpha,0,0}\in(0,\infty)$, $\alpha\geqq 0$, *satisfying the following.*

$$||(G_{i,q}\psi)(t)||_{p,0}\leqq C_{3,\alpha,0,0}q^{-((\alpha-1)\vee 0)}||\psi||_{p,\alpha,0},$$

and

$$||(G_{i,q}\psi)(t)-(G_{i,q}\psi)(s)||_{p,0}$$
$$\leqq C_{3,\alpha,0,0}(t-s)^{3/4}q^{-((\alpha-1/4)\vee 0)}||\psi||_{p,\alpha,0}$$

for any $\psi\in\mathscr{C}^0_{p,\alpha}$, $p=1,\infty$, $q\in(0,1/2)$, *and* $s,t\in(2q,1)$ *with* $s<t$.

(2) *There is a* $C_{3,\alpha,\gamma,1}\in(0,\infty)$, $\alpha\geqq 0$, $\gamma=0,1/4,1/2$, *satisfying the following.*

$$||D_x(G_{i,q}\psi)(t)||_{p,0}\leqq C_{3,\alpha,\gamma,0}q^{-((\alpha-1/2)\vee 0)}||\psi||_{p,\alpha,0},$$

and

$$||D_x(G_{i,q}\psi)(t)-D_x(G_{i,q}\psi)(s)||_{p,0}$$
$$\leqq C_{3,\alpha,\gamma,1}(t-s)^{\gamma+1/4}q^{-((\alpha+\gamma-1/4)\vee 0)}||\psi||_{p,\alpha,\gamma}$$

for any $\psi\in\mathscr{C}^\gamma_{p,\alpha}$, $p=1,\infty$, $q\in(0,1/2)$, *and* $s,t\in(2q,1)$ *with* $s<t$.

(3) *There is a* $C_{4,\alpha,1/2,0} \in (0,\infty)$, $\alpha \geqq 0$, *satisfying the following.*

$$||(B_q\psi)(t)||_{p,0} \leqq C_{4,\alpha,1/2,0}q^{-((\alpha-1/2)\vee 0)}||\psi||_{p,\alpha,1/2},$$

and

$$\begin{aligned}&||(B_q\psi)(t)-(B_q\psi)(s)||_{p,0}\\ &\leqq C_{4,\alpha,1/2,0}(t-s)^{3/4}q^{-(\alpha+1/4)}||\psi||_{p,\alpha,1/2}\end{aligned}$$

for any $\psi \in \mathscr{C}^{1/2}_{p,\alpha}$, $p=1,\infty$, $q\in(0,1/2)$, *and* $s,t\in(2q,1)$ *with* $s<t$.

Moreover, $B_q\psi : (2q,1) \to W^0_p(E)$ *is continuous for any* $\psi \in \mathscr{C}^{1/2}_{p,\alpha}$, $p=1,\infty$, $q\in(0,1/2)$.

(4) *There is a* $C_{3,\alpha,\gamma,2} \in (0,\infty)$, $\alpha \geqq 0$, $\gamma = 1/4, 1/2, 3/4, 1$, *satisfying the following.*

$$||D^2_x(G_{i,q}\psi)(t)||_{p,0} + ||\frac{d}{dt}(G_{i,q}\psi)(t)||_{p,0} \leqq C_{3,\alpha,1/4,2}q^{-\alpha}||\psi||_{p,\alpha,1/4}$$

for any $\psi \in \mathscr{C}^{\gamma}_{p,\alpha}$, $p=1,\infty$, $q\in(0,1/2)$, *and* $s,t\in(2q,1)$ *with* $s<t$. *Also,*

$$\begin{aligned}&||D^2_x(G_{i,q}\psi)(t)-D^2_x(G_{i,q}\psi)(s)||_{p,0}+||\frac{d}{dt}(G_{i,q}\psi)(t)-\frac{d}{dt}(G_{i,q}\psi)(s)||_{p,0}\\ &\leqq C_{3,\alpha,\gamma,2}(t-s)^{\gamma-1/4}q^{-(\alpha+\gamma-1/4)}||\psi||_{p,\alpha,\gamma}\end{aligned}$$

for any $\psi \in \mathscr{C}^{\gamma}_{p,\alpha}$, $\gamma = 1/2, 3/4, 1$ $p=1,\infty$, *and* $q\in(0,1/2)$.

(5) *There is a* $C_{4,\alpha,\gamma,1} \in (0,\infty)$, $\alpha \geqq 0$, $\gamma = 1/2, 3/4, 1$, *satisfying the following.*

$$||D_x(B_q\psi)(t)||_{p,0} \leqq C_{4,\alpha,\gamma,1}q^{-\alpha}||\psi||_{p,\alpha,1/2},$$

and

$$\begin{aligned}&||D_x(B_q\psi)(t)-D_x(B_q\psi)(s)||_{p,0}\\ &\leqq C_{4,\alpha,\gamma,1}(t-s)^{\gamma-1/4}q^{-(\alpha+\gamma-1/4)\vee 0}||\psi||_{p,\alpha,\gamma}\end{aligned}$$

for any $\psi \in \mathscr{C}^{\gamma}_{p,\alpha}$, $p=1,\infty$, $q\in(0,1/2)$, *and* $s,t\in(2q,1)$ *with* $s<t$.

(6) *There is a* $C_{4,\alpha,\gamma,2} \in (0,\infty)$, $\alpha \geqq 0$, $\gamma = 3/4, 1, 5/4, 3/2$, *satisfying the following.*

$$||D^2_x(B_q\psi)(t)||_{p,0} + ||\frac{d}{dt}(B_q\psi)(t)||_{p,0} \leqq C_{4,\alpha,\gamma,2}q^{-(\alpha+1/2)}||\psi||_{p,\alpha,3/4},$$

and $D_x^2 B_q\psi : (2q, 1) \to W_p^0(E)$ *is continuous for any* $\psi \in \mathscr{C}_{p,\alpha}^{3/4}$, $p = 1, \infty$, $q \in (0, 1/2)$. *Moreover,*

$$||D_x^2(B_q\psi)(t) - D_x^2(B_q\psi)(s)||_{p,0} + ||\frac{d}{dt}(B_q\psi)(t) - \frac{d}{dt}(B_q\psi)(s)||_{p,0}$$

$$\leqq C_{4,\alpha,\gamma,2}(t-s)^{\gamma-3/4}q^{-(\alpha+\gamma-1/4)}||\psi||_{p,\alpha,\gamma}$$

for any $\psi \in \mathscr{C}_{p,\alpha}^{\gamma}$, $\gamma = 1, 5/4, 3/2$, $p = 1, \infty$, $q \in (0, 1/2)$, *and* $s, t \in (2q, 1)$ *with* $s < t$.

(7) *There is a* $C_{4,\alpha,5/4,3} \in (0, \infty)$, $\alpha \geqq 0$, *satisfying the following.* $(B_q\psi)(t) \in W_p^3(E)$, $t \in (0, 1)$, $D_x^3 B_q\psi : (2q, 1) \to W_p^0(E)$ *is continuous, and*

$$||D_x^3(B_q\psi)(t)||_{p,0} \leqq C_{4,\alpha,5/4,3}q^{-(\alpha+1)}||\psi||_{p,\alpha,5/4}, \qquad t \in (2q, 1).$$

for any $\psi \in \mathscr{C}_{p,\alpha}^{5/4}$, $p = 1, \infty$, *and* $q \in (0, 1/2)$.

Proof Assertion (1) follows from Proposition 23. Assertion (2) follows from Propositions 23 and 26.

Assertion (3) follows from Propositions 24 and 26. Assertion (4) follows from Propositions 25, 27 and Assertions (2), (3). Assertion (5) follows from Propositions 24 and 26. Assertion (6) follows from Proposition 28.

The proof of Assertion (7) is similar to the proof of Proposition 28. Actually, if $p = \infty$,

$$D_x^3(B_q\psi)(t) = 2D_x B_q(\frac{d\psi}{dt})(t) + 2\frac{\partial g}{\partial x}(t-q, \cdot)\psi(q)(0),$$

and so by Proposition 26 we have our assertion. If $p = 1$, we see that

$$\int_q^t ||\frac{\partial^3 g}{\partial x^3}(t-r, \cdot)(\psi(r)(0) - \psi(t)(0))||_{1,0}dr$$

$$\leqq \int_q^t C_{1,0,3}(t-r)^{-3/2}(t-r)^{3/4}r^{-(\alpha+5/4)}||\psi||_{1,\alpha,5/4}dr$$

$$\leqq C_{1,0,3}C_{2,3/4,\alpha+5/4}q^{-(\alpha+1)}||\psi||_{1,\alpha,5/4}dr,$$

and so we have our assertion.

This completes the proof. ∎

Proposition 30 *Let* $\alpha \geqq 0$, $p = 1, \infty$, *and* $\gamma = \ell/4$, $\ell = 0, 1, \ldots$ *Then there is a* $C_{5,\alpha,\gamma} \in (0, \infty)$ *satisfying the following.*

If $\psi \in \mathscr{C}^{\gamma}_{p,\alpha}$, $\psi_{0,j} \in \mathscr{C}^{\gamma}_{p,\alpha}$, $j = 0, 1$, $\psi_{1,j} \in \mathscr{C}^{\gamma}_{p,\alpha+1}$, $j = 0, 1$, $\psi_{2,j} \in \mathscr{C}^{\gamma}_{p,\alpha+1/2}$, $j = 0, 1$, $\psi_3 \in \mathscr{C}^{\gamma+1/2}_{p,\alpha}$, $\psi_4 \in \mathscr{C}^{\gamma+1/2}_{p,\alpha}$, $\psi_5 \in \mathscr{C}^{\gamma+1/4}_{p,\alpha}$, *and* $\psi_{6,k} \in \mathscr{C}^{1/2}_{p,\alpha-(k+1)/2}$, $k = 0, 1, \dots, [2\alpha] - 1$, *and if*

$$\psi(t) = \sum_{j=0}^{1} Q_{j,t-q}\psi_{0,j}(q) + \sum_{j=0}^{1}(G_{j,q}\psi_{1,j})(t) + \sum_{j=0}^{1} D_x(G_{j,q}\psi_{2,j})(t)$$

$$+(B_q\psi_3)(t) + D_x(B_q\psi_4)(t) + \psi_5(t) + \sum_{k=0}^{[2\alpha]-1} \frac{\partial^k g}{\partial x^k}(t-q,\cdot)\psi_{6,k}(q)$$

for any $q \in (0, 1/2)$, $t \in (2q, 1)$, *then* $\psi \in \mathscr{C}^{\gamma+1/4}_{p,\alpha}$, *and*

$$||\psi||_{p,\alpha+1/4,\gamma+1/4}$$

$$\leqq C_{5,\alpha,\gamma}(||\psi_0||_{p,\alpha,\gamma} + \sum_{j=0}^{1}(||\psi_{1,j}||_{p,\alpha+1,\gamma}$$

$$+ \sum_{j=0}^{1}(||\psi_{2,j}||_{p,\alpha+1/2,\gamma} + ||\psi_3||_{p,\alpha,\gamma+1/2}$$

$$+ ||\psi_4||_{p,\alpha,\gamma+1/2} + ||\psi_5||_{p,\alpha,\gamma+1/4} + \sum_{k=0}^{[2\alpha]-1} ||\psi_{6,k}||_{p,\alpha-(k+1)/2,1/2}).$$

Proof First note that for $p = 1$

$$||\frac{\partial^{k+\ell} g}{\partial x^{k+\ell}}(t-q,\cdot)\psi_{6,k}(q)(0)||_{1,0}$$
$$\leqq C_{1,0,k}(t-q)^{-(k+\ell)/2}q^{-(\alpha-k/2)}||\psi_{6,k}||_{1,\alpha-(k+1)/2,1/2}$$
$$\leqq C_{1,0,k}q^{-(\alpha+\ell/2)}||\psi_{6,k}||_{1,\alpha-(k+1)/2,1/2},$$

and that for $p = \infty$

$$||\frac{\partial^{k+\ell} g}{\partial x^{k+\ell}}(t-q,\cdot)\psi_{6,k}(q)(0)||_{\infty,0}$$
$$\leqq C_{0,0,k}(t-q)^{-(k+\ell+1)/2}q^{-(\alpha-(k+1)/2)}||\psi_{6,k}||_{\infty,\alpha-(k+1)/2,0}$$
$$\leqq C_{0,0,k}q^{-(\alpha+\ell/2)}||\psi_{6,k}||_{\infty,\alpha-(k+1)/2,1/2}.$$

Also, note that

$$\frac{\partial}{\partial t}\frac{\partial^k g}{\partial x^k}(t-q,x) = \frac{1}{2}\frac{\partial^{k+2} g}{\partial x^{k+2}}(t-q,x), \qquad x > 0.$$

Then we see that

$$\begin{aligned}
&||\frac{\partial^k g}{\partial x^k}(t-q,\cdot)\psi_{6,k}(q)(0) - \frac{\partial^k g}{\partial x^k}(t-q,\cdot)\psi_{6,k}(q)(0)||_{p,0} \\
&\quad \leqq 2(C_{0,0,k} + C_{1,0,k} + C_{0,0,k+2} + C_{1,0,k+2})q^{-\alpha}(1 \wedge (t-s)q^{-1})||\psi_{6,k}||_{\infty,\alpha-(k+1)/2,1/2}.
\end{aligned}$$

We prove our assertion by induction in ℓ. Let $s, t \in (0,1)$ with $s < t$, and let $q = s/2$.

First, let $\gamma = 0$. Then by Propositions 21 and 29 we see that

$$\begin{aligned}
&||\psi(t) - \psi(s)||_{p,0} \\
&\quad \leqq \sum_{j=0}^{1}(C_{1,1,0} + 2C_{1,0,0})(t-s)^{1/4}(s-q)^{-1/4}q^{-\alpha}||\psi_{0,j}||_{p,\alpha,0} \\
&\quad + \sum_{j=0}^{1} C_{1,\alpha+1,0,0}(t-s)^{1/4}q^{-(\alpha+1/4)}||\psi_{1,j}||_{p,\alpha+1,0} \\
&\quad + \sum_{j=0}^{1} C_{3,\alpha+1/2,0,1}(t-s)^{1/4}q^{-(\alpha+1/4)}||\psi_{2,j}||_{p,\alpha+1/2,0} \\
&\quad + C_{4,\alpha,1/2,0}(t-s)^{3/4}q^{-(\alpha+1/4)}||\psi_3||_{p,\alpha,1/2} \\
&\quad + C_{4,\alpha,0,1}(t-s)^{1/4}q^{-(\alpha+1/4)}||\psi_4||_{p,\alpha,1/2} + (t-s)^{1/4}q^{-(\alpha+1/4)}||\psi_5||_{\infty,\alpha,1/4} \\
&\quad + \sum_{k=0}^{[2\alpha]-1} 2(C_{0,0,k} + C_{1,0,k} + C_{0,0,k+2} + C_{1,0,k+2})(t-s)^{1/4}q^{-(\alpha+1/4)} \\
&\quad \times ||\psi_{6,k}||_{\infty,\alpha-(k+1)/2,1/2}.
\end{aligned}$$

This implies that $\psi \in \mathscr{C}^{1/4}_{p,\alpha}$.

Let $\gamma = 1/4$. Then estimating $||\psi(t) - \psi(s)||_{p,0}$, and $||D_x\psi(t)||_{p,0}$ by Propositions 21 and 29, we see that $\psi \in \mathscr{C}^{1/2}_{p,\alpha}$.

Let $\gamma = 1/2$. Then by Propositions 20 and 25, we see that

$$\begin{aligned}
&D_x\psi(t) \\
&\quad = \sum_{j=0}^{1} Q_{1-j,t-q}\psi_{0,j}(q) + 2g(t-q,\cdot)\psi_{0,0}(q)(0)
\end{aligned}$$

$$+\sum_{j=0}^{1} G_{1-j,q} D_x \psi_{1,j})(t) + 2(B_q \psi_{1,0})(t)$$

$$+\sum_{j=0}^{1}(D_x G_{j,q} D_x \psi_{2,j})(t) + 2(D_x B_q \psi_{2,0})(t) + (D_x B \psi_3)(t)$$

$$+ D_x^2 (B_q \psi_4)(t) + (D_x \psi_5)(t) + \sum_{k=0}^{[2\alpha]-1} \frac{\partial^{k+1} g}{\partial x^{k+1}}(t-q,\cdot)\psi_{6,k}(q)(0).$$

Then estimating $||\psi(t)-\psi(s)||_{p,0}$, and $||D_x\psi(t)-D_x\psi(s)||_{p,0}$ by Propositions 21 and 29, we see that $\psi \in \mathscr{C}^{3/4}_{p.\alpha}$.

Let $\gamma = 3/4$. Then by Propositions 20 and 25, we see that

$$\begin{aligned} D_x\psi(t) &= \sum_{j=0}^{1} Q_{1-j,t-q}\psi_{0,j}(q) + 2g(t-q,\cdot)\psi_{0,0}(q)(0) \\ &\quad + \sum_{j=0}^{1}(G_{1-j,q} D_x \psi_{1,j})(t) + 2(B_q \psi_{1,0})(t) \\ &\quad + \sum_{j=0}^{1}(D_x G_{j,q} D_x \psi_{2,j})(t) + 2(D_x B_q \psi_{2,0})(t) + (D_x B \psi_3)(t) \\ &\quad + (B_q \frac{d\psi_4}{dt})(t) + g(t-q,\cdot)\psi_4(q)(0) + (D_x\psi_5)(t) \\ &\quad + \sum_{k=0}^{[2\alpha]-1} \frac{\partial^{k+1} g}{\partial x^{k+1}}(t-q,\cdot)\psi_{6,k}(q)(0). \end{aligned} \tag{20}$$

So from the assumption of induction, we see that $D_x\psi \in \mathscr{C}^{\gamma-1/2}_{p,\alpha+1/2}$.

Also, we see that

$$\begin{aligned} \frac{d\psi}{dt}(t) &= \frac{1}{2}\sum_{j=0}^{1}(D_x^2 Q_{j,t-q}\psi_{0,j})(q) + \sum_{j=0}^{1}(\frac{1}{2}(D_x^2 G_{j,q}\psi_{1,j})(t) + \psi_{1,j}(t)) \\ &\quad + \sum_{j=0}^{1}(\frac{1}{2}(D_x^2 G_{j,q} D_x \psi_{2,j})(t) + D_x\psi_{2,j}(t)) \end{aligned}$$

$$+ (D_x^2 B_q \psi_{2,0})(t) + \frac{1}{2}(D_x^2 B \psi_3)(t)$$
$$+ (D_x B_q \frac{d\psi_4}{dt})(t) + \frac{d\psi_5}{dt}(t) + \frac{\partial g}{\partial x}(t-q,\cdot)\psi_4(q)(0)$$
$$+ \frac{1}{2}\sum_{k=0}^{[2\alpha]-1} \frac{\partial^{k+2} g}{\partial x^{k+1}}(t-q,\cdot)\psi_{6,k}(q)(0).$$

Then estimating $||\frac{d\psi}{dt}(t)||_{p,0}$, by Propositions 21 and 29, we see that $\psi \in \mathscr{C}^1_{p.\alpha}$. Let $\gamma \geqq 1$. Then we see that

$$\frac{d\psi}{dt}(t)$$
$$= \frac{1}{2}\sum_{j=0}^{1}(Q_{j,t-q} D_x^2 \psi_{0,j})(q) + g(t-q,\cdot)\psi_{0,1}(q)(0)$$
$$+ \frac{\partial g}{\partial x}(t-q,\cdot)\psi_{0,0}(q)(0) + \sum_{j=0}^{1}(\frac{1}{2}(G_{j,q} D_x^2 \psi_{1,j})(t) + \psi_{1,j}(t))$$
$$+ (B_q \psi_{1,1})(t) + (D_x B_q \psi_{1,0})(t) + \sum_{j=0}^{1}(\frac{1}{2} D_x G_{j,q} D_x^2 \psi_{2,j})(t)$$
$$+ D_x^2 \psi_{2,j}(t)) + D_x B_q \frac{d\pi_{2,0}}{dt} + \frac{\partial g}{\partial x}(t-q,\cdot)\psi_{2,0}(q)(0)$$
$$+ (B_q \psi_{2,1})(t) + g(t-q,\cdot)\psi_{2,1}(q)(0) + B_q \frac{d\psi_3}{dt}(t)$$
$$+ g(t-q,\cdot)\psi_3(q)(0) + (D_x B_q \frac{d\psi_4}{dt})(t) + \frac{\partial g}{\partial x}(t-q,\cdot)\psi_4(q)(0) + \frac{d\psi_5}{dt}(t)$$
$$+ \frac{1}{2}\sum_{k=0}^{[2\alpha]-1} \frac{\partial^{k+2} g}{\partial x^{k+2}}(t-q,\cdot)\psi_{6,k}(q)(0).$$

So this equation, Eq. (20) and the assumption of induction imply that $D_x\psi \in \mathscr{C}^{\gamma-1/4}_{p,\alpha+1/2}$ and that $\frac{d}{dt}\psi \in \mathscr{C}^{\gamma-3/4}_{p,\alpha+1}$. So we see that $\psi \in \mathscr{C}^{\gamma+1/4}_{\alpha}$.

This completes the proof. ■

3.3 Differential Operators and Basic Equality

Let us remind the definition of $\mathscr{D}_n$, $n \geqq 0$, and their property in Sect. 2.5.

The following is an easy consequence of Theorem 3.

Proposition 31 *Let $n_1, n_2, m \geqq 0$ and $A_i \in \mathscr{D}_{n_i}$, $i = 1, 2$. Then there is a $C \in (0, \infty)$ such that*

$$\sup_{x\in(0,\infty)\times\mathbf{R}^{N-1}} |(A_1\, adj^m(V_0)(P_t^0)A_2 f)(x)| \leqq Ct^{-m-(n_1+n_2)/2} \sup_{x\in(0,\infty)\times\mathbf{R}^{N-1}} |f(x)|$$

and

$$\int_{(0,\infty)\times\mathbf{R}^{N-1}} |(A_1\, adj^m(V_0)(P_t^0)A_2 f)(x)|\, dx \leqq Ct^{-m-(n_1+n_2)/2} \int_{(0,\infty)\times\mathbf{R}^{N-1}} |f(x)|\, dx$$

for any $t \in (0, 1]$ and $f \in C_0^\infty(\mathbf{R}^N)$.

Also, we have the following.

Proposition 32 *Let $n \geqq 1$. Then we have the following.*

(1) *There are $A_{1,k} \in \mathscr{D}_{2(n-k)+1}$, $k = 0, \ldots, n-1$, and $A_{0,k} \in \mathscr{D}_{2(n-k)+2}$, $k = 0, \ldots, n-2$, such that*

$$adj(V_0)^n(V_1^2 B) = V_1^2 adj(V_0)^n(B) + \sum_{k=0}^{n-1} V_1 A_{1,k} adj(V_0)^k(B) + \sum_{k=0}^{n-2} A_{0,k} adj(V_0)^k(B)$$

for any linear operator B defined in $C_b^\infty(\mathbf{R}^N)$.

(2) *There are $C_k \in \mathscr{D}_{2(n-k)+1}$, $k = 0, \ldots, n-1$, such that*

$$adj(V_0)^n(b_1 V_1 B) = b_1 V_1 adj(V_0)^n(B) + \sum_{k=0}^{n-1} C_k adj(V_0)^k(B)$$

for any linear operator B defined in $C_b^\infty(\mathbf{R}^N)$. Here b_1 is a function in the definition of P_t^0.

Proof Note that

$$\begin{aligned} &adj(V_0)^n(V_1^2 B) \\ &= \sum_{k=0}^{n}\sum_{\ell=0}^{n-k} \binom{n}{k}\binom{n-k}{\ell} adj(V_0)^k(V_1) adj(V_0)^\ell(V_1) adj(V_0)^{n-k-\ell}(B) \\ &= V_1^2 adj(V_0)^n(B) + \sum_{\ell=1}^{n} \binom{n}{\ell} V_1 adj(V_0)^\ell(V_1) adj(V_0)^{n-\ell}(B) \end{aligned}$$

$$+\sum_{k=1}^{n}\binom{n}{k}V_1 adj(V_0)^k(V_1)adj(V_0)^{n-k-\ell}(B)$$

$$-\sum_{k=1}^{n}\binom{n}{k}[V_1, adj(V_0)^k(V_1)]adj(V_0)^{n-k-\ell}(B)$$

$$+\sum_{k=1}^{n}\sum_{\ell=1}^{n-k}\binom{n}{k}\binom{n-k}{\ell}adj(V_0)^k(V_1)adj(V_0)^\ell(V_1)adj(V_0)^{n-k-\ell}(B)$$

So we have Assertion (1). The proof of Assertion (2) is similar. ■

Also, we have the following.

Proposition 33 *Let $n \geqq 1$, and $A \in \mathscr{D}_m$, $m \geqq 0$.*

(1) *There are $C_{1,k} \in \mathscr{D}_{m+2(n-k)+1}$, $k = 0, \ldots, n-1$, and $C_{0,k} \in \mathscr{D}_{m+2(n-k)+2}$, $k = 0, \ldots, n-2$, such that*

$$A\ adj(V_0)^n((\frac{1}{2}V_1^2 + b_1 V_1)B)$$

$$= \frac{1}{2}V_1^2 A\ adj(V_0)^n(B) + \sum_{k=0}^{n-1} V_1 C_{1,k} adj(V_0)^k(B) + \sum_{k=0}^{n-2} C_{0,k} adj(V_0)^k(B)$$

for any linear operator B in $C_b^\infty(\mathbf{R}^N)$.

(2) *Let $A' \in \mathscr{D}_{m'}$, $m' \geqq 0$. Then for any $n \geqq 1$, there are $C'_k \in \mathscr{D}_{m+m'+2(n-k)}$, $k = 0, \ldots, n$, such that*

$$A\ adj(V_0)^n(A'B) = \sum_{k=0}^{n} C'_k adj(V_0)^k(B)$$

for any linear operator B in $C_b^\infty(\mathbf{R}^N)$.

Now let $U \in C_b^\infty(\mathbf{R})$ such that $U(z) = 0$ for $z > 0$, $U(z) > 0$ for $z < 0$, and $U(z) = -1$ for $z \leqq -1$. We define a semigroup of linear operators $P_t^{(\lambda)}$, $t \in [0, \infty)$, $\lambda \in (0, \infty)$, in $C_b^\infty(\mathbf{R}^N)$ by

$$(P_t^{(\lambda)} f)(x)$$

$$= E^\mu[\exp(-\lambda \int_0^t U(\tilde{X}^1(s, x))ds) \exp(\sum_{k=0}^{d} \int_0^t b_k(X(s, x)) \circ dB^k(s)) f(X(t, x))],$$

$x \in \mathbf{R}^N$, $f \in C_b^\infty(\mathbf{R}^N)$. Note that the infinitesimal generator $L^{(\lambda)}$ of $P_t^{(\lambda)}$ is given by

$$L^{(\lambda)} = V_0 + \frac{1}{2}\sum_{k=1}^d V_k^2 + \sum_{k=1}^d b_k V_k + b_0 + \frac{1}{2}\sum_{k=1}^d (b_k^2 + (V_k b_k)) - \lambda U.$$

Then we see that for any $A \in \mathscr{D}_m$, $m \geqq 0$, $n \geqq 0$, and $f \in C_0^\infty(\mathbf{R}^N)$

$$\begin{aligned}
&\frac{d}{dt}(A\, adj(V_0)^n (P_t^{(\lambda)}) Exp(-tV_0) f)(x) \\
&\quad = (A\, adj(V_0)^n ((\frac{1}{2}V_1^2 + b_1 V_1 + L_0) P_t^{(\lambda)}) Exp(-tV_0) f))(x)) \\
&\quad\quad + (A\, adj(V_0)^{n+1} (P_t^{(\lambda)}) Exp(-tV_0) f)(x)
\end{aligned}$$

for $x \in (0, \infty) \times \mathbf{R}^{N-1}$. Here

$$L_0 = \frac{1}{2}\sum_{k=2}^d V_k^2 + \sum_{k=2}^d b_k V_k + b_0 + \frac{1}{2}\sum_{k=1}^d (b_k^2 + (V_k b_k)) \in \mathscr{D}_2.$$

So we see have the following by Proposition 33.

Proposition 34 *For any $A \in \mathscr{D}_m$, $m \geqq 0$, and $n \geqq 0$, there are $C_{1,k} \in \mathscr{D}_{m+2(n-k)+1}$, $k = 0, \ldots, n-1$, and $C_{0,k} \in \mathscr{D}_{2(n-k)+2}$, $k = 0, \ldots, n$, such that*

$$\begin{aligned}
&(\frac{d}{dt} - \frac{1}{2}V_1^2)(A\, adj(V_0)^n (P_t^{(\lambda)}) Exp(-tV_0) f)(x) \\
&\quad = \sum_{k=0}^n (V_1 C_{1,k}\, adj(V_0)^k (P_t^{(\lambda)}) Exp(-tV_0) f)(x) \\
&\quad\quad + \sum_{k=0}^n (C_{0,k}\, adj(V_0)^k (P_t^{(\lambda)}) Exp(-tV_0) f)(x) \\
&\quad\quad + (A\, adj(V_0)^{n+1} (P_t^{(\lambda)}) Exp(-tV_0) f)(x) \qquad (21)
\end{aligned}$$

for any $f \in C_0(\mathbf{R}^N)$, $t > 0$, and $x \in (0, \infty) \times \mathbf{R}^{N-1}$.

Here $C_{1,k} \in \mathscr{D}_{m+2(n-k)+1}$, $k = 0, \ldots, n-1$, and $C_{0,k} \in \mathscr{D}_{m+2(n-k)+2}$, $k = 0, \ldots, n-2$, are those in Proposition 33.

For any $f \in C_0^\infty(\mathbf{R}^N)$, $n, m \geqq 0$, and $A \in \mathscr{D}_m$, let $\hat{\psi}_{n,A,f} : (0,1) \times (0,\infty) \to C_b^\infty(\mathbf{R}^{N-1})$, be given by

$$\hat{\psi}_{n,A,f}(t)(x)(\tilde{x}) = (A\, adj(\tilde{V}_0)^n(P_t^0 Exp(-\tilde{V}_0))f)(x,\tilde{x}), \qquad t, x > 0,\ \tilde{x} \in \mathbf{R}^{N-1}.$$

Notice that $\hat{\psi}_{n,A,f}(t)(x)(\tilde{x}) \to 0$ as $x + |\tilde{x}| \to \infty$, and

$$\int_{(0,\infty)} dx \int_{\mathbf{R}^{N-1}} d\tilde{x} |\psi_{n,A,f}(t)(x)(\tilde{x})| < \infty.$$

Then we have the following.

Proposition 35 *Let $n, m \geqq 0$, and $A \in \mathscr{D}_m$. Then for any $q \in (0,1/2)$, $f \in C_0^\infty(\mathbf{R}^N)$, $t \in (2q,1)$, $x \in (0,\infty)$, and $\tilde{x}$,*

$$\begin{aligned}
&\hat{\psi}_{n,A,f}(t)(x)(\tilde{x}) \\
&= \int_0^\infty g_0(t-q,x,y)\hat{\psi}_{n,A,f}(q)(y)dy \\
&\quad - \sum_{k=0}^n \int_q^t ds \int_0^\infty dy \frac{\partial g_1}{\partial x}(t-s,x,y)\frac{\partial}{\partial y}\hat{\psi}_{k,C_{1,k},f}(s)(y)(\tilde{x}) \\
&\quad + \sum_{k=0}^n \int_q^t ds \int_0^\infty dy g_0(t-s,x,y)\hat{\psi}_{k,C_{0,k},f}(s)(y)(\tilde{x}) \\
&\quad + \int_q^t ds \int_0^\infty dy g_0(t-s,x,y)\hat{\psi}_{n+1,A,f}(s)(y).(\tilde{x}).
\end{aligned}$$

Here $C_{1,k} \in \mathscr{D}_{m+2(n-k)+1}$, $k = 0,\dots,n-1$, and $C_{0,k} \in \mathscr{D}_{m+2(n-k)+2}$, $k = 0,\dots,n-2$, are those in Proposition 34.

Proof For any $f \in C_0^\infty(\mathbf{R}^N)$, $n, m \geqq 0$, $A \in \mathscr{D}_m$, and $\lambda > 0$, let $\hat{\psi}^{(\lambda)} = \hat{\psi}^{(\lambda)}_{n,A,f} :$ $(0,\infty) \times ([0,\infty) \times \mathbf{R}^{N-1}) \to \mathbf{R}$ be given by

$$\hat{\psi}^{(\lambda)}(t,x,\tilde{x}) = (A\, adj(V_0)^n(P_t^{(\lambda)} Exp(-tV_0))f)(x,\tilde{x}), \qquad t \in (0,1),\ x \geqq 0,\ \tilde{x} \in \mathbf{R}^{N-1}.$$

Note that $\hat{\psi}^{(\lambda)}_{n,A,f}(t,x,\tilde{x}) \to 0$, as $x + |\tilde{x}| \to \infty$, for any $t \in [0,\infty)$,

$$\sup_{\lambda>0} \int_{(0,\infty)\times\mathbf{R}^{N-1}} |\hat{\psi}^{(\lambda)}(t,x,\tilde{x})|dxd\tilde{x} < \infty, \qquad t \in [0,\infty),$$

and

$$\sup_{\lambda>0} \sup_{x\in(0,\infty),\tilde{x}\in\mathbf{R}^{N-1}} |\hat{\psi}^{(\lambda)}(t,x,\tilde{x})| < \infty, \qquad t \in [0,\infty).$$

Let $q \in (0, 1/2)$. Then for any $x > 0$, $t \in (2q, 1)$, and $\varepsilon \in (0, q)$,

$$\begin{aligned}
0 &= \int_q^{t-\varepsilon} ds \int_0^\infty dy((\frac{\partial}{\partial s} + \frac{1}{2}\frac{\partial^2}{\partial y^2})g_0(t-s,x,y))\hat{\psi}^{(\lambda)}(s,y,\tilde{x}) \\
&= \int_0^\infty dy(g_0(\varepsilon,x,y)\hat{\psi}^{(\lambda)}(t-\varepsilon,y,\tilde{x}) - g_0(t-q,x,y)\hat{\psi}^{(\lambda)}(q,y,\tilde{x})) \\
&\quad - \int_q^{t-\varepsilon} ds\frac{\partial\hat{\psi}^{(\lambda)}}{\partial s}(s,y)g_0(t-s,x,y) + \frac{1}{2}\int_q^{t-\varepsilon} ds(-\frac{\partial g_0}{\partial y}(t-s,x,0)\hat{\psi}^{(\lambda)}(s,0,\tilde{x}) \\
&\quad + \int_0^\infty dy g_0(t-s,x,y)\frac{\partial^2\hat{\psi}^{(\lambda)}}{\partial y^2}(s,y,\tilde{x})) \\
&= \int_0^\infty dy(g_0(\varepsilon,x,y)\hat{\psi}(t-\varepsilon,y) - g_0(t-q,x,y)\hat{\psi}^{(\lambda)}(q,y,\tilde{x})) \\
&\quad - \int_q^{t-\varepsilon} \frac{\partial g_0}{\partial y}(t-s,x,0)\hat{\psi}^{(\lambda)}(s,0,\tilde{x})ds \\
&\quad + \int_q^{t-\varepsilon} ds\int_0^\infty dy g_0(t-s,x,y)(-\frac{\partial}{\partial s} + \frac{1}{2}\frac{\partial^2}{\partial y^2})\hat{\psi}^{(\lambda)}(s,y,\tilde{x}).
\end{aligned}$$

So we have

$$\begin{aligned}
&\int_0^\infty g_0(\varepsilon,x,y)\hat{\psi}^{(\lambda)}(t-\varepsilon,y,\tilde{x})dy \\
&\quad = \int_0^\infty g_0(t-q,x,y)\hat{\psi}^{(\lambda)}(q,y,\tilde{x})dy \\
&\qquad - \int_q^{t-\varepsilon} \frac{\partial g_0}{\partial y}(t-s,x,0)\hat{\psi}^{(\lambda)}(s,0,\tilde{x})ds \\
&\qquad + \int_q^{t-\varepsilon} ds\int_0^\infty dy g_0(t-s,x,y)(\frac{\partial}{\partial s} - \frac{1}{2}\frac{\partial^2}{\partial y^2})\hat{\psi}^{(\lambda)}(s,y,\tilde{x}).
\end{aligned}$$

Letting $\varepsilon \downarrow 0$, we see that

$$\begin{aligned}
&\hat{\psi}^{(\lambda)}(t,x,\tilde{x}) \\
&\quad = \int_0^\infty g_0(t-q,x,y)\hat{\psi}^{(\lambda)}(q,y,\tilde{x})dy - \int_q^t \frac{\partial g_0}{\partial y}(t-s,x,0)\hat{\psi}^{(\lambda)}(s,0,\tilde{x})ds \\
&\qquad + \int_q^t ds\int_0^\infty dy g_0(t-s,x,y)(\frac{\partial}{\partial s} - \frac{1}{2}\frac{\partial^2}{\partial y^2})\hat{\psi}^{(\lambda)}(s,y,\tilde{x}).
\end{aligned}$$

Therefore by Proposition 34, we see that

$$
\begin{aligned}
&\hat{\psi}^{(\lambda)}_{n,A,f}(t,x,\tilde{x}) \\
&= \int_0^\infty g_0(t-q,x,y)\hat{\psi}^{(\lambda)}_{n,A,f}(q,y,\tilde{x})dy \\
&\quad - \int_q^t \frac{\partial g_0}{\partial y}(t-s,x,0)\hat{\psi}^{(\lambda)}_{n,A,f}(s,0,\tilde{x})ds \\
&\quad + \sum_{k=0}^n \int_q^t ds \int_0^\infty dy g_0(t-s,x,y)\frac{\partial}{\partial y}\hat{\psi}^{(\lambda)}_{k,C_{1,k},f}(s,y,\tilde{x}) \\
&\quad + \sum_{k=0}^n \int_q^t ds \int_0^\infty dy g_0(t-s,x,y)\hat{\psi}^{(\lambda)}_{k,C_{0,k},f}(s,y,\tilde{x}) \\
&\quad + \int_q^t ds \int_0^\infty dy g_0(t-s,x,y)\hat{\psi}^{(\lambda)}_{n+1,A,f}(s,y,\tilde{x}) \\
&= \int_0^\infty g_0(t-q,x,y)\hat{\psi}^{(\lambda)}_{n,A,f}(q,y,\tilde{x})dy \\
&\quad - \sum_{k=0}^n \int_q^t ds \int_0^\infty dy \frac{\partial g_1}{\partial x}(t-s,x,y)\hat{\psi}^{(\lambda)}_{k,C_{1,k},f}(s,y,\tilde{x}) \\
&\quad + \sum_{k=0}^n \int_q^t ds \int_0^\infty dy g_0(t-s,x,y)\hat{\psi}^{(\lambda)}_{k,C_{0,k},f}(s,y,\tilde{x}) \\
&\quad + \int_q^t ds \int_0^\infty dy g_0(t-s,x,y)\hat{\psi}^{(\lambda)}_{n+1,A,f}(s,y,\tilde{x}) \\
&\quad - \int_q^t \frac{\partial g_0}{\partial y}(t-s,x,0)\hat{\psi}^{(\lambda)}_{n,A,f}(s,0,\tilde{x})ds
\end{aligned}
$$

It is easy to see that $\hat{\psi}^{(\lambda)}_{n,A,f}(t)(x,\tilde{x}) \to \hat{\psi}_{n,A,f}(t)(x,\tilde{x})$, as $\lambda \to \infty$, and $\hat{\psi}^{(\lambda)}_{n,A,f}(t)(0,\tilde{x}) \to 0$, as $\lambda \to \infty$, for any $t \in (0,1)$, $x \in (0,\infty)$, $\tilde{x} \in \mathbf{R}^{N-1}$. So letting $\lambda \to \infty$, we have our assertion. ■

3.4 L^∞-Estimate

Now let $E = E_\infty = C_\infty(\mathbf{R}^{N-1})$ be a separable Banach space consisting of continuous functions defined on $\mathbf{R}^{N-1}$ such that $f(\tilde{x}) \to 0$ as $|\tilde{x}| \to \infty$ with a norm

$$||f||_{E_\infty} = \sup\{|f(\tilde{x})|;\ \tilde{x} \in \mathbf{R}^{N-1}\}.$$

Note that $C_0^\infty(\mathbf{R}^{N-1})$ is a dense subset in E_∞. So we see that $\{f|_{(0,\infty)\times\mathbf{R}^{N-1}};\ f \in C_0^\infty(\mathbf{R}^N)\}$ is a dense subset of $W_\infty^0(E_\infty)$.

For any $f \in W_\infty^0(E_\infty)$, $n, m \geqq 0$, and $A \in \mathscr{D}_m$, let $\hat{\psi}_{n,A,f} : (0,1)\times(0,\infty) \to C_b^\infty(\mathbf{R}^{N-1})$, be given by

$$\psi_{n,A,f}(t)(x)(\tilde{x}) = (A\, adj(\tilde{V}_0)^n (P_t^0 Exp(-\tilde{V}_0))f)(x,\tilde{x}), \qquad t, x > 0,\ \tilde{x} \in \mathbf{R}^{N-1}.$$

Then we have the following

Proposition 36

(1) $\psi_{n,A,f} \in \mathscr{C}^0_{\infty,m/2+n}$ *for any* $n, m \geqq 0$, $A \in \mathscr{D}_m$, *and* $f \in C_0^\infty(\mathbf{R}^N)$. *Moreover, for any* $n, m \geqq 0$, *and* $A \in \mathscr{D}_m$, *there is a* $C \in (0,\infty)$ *such that*

$$||\psi_{n,A,f}||_{\infty,m/2+n,0} \leqq C \sup_{x\in(0,\infty)\times\mathbf{R}^N} |f(x)|$$

for any $f \in C_0^\infty(\mathbf{R}^N)$.

(2) *For any* $n, m \geqq 0$, *and* $A \in \mathscr{D}_m$, *there are* $C_{1,k} \in \mathscr{D}_{m+2(n-k)+1}$, $k = 0, \ldots, n-1$, *and* $C_{0,k} \in \mathscr{D}_{m+2(n-k)+2}$, $k = 0, \ldots, n-2$, *satisfying the following. For any* $q \in (0,1/2)$, $t \in (2q,1)$, *and* $f \in C_0^\infty(\mathbf{R}^N)$,

$$\begin{aligned}&\psi_{n,A,f}(t)\\ &\quad = Q_{0,t-q}\psi_{n,A,f}(q) + \sum_{k=0}^{n} D_x(G_{1,Q}\psi_{k,C_{1,k},f})(t)\\ &\qquad + \sum_{k=0}^{n}(G_{0,q}\psi_{k,C_{0,k},f})(t) + (G_{0,q}\psi_{n+1,A,f})(t).\end{aligned}$$

(3) $\psi_{n,A,f} \in \mathscr{C}^{k/4}_{\infty,m/2+n}$ *for any* $n, m, k \geqq 0$, $A \in \mathscr{D}_m$, *and* $f \in C_0^\infty(\mathbf{R}^N)$. *Moreover, for any* $n, m, k \geqq 0$, *and* $A \in \mathscr{D}_m$, *there is a* $C \in (0,\infty)$ *such that*

$$||\psi_{n,A,f}||_{\infty,m/2+n,k/4} \leqq C \sup_{x\in(0,\infty)\times\mathbf{R}^N} |f(x)|$$

for any $f \in C_0^\infty(\mathbf{R}^N)$.

Proof Assertions (1) and (2) follow from Propositions 31 and 35.
Then by Proposition 30, we have Assertion (3) by induction in k. ∎

As a corollary to Proposition 36, we have the following.

Corollary 3 *For for any* $n, m, k \geqq 0$, $A \in \mathscr{D}_m$, *and* $f \in C_0^\infty(\mathbf{R}^N)$, $(AP_t^0 f)(\cdot,\tilde{x}) : (0,\infty) \to \mathbf{R}$ *is smooth for any* $t \in (0,1)$ *and* $\tilde{x} \in \mathbf{R}^{N-1}$.

Moreover, for any $n, m \geqq 0$, and $A \in \mathscr{D}_m$, there is a $C \in (0, \infty)$ such that

$$\sup_{x \in (0,\infty)\times \mathbf{R}^{N-1}} |V_1^k A\, adj(V_0)^n (P_t^0) f(x)| \leqq C t^{-(2n+m+k)/2} \sup_{x \in (0,\infty)\times \mathbf{R}^{N-1}} |f(x)|$$

for any $t \in (0, 1)$ and $f \in C_0^\infty(\mathbf{R}^N)$.

3.5 L^1-Estimate

Let $E = E_1 = L^1(\mathbf{R}^{N-1}, d\tilde{x})$. Then E ie a separable Banach space.

For any $f \in W_1^0(E_1)$, $n, m \geqq 0$, and $A \in \mathscr{D}_m$, let $\psi_{n,A,f} : (0, 1) \times (0, \infty) \to C_b^\infty(\mathbf{R}^{N-1})$, be given by

$$\psi_{n,A,f}(t)(x)(\tilde{x}) = (A\, adj(\tilde{V}_0)^n (P_t^0 Exp(-\tilde{V}_0)) f)(x, \tilde{x}), \qquad t, x > 0,\ \tilde{x} \in \mathbf{R}^{N-1}.$$

Proposition 31 guarantees that $\psi_{n,A,f}$ is well-defined.

By a similar argument in the previous Section, we have the following.

Proposition 37

(1) $\psi_{n,A,f} \in \mathscr{C}^0_{1,m/2+n}$ *for any $n, m \geqq 0$, $A \in \mathscr{D}_m$, and $f \in C_0^\infty(\mathbf{R}^N)$. Moreover, for any $n, m \geqq 0$, and $A \in \mathscr{D}_m$, there is a $C \in (0, \infty)$ such that*

$$||\psi_{n,A,f}||_{1,m/2+n,0} \leqq C \int_{(0,\infty)\times \mathbf{R}\ N-1} |f(x)|$$

for any $f \in C_0^\infty(\mathbf{R}^N)$.

(2) *For any $n, m \geqq 0$, and $A \in \mathscr{D}_m$, there are $C_{1,k} \in \mathscr{D}_{m+2(n-k)+1}$, $k = 0, \ldots, n-1$, and $C_{0,k} \in \mathscr{D}_{m+2(n-k)+2}$, $k = 0, \ldots, n-2$, satisfying the following. For any $q \in (0, 1/2)$, $t \in (2q, 1)$, and $f \in C_0^\infty(\mathbf{R}^N)$,*

$$\begin{aligned}
&\psi_{n,A,f}(t) \\
&= Q_{0,t-q}\psi_{n,A,f}(q) + \sum_{k=0}^{n} D_x(G_{1,Q}\psi_{k,C_{1,k},f})(t) \\
&+ \sum_{k=0}^{n} (G_{0,q}\psi_{k,C_{0,k},f})(t) + (G_{0,q}\psi_{n+1,A,f})(t).
\end{aligned}$$

(3) $\psi_{n,A,f} \in \mathscr{C}^{k/4}_{1,m/2+n}$ *for any* $n, m, k \geqq 0$, $A \in \mathscr{D}_m$, *and* $f \in C_0^\infty(\mathbf{R}^N)$. *Moreover, for any* $n, m, k \geqq 0$, *and* $A \in \mathscr{D}_m$, *there is a* $C \in (0, \infty)$ *such that*

$$||\psi_{n,A,f}||_{1,m/2+n,k/4} \leqq C \int_{(0,\infty)\times\mathbf{R}^{N-1}} |f(x)|dx$$

So we have the following.

Corollary 4 *For for any* $n, m, k \geqq 0$, $A \in \mathscr{D}_m$, *and* $f \in C_0^\infty(\mathbf{R}^N)$, $(A\hat{P}_t^0 f)(\cdot, \tilde{x}) : (0, \infty) \to \mathbf{R}$ *is smooth for any* $t \in (0, 1)$ *and* $\tilde{x} \in \mathbf{R}^{N-1}$.

Moreover, for any $n, m \geqq 0$, *and* $A \in \mathscr{D}_m$, *there is a* $C \in (0, \infty)$ *such that*

$$\int_{(0,\infty)\times\mathbf{R}^{N-1}} |V_1^k A\, adj(V_0)^n(\hat{P}_t^0) f(x)|dx \leqq Ct^{-(2n+m+k)/2} \int_{(0,\infty)\times\mathbf{R}^{N-1}} |f(x)|dx$$

for any $t \in (0, 1)$ *and* $f \in C_0^\infty(\mathbf{R}^N)$.

3.6 Proof of Theorem 2

Let $\bar{P}_t^0$, $t \geqq 0$, be the diffusion semi-group given in Sect. 2.5. Then by Eq. (15) and arguments in Sect. 2.5, we see that

$$\int_{(0,\infty)\times\mathbf{R}^{N-1}} g(x)(P_t^0 f)(x)dx = \int_{(0,\infty)\times\mathbf{R}^{N-1}} f(x)(\bar{P}_t^0 g)(x)dx$$

for $t > 0$, and $f, g \in C_0^\infty((0, \infty) \times \mathbf{R}^{N-1})$, and so by Corollaries 3 and 4, we have the following.

Proposition 38 *For any* $n, m, k \geqq 0$, *and* $A \in \mathscr{D}_m$, *there is a* $C \in (0, \infty)$ *such that*

$$\sup_{x\in(0,\infty)\times\mathbf{R}^{N-1}} |adj(V_0)^n(P_t^0)AV_1^k f(x)| \leqq Ct^{-(2n+m+k)/2} \sup_{x\in(0,\infty)\times\mathbf{R}^{N-1}} |f(x)|$$

and

$$\int_{(0,\infty)\times\mathbf{R}^{N-1}} |adj(V_0)^n(P_t^0)AV_1^k f(x)|dx \leqq Ct^{-(2n+m+k)/2} \int_{(0,\infty)\times\mathbf{R}^{N-1}} |f(x)|dx$$

for any $t \in (0, 1)$ *and* $f \in C_0^\infty((0, \infty) \times \mathbf{R}^{N-1})$.

Combining Corollaries 3, 4 and Proposition 38, we have Theorem 2.

References

1. Bergh J, Löfstrom J (1976) Interpolation spaces. An introduction. Springer, Berlin
2. Kunita H (1997) Stochastic flows and stochastic differential equations Cambridge studies in advanced mathematics. Cambridge University Press, Cambridge
3. Kusuoka S (2003) Malliavin calculus revisited. J Math Sci Univ Tokyo 10:261–277
4. Kusuoka S (2012) A remark on Malliavin calculus: uniform estimate and localization. J Math Sci Univ Tokyo 19:533–558
5. Kusuoka S, Stroock DW (1987) Applications of Malliavin calculus III. J Fac Sci Univ Tokyo Sect IA Math 34:391–442
6. Shigekawa I (2000) Stochastic analysis. Translation of mathematical monographs, vol. 224. American Mathematical Society, Providence
7. Watanabe S (1997) Wiener functionals with the regularity of fractional order. New trends in stochastic analysis (Charingworth, 1994). World Scientific Publishing, Singapore, pp 416–429

Disintegration of Young Measures and Nonlinear Analysis

Toru Maruyama

Abstract Selected fundamental theories of Young measures are systematically presented together with preparatory materials.
The main topics to be discussed here are the following three.

1. The existence and uniqueness of a representation (disintegration) of a Young measure by means of a measurable family of Radon probability measures.
2. The topological structure of the space of Young measures.
3. The lower semi-continuity of nonlinear integral functionals: a crucial topic in the existence theory of calculus of variations.

Several applications are also shown, which include

1. the relation between the concept of disintegration and that of conditional expectation, as well as
2. the existence of solutions for simple variational problems and their "purifications".

Keywords Young measure · Disintegration · Narrow topology · Purification · Continuity of nonlinear integral functional

Article type: Research Article
Received: May 10, 2019
Revised: July 19, 2019

JEL Classification: C61
Mathematics Subject Classification: 28A33, 49J24

T. Maruyama (✉)
Keio University, Tokyo, Japan
e-mail: maruyama@econ.keio.ac.jp

T. Maruyama (ed.), *Advances in Mathematical Economics*, Advances in Mathematical Economics 23, https://doi.org/10.1007/978-981-15-0713-7_6

1 Introduction

Let $I = [0, 1]$ be the unit interval of the real axis $\mathbb{R}$ endowed with the Lebesgue σ-field $\mathscr{L}$ and the Lebesgue measure m. $\{\nu_\omega | \omega \in I\}$ denotes a family of finite measures on $(\mathbb{R}^l, \mathscr{B}(\mathbb{R}^l))$, where $\mathscr{B}(\mathbb{R}^l)$ is the Borel σ- field on $\mathbb{R}^l$. $\{\nu_\omega | \omega \in I\}$ is called a measurable family if the function $\omega \mapsto \nu_\omega(B)$ of I into $\mathbb{R}$ is measurable for all $B \in \mathscr{B}(\mathbb{R}^l)$.

The function

$$\omega \mapsto \int_{\mathbb{R}^l} \chi_A(\omega, x) d\nu_\omega$$

of I into $\mathbb{R}$ is shown to be measurable for any $A \in \mathscr{L} \otimes \mathscr{B}(\mathbb{R}^l)$. ($\chi_A$ is the characteristic function of A.) Hence we can define the integration of this function on I. When a measure γ on $(I \times \mathbb{R}^l, \mathscr{L} \otimes \mathscr{B}(\mathbb{R}^l))$ is expressed in the form

$$\gamma(A) = \int_I \{\int_{\mathbb{R}^l} \chi_A(\omega, x) d\nu_\omega\} dm, \tag{1.1}$$

this expression is called the disintegration of γ by means of the measurable family $\{\nu_\omega | \omega \in I\}$.[1] Equation (1.1) can also be expressed as

$$\gamma(A) = \int_I \{\int_{I \times \mathbb{R}^l} \chi_A(\omega, x) d(\delta_\omega \otimes \nu_\omega)\} dm. \tag{1.1'}$$

When γ satisfies the relation (1.1) ($\Leftrightarrow$(1.1')), we symbolically express γ as

$$\gamma = \int_I \delta_\omega \otimes \nu_\omega dm. \tag{1.2}$$

δ_ω is the Dirac measure which assigns mass 1 at ω.

As a special case of a measurable family, we often consider the one defined by some $x(\cdot) \in \mathfrak{L}^1(I, \mathbb{R}^l)$:

$$\{\nu_\omega = \delta_{x(\omega)} | \omega \in \Omega\}.$$

The measure γ on $I \times \mathbb{R}^l$ corresponding to this measurable family is

$$\gamma = \int_I \delta_\omega \otimes \delta_{x(\omega)} dm. \tag{1.3}$$

[1]Let ζ be a finite measure on $(\mathbb{R}^l, \mathscr{B}(\mathbb{R}^l))$. If $\nu_\omega = \zeta$ for all $\omega \in I$, $\{\nu_\omega | \omega \in I\}$ is a measurable family and γ defined by (1.1) is nothing other than the product measure $m \otimes \zeta$.

This is an elementary but important example of a measure which has an expression by disintegration. $\Delta(\mathfrak{L}^1(I, \mathbb{R}^l))$ is the set of all the measures on $I \times \mathbb{R}^l$ that can be expressed in the form (1.3) for some $x(\cdot) \in \mathfrak{L}^1(I, \mathbb{R}^l)$.

Suppose that a certain function $f : I \times \mathbb{R}^l \to \mathbb{R}$ is given. We define an integral functional $J : \mathfrak{L}^1(I, \mathbb{R}^l) \to \mathbb{R}$ by

$$J : x(\cdot) \mapsto \int_I f(\omega, x(\omega))dm. \tag{1.4}$$

On the other hand, we also define a functional $H : \Delta(\mathfrak{L}^1(I, \mathbb{R}^l)) \to \mathbb{R}$ by

$$H : \gamma = \int_I \delta_\omega \otimes \delta_{x(\omega)} dm \mapsto \int_{I \times \mathbb{R}^l} f(\omega, x)d\gamma. \tag{1.5}$$

Then it is obvious that (1.4) and (1.5) are different expressions of the same object. J is a nonlinear functional on $\mathfrak{L}^1(I, \mathbb{R}^l)$. Whereas, roughly speaking, H is a "linear" functional of γ. Such a transformation from nonlinear to linear is made possible by changing variable from $x(\cdot)$ to γ interpreted as an operator acting on f.[2]

In classical calculus of variations, a basic integral functional

$$x(\cdot) \mapsto \int_I f(\omega, x(\omega), \dot{x}(\omega))dm$$

on $\mathfrak{C}^1(I, \mathbb{R}^l)$[3] appears frequently. Although this functional is nonlinear, the concept of disintegration of measures can be effectively used for paving the way to "linearization" of the nonlinear problem.

Thus a transparent route can be developed for calculus of variations,[4] nonlinear partial differential equations[5] and etc. by means of disintegration of measures. However we have to prepare exact answers to the following two questions.

First of all, under what conditions, is a measure γ on $(I \times \mathbb{R}^l, \mathfrak{L} \otimes \mathscr{B}(\mathbb{R}^l))$ expressible in the form of disintegration?

The second question to be answered is a detailed examination from the viewpoint of functional analysis of the space of disintegrable measures.

We discuss these problems in the first two parts of this article. Being based upon these basic studies, we devote the remaining part to some selected applications. The continuity of nonlinear integral functionals is our summit target. It is an

[2]Of course, H is not a linear functional in an exact sense because $\Delta(\mathfrak{L}^1(I, \mathbb{R}^l))$ is not a vector space.

[3]$\mathfrak{C}^1(I, \mathbb{R}^l)$ is the set of all continuously differentiable functions of I into $\mathbb{R}^l$. We also adopt more generally the Sobolev space, say $\mathfrak{W}^{1,2}(I, \mathbb{R}^l)$, instead of $\mathfrak{C}^1(I, \mathbb{R}^l)$.

[4]Berliocchi–Lasry [7] is a forerunner in this field. See also Maruyama [33].

[5]For instance, Evans [22] provides a lucid explanation.

indispensable preparation for the existence theory in calculus of variations or optimal controls.

The purpose of this work is to give a systematic overview of the theory of disintegration including my own results. In the course of my research, I have learned a lot about this discipline from mathematicians of Montpelier School, including Professor C. Castaing and Professor M. Valadier. In particular, I have to appreciate what I learned from Professor Valadier's survey articles [48, 49] for writing up this work. I also recommend the readers to consult Castaing et al. [14] as a reliable treatise. See Bourbaki [11] for functional analysis, in general.

2 Disintegrations

2.1 Measurable Family of Measures

The concept of a measurable family of measures was introduced already in the preface. We recapitulate this concept in a more general framework.

Definition 2.1 Let $(\Omega_1, \mathscr{E}_1)$ and $(\Omega_2, \mathscr{E}_2)$ be measurable spaces. A measure μ_1 on $(\Omega_1, \mathscr{E}_1)$ and a measurable function $\theta : \Omega_1 \to \Omega_2$ are given. Then the measure μ_2 on $(\Omega_2, \mathscr{E}_2)$ defined by

$$\mu_2(E) = (\mu_1 \circ \theta^{-1})(E) = \mu_1(\theta^{-1}(E)), \quad E \in \mathscr{E}_2$$

is called the **image measure of μ_1 via θ**.

The following properties of image measures can be checked easily.

1° If a measurable function $f : \Omega_2 \to \bar{\mathbb{R}}$ is either nonnegative or μ_2-integrable, then

$$\int_{\Omega_2} f d\mu_2 = \int_{\Omega_1} f \circ \theta d\mu_1.$$

2° If $(\Omega_3, \mathscr{E}_3)$ is also a measurable space and $\tau : \Omega_2 \to \Omega_3$ is a measurable function, then

$$\mu_1 \circ (\tau \circ \theta)^{-1} = \mu_2 \circ \tau^{-1}.$$

In the following, $(\Omega, \mathscr{E}, \mu)$ stands for a finite measure space, and X a Hausdorff topological space endowed with the Borel σ-field $\mathscr{B}(X)$. The projection of the product space $\Omega \times X$ into Ω (resp. X) is denoted by π_Ω (resp. π_X).

Definition 2.2 A measure γ on $(\Omega \times X, \mathscr{E} \otimes \mathscr{B}(X))$ which satisfies

$$\gamma \circ \pi_\Omega^{-1} = \mu \quad (\text{resp.} \gamma \circ \pi_\Omega^{-1} \leqq \mu)$$

is called a **Young measure** (resp. **sub-Young measure**). The set of all Young measures (resp. sub-Young measures) is denoted by $\mathfrak{Y}(\Omega, \mu; X)$ (resp. $\mathfrak{Y}_s(\Omega, \mu; X)$).

Definition 2.3 A set $\{\nu_\omega | \omega \in \Omega\}$ of finite measures on $(X, \mathscr{B}(X))$ is called a **measurable family** if the mapping

$$\omega \mapsto \nu_\omega(B)$$

is measurable for all $B \in \mathscr{B}(X)$.

The above definition does not cover the case of signed measures. A family $\{\nu_\omega | \omega \in \Omega\}$ of signed measures with $|\nu_\omega| < \infty$ is said to be a measurable family if both of $\{\nu_\omega^+ | \omega \in \Omega\}$ and $\{\nu_\omega^- | \omega \in \Omega\}$ form measurable families, where ν_ω^+ (resp. ν_ω^-) is the positive (resp. negative) part of the Jordan decomposition of ν_ω.

Theorem 2.1 *Let (X, ρ) be a metric space. The following two statements are equivalent for a family $\{\nu_\omega | \omega \in \Omega\}$ of finite measures on $(X, \mathscr{B}(X))$.*

(i) $\{\nu_\omega | \omega \in \Omega\}$ is a measurable family.
(ii) The function

$$\omega \mapsto \int_X f(x) d\nu_\omega$$

is measurable for any $f \in \mathfrak{C}^b(X, \mathbb{R})$.[6]

If, in addition, (X, ρ) is a locally compact and σ-compact metric space, (i) and (ii) are equivalent to (iii).

(iii) The function

$$\omega \mapsto \int_X f(x) d\nu_\omega$$

is measurable for any $f \in \mathfrak{C}_\infty(X, \mathbb{R})$.[7]

Proof (i)⇒ (ii): We may assume that $f \in \mathfrak{C}^b(X, \mathbb{R})$ is nonnegative, without loss of generality. As is well-known, there exists a sequence $\{\varphi_n\}$ of nonnegative simple functions such that

$$\varphi_n(x) \to f(x) \quad \text{pointwise as} \quad n \to \infty,$$
$$\varphi_n(x) \leqq \varphi_{n+1}(x) \quad ; \quad n = 1, 2, \cdots .$$

Since

$$\omega \mapsto \int_X \varphi_n(x) d\nu_\omega; \quad n = 1, 2, \cdots$$

[6] $\mathfrak{C}^b(X, \mathbb{R})$ is the set of all bounded continuous real-valued functions defined on X.

[7] $\mathfrak{C}_\infty(X, \mathbb{R})$ is the set of all continuous real-valued functions which vanish at infinity.

is measurable by (i),

$$\omega \mapsto \int_X f(x)d\nu_\omega = \int_X \lim_{n\to\infty} \varphi_n(x)d\nu_\omega$$
$$= \lim_{n\to\infty} \int_X \varphi_n(x)d\nu_\omega$$

(monotone convergence theorem)

is also measurable.

(ii)⇒ (i): Let me start by examining a closed set F. If we define a sequence $f_n : X \to \mathbb{R}$ $(n = 1, 2, \cdots)$ by

$$f_n(x) = \mathrm{Max}\{1 - n\rho(x, F), 0\}$$

(cf. Fig. 1), then $f_n \in \mathfrak{C}^b(X, \mathbb{R})$ and

$$f_n(x) \to \chi_F(x) \quad \text{as} \quad n \to \infty,$$
$$f_n(x) \geqq f_{n+1}(x)\,; \quad n = 1, 2, \cdots .$$

The function

$$\omega \mapsto \int_X f_n(x)d\nu_\omega\,; \quad n = 1, 2, \cdots \tag{2.1}$$

is measurable by (ii).

Since

$$\nu_\omega(F) = \int_X \chi_F(x)d\nu_\omega = \int_X \lim_{n\to\infty} f_n(x)d\nu_\omega = \lim_{n\to\infty} \int_X f_n(x)d\nu_\omega$$

(monotone convergence theorem),

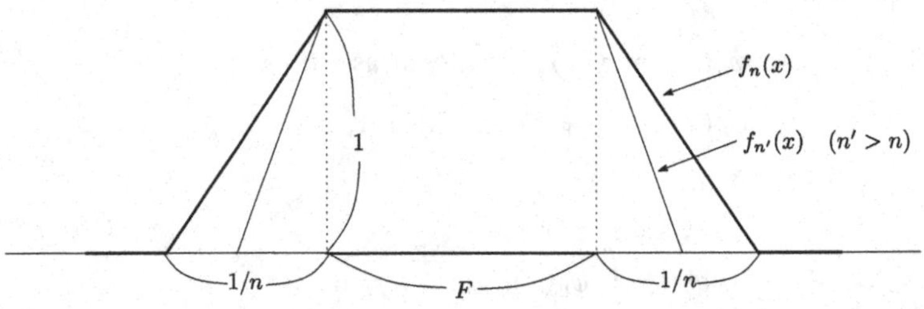

Fig. 1 Graph of $f_n(x)$

the function $\omega \mapsto \nu_\omega(F)$ is the limit of a sequence (2.1) of measurable functions. So it is measurable.

We now consider a general Borel set B. The family

$$\mathscr{B}' = \{B \in \mathscr{B}(X) | \omega \mapsto \nu_\omega(B) \text{is measurable}\}$$

is a Dynkin family[8] which contains all the closed sets. On the other hand, the class $\mathscr{F}$ of all the closed sets of X is multiplicative, the Dynkin class $\mathscr{F}$ generates and $\mathscr{B}(X)$ coincide by Dynkin class theorem. Consequently, we obtain $\mathscr{B}' \supset \mathscr{B}(X)$. Thus (i) holds good.

Finally, we show the equivalence of (i), (ii) and (iii) under the additional assumptions. Since (i)⇒ (iii) can be proved quite similarly as (i)⇒(ii), we have only to show (iii)⇒ (i).

It is enough to show that the function $\omega \mapsto \nu_\omega(F)$ is measurable for any closed set F in X.[9] Since X is σ-compact, there exists a countable compact sets $K_n (n = 1, 2, \cdots)$ such that $X = \bigcup_{n=1}^{\infty} K_n$. Clearly $F = \bigcup_{n=1}^{\infty} (F \cap K_n)$. We can prove the measurability of

$$\omega \mapsto \nu_\omega(F \cap (\bigcup_{j=1}^{n} K_j)) = \int_X \sup_{1 \leqq j \leqq n} \chi_{F \cap K_j}(x) d\nu_\omega$$

by a similarly reasoning as the proof of (ii)⇒ (i).[10] Hence

$$\begin{aligned}
\omega \mapsto \nu_\omega(F) &= \int_X \sup_n \chi_{F \cap K_n}(x) d\nu_\omega \\
&= \int_X \lim_{n \to \infty} \sup_{1 \leqq j \leqq n} \chi_{F \cap K_j}(x) d\nu_\omega \\
&= \lim_{n \to \infty} \int_X \sup_{1 \leqq j \leqq n} \chi_{F \cap K_j}(x) d\nu_\omega
\end{aligned}$$

(monotone convergence theorem)

is measurable as the limit of a sequence of measurable functions. □

---

[8]Let $\mathscr{D}$ be a class of subsets of a set Z. $\mathscr{D}$ is called a **Dynkin class** if it satisfies the following three conditions. (a) $A_n \in \mathscr{D}$, $A_n \cap A_m = \emptyset (m \neq n) \Rightarrow \cup_{n=1}^{\infty} A_n \in \mathscr{D}$. (b) $A_1, A_2 \in \mathscr{D}$, $A_1 \subset A_2 \Rightarrow A_2 \setminus A_1 \in \mathscr{D}$. (c) $Z \in \mathscr{D}$. If a family of subsets of Z is closed under the operation of finite intersection, this family is said to be multiplicative.

Dynkin Class Theorem *If a family of subsets of Z is multiplicative, then the Dynkin class and the σ-field it generates coincide.* cf. Itô [27, p. 50].

[9]It is easy to prove that $\{C \in \mathscr{B}(X) | \omega \mapsto \nu_\omega(C) \text{is measurable}\}$ forms a σ-field. If all the closed sets are contained in this family, so is every member of $\mathscr{B}(X)$.

[10]Approximate the characteristic function of $F \cap (\cup_{j=1}^{n} K_j)$ by a decreasing sequence in $\mathscr{C}_\infty(X, \mathbb{R})$.

We now proceed to prove that the function

$$\omega \mapsto \int_X f(\omega, x) d\nu_\omega$$

is measurable for a measurable function $f : \Omega \times X \to \mathbb{R}$. We prepare a couple of lemmata.

Lemma 2.1 *The σ-field generated by the family $\mathscr{R} \equiv \{E \times F | E \in \mathscr{E}, F$ is a closed set in $X\}$ coincides with $\mathscr{E} \otimes \mathscr{B}(X)$.*

Proof Since it is clear that the σ-field generated by $\mathscr{R}$ is contained in $\mathscr{E} \otimes \mathscr{B}(X)$, we have only to show the converse inclusion.

Let $\mathscr{B}'$ be the family of sets $B \in \mathscr{B}(X)$ such that $E \times B (E \in \mathscr{E})$ is contained in the σ-field generated by $\mathscr{R}$. Then $\mathscr{B}'$ is a Dynkin class. The family $\mathscr{F}$ of all the closed sets in X is multiplicative and contained in $\mathscr{B}'$. Since the Dynkin class generated by $\mathscr{F}$ is equal to $\mathscr{B}(X)$ (Dynkin class theorem, cf. footnote 8), we obtain $\mathscr{B}(X) \subset \mathscr{B}'$. Thus the lemma follows. □

Lemma 2.2 *Let $\{\nu_\omega | \omega \in \Omega\}$ be a measurable family of finite measures on $(X, \mathscr{B}(X))$. Then the function*

$$\omega \mapsto \int_X \chi_A(\omega, x) d\nu_\omega \tag{2.2}$$

is measurable for any $A \in \mathscr{E} \otimes \mathscr{B}(X)$.

Proof The function

$$\begin{aligned} \omega \mapsto \int_X \chi_{E\times F}(\omega, x) d\nu_\omega &= \int_X \chi_E(\omega)\chi_F(x) d\nu_\omega \\ &= \chi_E(\omega) \int_X \chi_F(x) d\nu_\omega = \chi_E(\omega)\nu_\omega(F) \end{aligned}$$

is measurable for any $E \in \mathscr{E}$ and any closed set F in X, since $\{\nu_\omega | \omega \in \Omega\}$ is a measurable family. Consequently the function (2.2) is measurable for any set $A \in \mathscr{R} \equiv \{E \times F | E \in \mathscr{E}, F$ is a closed set in $X\}$.

Let $\mathscr{R}'$ be the set of $A \in \mathscr{E} \otimes \mathscr{B}(X)$ such that the function (2.2) is measurable for A. Then $\mathscr{R}'$ is a Dynkin class.[11]

---

[11]If $A_n \in \mathscr{R}'$ and $A_n \cap A_m = \emptyset \ (n \neq m)$, then the function

$$\omega \mapsto \int_X \chi_{\cup_{n=1}^\infty A_n} d\nu_\omega = \sum_{n=1}^\infty \underbrace{\int_X \chi_{A_n} d\nu_\omega}_{\text{measurable}}$$

is measurable. Furthermore if $A_1, A_2 \in \mathscr{R}'$ and $A_1 \subset A_2$, then

Hence the Dynkin class generated by $\mathscr{R}$ is contained in $\mathscr{R}'$. However, the Dynkin class generated by $\mathscr{R}$ coincides with the σ-field generated by $\mathscr{R}$ (Dynkin class theorem), which is equal to $\mathscr{E} \otimes \mathscr{B}(X)$, by Lemma 2.1. Thus we obtain $\mathscr{E} \otimes \mathscr{B}(X) \subset \mathscr{R}'$, which implies that $\mathscr{E} \otimes \mathscr{B}(X) = \mathscr{R}'$. We conclude that the function (2.2) is measurable for any $A \in \mathscr{E} \otimes \mathscr{B}(X)$. □

Theorem 2.2 *Let $\{\nu_\omega | \omega \in \Omega\}$ be a measurable family of finite measures on $(X, \mathscr{B}(X))$. If $f : \Omega \times X \to \bar{\mathbb{R}}$ is a nonnegative measurable function, then the function*

$$\omega \mapsto \int_X f(\omega, x) d\nu_\omega$$

is measurable.

The above theorem can be proved easily by approximating f by a sequence of simple functions and applying Lemma 2.2. Furthermore the following theorem is an easy consequence of Theorem 2.2.

Theorem 2.3 *Let $\{\nu_\omega | \omega \in \Omega\}$ is a measurable family of finite measures on $(X, \mathscr{B}(X))$. We define*

$$\gamma(A) = \int_\Omega \{\int_X \chi_A(\omega, x) d\nu_\omega\} d\mu$$

for each $A \in \mathscr{E} \otimes \mathscr{B}(X)$.

(i) The set function γ is a measure on $(\Omega \times X, \mathscr{E} \otimes \mathscr{B}(X))$. If $\sup\{\nu_\omega(X) | \omega \in \Omega\} < \infty$, then γ is a finite measure.
(ii) If a measurable function $f : \Omega \times X \to \bar{\mathbb{R}}$ is either nonnegative or γ-integrable, then the formula

$$\int_{\Omega \times X} f(\omega, x) d\gamma = \int_\Omega \{\int_X f(\omega, x) d\nu_\omega\} d\mu \tag{2.3}$$

holds good.

The formula (2.3) is an analogue of the Fubini theorem for the usual product measures.

$$\omega \mapsto \int_X \chi_{A_2 \setminus A_1} d\nu_\omega = \int_X (\chi_{A_2} - \chi_{A_1}) d\nu_\omega$$

is also measurable. Finally it is clear that $X \in \mathscr{R}'$.

2.2 The Dual of $\mathfrak{L}^1(\Omega, \mathfrak{C}_\infty(X, \mathbb{R}))$

Definition 2.4 Let $(\Omega, \mathscr{E})$ be a measurable space. X and Y are given topological spaces. A function $f : \Omega \times X \to Y$ is called a **Carathéodory function** if it satisfies the following conditions.

(i) $\omega \mapsto f(\omega, x)$ is $(\mathscr{E}, \mathscr{B}(X))$-measurable for any fixed $x \in X$.
(ii) $x \mapsto f(\omega, x)$ is continuous for any fixed $\omega \in \Omega$.[12]

The measurability of Carathéodory functions are well-known.[13]

Proposition 2.1 *Let $(\Omega, \mathscr{E})$ be a measurable space. Suppose that (X, ρ) and (Y, ρ') are metric spaces, and, in particular, X is separable. Then a Carathéodory function $f : \Omega \times X \to Y$ is $(\mathscr{E} \otimes \mathscr{B}(X), \mathscr{B}(Y))$-measurable.*

In our discussion below, we are mainly interested in a real-valued Carathéodory function such that $\omega \mapsto f(\omega, \cdot)$ is integrable. So I would like to examine Carathéodory functions of this kind more in detail. For that purpose, **we assume that the metric space** (X, ρ) **is locally compact and separable.**[14]

A Carathéodory function $f : \Omega \times X \to \mathbb{R}$ defines a continuous function $f(\omega, \cdot) \equiv h(\omega)$ for each $\omega \in \Omega$. Assume that $h(\omega)$ is a continuous function which vanishes at infinity for each $\omega \in \Omega$; i.e. $h(\omega) \in \mathfrak{C}_\infty(X, \mathbb{R})$.

$\mathfrak{C}_\infty(X, \mathbb{R})$ is endowed with the uniform convergence norm $||\cdot||_\infty$. $\mathscr{B}(\mathfrak{C}_\infty(X, \mathbb{R}))$ is the Borel σ-field on this space. The function

$$h : \Omega \to \mathfrak{C}_\infty(X, \mathbb{R})$$

is $(\mathscr{E}, \mathscr{B}(\mathfrak{C}_\infty(X, \mathbb{R})))$-measurable.

In fact, this can be proved as follows. Let $\{\xi_1, \xi_2, \cdots\}$ be a countable dense set in X. Then we have, for any $g \in \mathfrak{C}_\infty(X, \mathbb{R})$,

$$\begin{aligned}\|h(\omega) - g\|_\infty &= \sup_{x \in X} |h(\omega)(x) - g(x)| \\ &= \sup_n |f(\omega, \xi_n) - g(\xi_n)|.\end{aligned}$$

[12] Assume that Ω is a measure space. Then there occurs no essential difference by changing "for any fixed $\omega \in \Omega$" to "for a.e. fixed $\omega \in \Omega$".

[13] See Maruyama [36, pp. 412–413] for a proof.

[14] We should remind of several important facts in general topology (cf. Boubaki [9, Part 1. pp. 90–94]).

1° A σ-compact topological space is Lindelöf.
2° In a metric space, the second countability, separability and Lindelöfness are all equivalent.
3° A metric space which is separable and locally compact is σ-compact.

Since the function $\omega \mapsto |f(\omega, \xi_n) - g(\xi_n)|$ is measurable, $\omega \mapsto \|h(\omega) - g\|_\infty$ is also measurable. We denote by $B_r(g)$ the open ball with center g and radius $r > 0$ in $\mathfrak{C}_\infty(X, \mathbb{R})$. Then $h^{-1}(B_r(g)) \in \mathscr{E}$ because

$$h^{-1}(B_r(g)) = \{\omega \in \Omega \,|\, \|h(\omega) - g\|_\infty < r\}$$

and $\omega \mapsto \|h(\omega) - g\|_\infty$ is measurable. Taking account of the separability of $\mathfrak{C}_\infty(X, \mathbb{R})$ (with sup-norm),[15] we obtain $h^{-1}(B) \in \mathscr{E}$ for any Borel set B in $\mathfrak{C}_\infty(X, \mathbb{R})$.

Thus we proved that h is $(\mathscr{E}, \mathscr{B}(\mathfrak{C}_\infty(X, \mathbb{R})))$-measurable. This justifies the integration of $\|h(\omega\|_\infty = \|f(\omega, \cdot)\|_\infty$ when a measure μ is given on $(\Omega, \mathscr{E})$.

Definition 2.5 Let $(\Omega, \mathscr{E}, \mu)$ be a finite measure space and (X, ρ) a locally compact and separable metric space. A Carathéodory function $f : \Omega \times X \to \mathbb{R}$ is called **integrable** if the following two conditions are satisfied.

(i) The continuous function $h(\omega) = f(\omega, \cdot)$ vanishes at infinity.

(ii) $\displaystyle\int_\Omega \|h(\omega)\|_\infty d\mu < \infty.$

The set of all integrable Carathéodory functions is denoted by $\mathfrak{G}_{\mathfrak{C}_\infty}(\Omega, \mu; X)$.

When X is a metric space which is not locally compact, the concept "vanishing at infinity" does not make sense. In such a case, we consider the set of all Carathéodory functions that satisfy (ii), without mentioning (i). The space of functions of this kind is denoted by $\mathfrak{G}_{\mathfrak{C}}(\Omega, \mu; X)$.[16]

The following theorem gives a justification for identifying $\mathfrak{G}_{\mathfrak{C}_\infty}(\Omega, \mu; X)$ and $\mathfrak{L}^1(\Omega, \mathfrak{C}_\infty(X, \mathbb{R}))$, the space of Bochner integrable functions of Ω into $\mathfrak{C}_\infty(X)$.

Theorem 2.4 **($\mathfrak{G}_{\mathfrak{C}_\infty} \cong \mathfrak{L}^1(\Omega, \mathfrak{C}_\infty)$)** *Let $(\Omega, \mathscr{E}, \mu)$ be a finite measure space and (X, ρ) a locally compact and separable metric space.*[17] *Define an operator T : $\mathfrak{G}_{\mathfrak{C}_\infty}(\Omega, \mu; X) \to \mathfrak{L}^1(\Omega, \mathfrak{C}_\infty(X, \mathbb{R}))$ by*

$$\begin{array}{ccc} T : f(\omega, x) & \longmapsto & f(\omega, \cdot) \\ \in \mathfrak{G}_{\mathfrak{C}_\infty}(\Omega, \mu; X) & & \in \mathfrak{L}^1(\Omega, \mathfrak{C}_\infty(X, \mathbb{R})). \end{array}$$

Then T is a bijection.[18]

---

[15]Since X is a separable and locally compact metric space, its Alexandrov compactification $\hat{X} = X \cup \{\infty\}$ is compact and metrizable (and so separable). As is well-known, $\mathfrak{C}(\hat{X}, \mathbb{R})$ is separable (cf. Dunford–Schwartz [21, p. 340], Maruyama [34, pp. 155–157]). Therefore $\mathfrak{C}_\infty(X, \mathbb{R})$, which is identified with $\{g \in \mathfrak{C}(\hat{X}, \mathbb{R}) | g(\infty) = 0\}$, is also separable.

[16]In case X is compact, there is no distinction between $\mathfrak{G}_{\mathfrak{C}_\infty}$ and $\mathfrak{G}_{\mathfrak{C}}$.

[17]By the separability of X and Proposition 2.1, any element of $\mathfrak{G}_{\mathfrak{C}_\infty}(\Omega, \mu; X)$ is measurable with respect to $\mathscr{E} \otimes \mathscr{B}(X)$.

[18]We identify f_1 and $f_2 \in \mathfrak{G}_{\mathfrak{C}_\infty}$ if $\mu\{\omega \in \Omega | f_1(\omega, \cdot) \neq f_2(\omega, \cdot)\} = 0$.

Proof As explained already, the operator T which associates each $f(\omega, x) \in \mathfrak{G}_{\mathfrak{C}_\infty}(\Omega, \mu; X)$ with a continuous function $f(\omega, \cdot) \in \mathfrak{C}_\infty(X, \mathbb{R})$ has a form $T : \mathfrak{G}_{\mathfrak{C}_\infty}(\Omega, \mu; X) \to \mathfrak{L}^1(\Omega, \mathfrak{C}_\infty(X, \mathbb{R}))$. It is clearly injective.

We now show that T is surjective. Corresponding to $h \in \mathfrak{L}^1(\Omega, \mathfrak{C}_\infty(X, \mathbb{R}))$, a function $f : \Omega \times X \to \mathbb{R}$ is defined by

$$f(\omega, x) = h(\omega)(x)$$

(abbreviated as $h(\omega)x$).

Then the function $h(\omega) : x \mapsto f(\omega, x)$ is continuous and vanishes at infinity (for each fixed $\omega \in \Omega$). It also satisfies

$$\int_\Omega \|f(\omega, \cdot)\|_\infty d\mu = \int_\Omega \|h(\omega)\|_\infty d\mu < \infty.$$

Furthermore the function $\omega \mapsto f(\omega, x)$ is measurable (for each $x \in X$). In fact, the set $\{g \in \mathfrak{C}_\infty(X, \mathbb{R}) | g(x) < \alpha\}$ is open (with respect to $\| \cdot \|_\infty$) for any $\alpha \in \mathbb{R}$, x being fixed. Hence

$$\begin{aligned}\{\omega \in \Omega | f(\omega, x) < \alpha\} &= \{\omega \in \Omega | h(\omega)x < \alpha\} \\ &= h^{-1}[\{g \in \mathfrak{C}_\infty(X, \mathbb{R}) | g(x) < \alpha\}] \in \mathscr{E}\end{aligned}$$

by the measurability of h.

Thus we conclude that $f \in \mathfrak{G}_{\mathfrak{C}_\infty}(\Omega, \mu; X)$ and $Tf = h$. □

We next discuss the dual space of $\mathfrak{L}^1(\Omega, \mathfrak{C}_\infty(X, \mathbb{R}))$, keeping in mind its relation with the concept of measurable families.

Definition 2.6 Let X be a Hausdorff topological space. A (positive) finite measure λ on $(X, \mathscr{B}(X))$ is called a **Radon measure** if it is inner regular.[19]

A signed measure λ on $(X, \mathscr{B}(X))$ is called a **Radon signed measure** if both components, λ^+ and λ^-, of the Jordan decomposition $\lambda = \lambda^+ - \lambda^-$ are Radon measures.

The set of all the Radon signed measures is denoted by $\mathfrak{M}(X)$. The set of (positive) Radon measures is denoted by $\mathfrak{M}_+(X)$ and the set of Radon probability measures by $\mathfrak{M}_+^1(X)$.

$\mathfrak{M}(X)$ is a normed vector space with the norm $\|\lambda\| = |\lambda|(X)$ (total variation). The following result due to Riesz–Markov–Kakutani is well-known.[20]

Proposition 2.2 *Let X be a locally compact Hausdorff topological space. Then the dual space of $\mathfrak{C}_\infty(X, \mathbb{R})$ is isomorphic to $\mathfrak{M}(X)$; i.e.*

$$\mathfrak{C}_\infty(X, \mathbb{R})' \cong \mathfrak{M}(X).$$

[19] That is, for any $B \in \mathscr{B}(X)$ and $\varepsilon > 0$, there exists a compact set $K \subset B$ such that $\lambda(B \setminus K) < \varepsilon$.
[20] For a proof, see Malliavin [32, p. 97].

By the duality, we can introduce the weak*-topology for $\mathfrak{M}(X)$; i.e. $\sigma(\mathfrak{M}(X), \mathfrak{C}_\infty(X, \mathbb{R}))$. The set of all measurable functions $\theta : \Omega \to \mathfrak{M}(X)$ such that

$$\operatorname*{ess\,sup}_{\omega\in\Omega} \|\theta(\omega)\| = \operatorname*{ess\,sup}_{\omega\in\Omega} |\theta(\omega)| < \infty$$

is denoted by $\mathfrak{L}^\infty(\Omega, \mathfrak{M}(X))$, where **the σ-field on $\mathfrak{M}(X)$ is the Borel σ-field generated by the w^*-topology**.

Definition 2.7 Let X be a Hausdorff topological space. X is called a **Radon space** if any finite (positive) measure on $(X, \mathscr{B}(X))$ is a Radon measure.

Proposition 2.3 (P. A. Meyer) *Any Souslin space is a Radon space.*[21]

Consequently, a complete separable metric space is a Radon space. If X is a locally compact and separable metric space, its Alexandrov's compactification $\hat{X} = X \cup \{\infty\}$ is metrizable (and so Polish). Since X is an open subset of $\hat{X}$, it is also a Polish space (and so a Souslin space). It follows that any finite (positive) Borel measure on X is a Radon measure, and the set of all the Borel signed measures with finite total variations is just equal to $\mathfrak{M}(X)$.

Taking account of Proposition 2.2, we naturally reach at a conjecture that the dual space of $\mathfrak{L}^1(\Omega, \mathfrak{C}_\infty(X, \mathbb{R}))$ is isomorphic to $\mathfrak{L}^\infty(\Omega, \mathfrak{M}(X))$ when X is a locally compact Hausdorff topological space. The exact answer to this conjecture is given by following Proposition.

Proposition 2.4[22] *Let $(\Omega, \mathscr{E}, \mu)$ be a finite measure space, and (X, ρ) a locally compact and separable metric space.*[23] *For each $\nu \in \mathfrak{L}^\infty(\Omega, \mathfrak{M}(X))$, define a functional Λ_ν on $\mathfrak{L}^1(\Omega, \mathfrak{C}_\infty(X, \mathbb{R}))$ by*

$$\Lambda_\nu h = \int_\Omega \{\int_X h(\omega)x d\nu(\omega)\} d\mu, \; h \in \mathfrak{L}^1(\Omega, \mathfrak{C}_\infty(X, \mathbb{R})). \tag{2.4}$$

Then $\Lambda_\nu \in \mathfrak{L}^1(\Omega, \mathfrak{C}_\infty(X, \mathbb{R}))'$. Under the mapping $\nu \mapsto \Lambda_\nu$, $\mathfrak{L}^1(\Omega, \mathfrak{C}_\infty(X, \mathbb{R}))'$ and $\mathfrak{L}^\infty(\Omega, \mathfrak{M}(X))$ are isomorphic.

$$\mathfrak{L}^1(\Omega, \mathfrak{C}_\infty(X, \mathbb{R}))' \cong \mathfrak{L}^\infty(\Omega, \mathfrak{M}(X)). \tag{2.5}$$

Remark 1 If we write $f(\omega, x) = h(\omega)x$, f is a Carathéodory function. Since f is $(\mathscr{E} \otimes \mathscr{B}(X), \mathscr{B}(\mathbb{R}))$-measurable by Proposition 2.1, the integration (2.4) makes

[21] A separable topological space is called a **Polish space** if it is completely metrizable. A metrizable topological space X is called a **Souslin space** if there exists a Polish space P and a continuous function $f : P \to X$ such that $f(P) = X$. cf. Bourbaki [9, Part 2, pp. 195–200], Schwartz [43, pp. 122–124] and Maruyama [36, pp. 392–395].

[22] cf. Bourbaki [10, Chap. VI], Warga [51, Chap. IV] for the details of this delicate theory.

[23] Hence X is σ-compact.

sense. By Theorem 2.4, we can identify $\mathfrak{G}_{\mathfrak{C}_\infty}(\Omega, \mu; X)$ and $\mathfrak{L}^1(\Omega, \mathfrak{C}_\infty(X, \mathbb{R}))$. Hence (2.4) has an alternative expression:

$$\Lambda_\nu f = \int_\Omega \{ \int_X f(\omega, x) d\nu(\omega) \} d\mu, \quad f \in \mathfrak{G}_{\mathfrak{C}_\infty}(\Omega, \mu; X). \tag{2.4'}$$

Theorem 2.5 *Let (X, ρ) be a locally compact and separable metric space. Assume that a family $\{\nu_\omega | \omega \in \Omega\}$ of finite measures on $(X, \mathscr{B}(X))$ satisfies*

$$\sup_{\omega \in \Omega} \|\nu_\omega\| < \infty.$$

Then the following three statements are equivalent.

(i) *$\{\nu_\omega | \omega \in \Omega\}$ is a measurable family.*
(ii) *For any $f \in \mathfrak{C}_\infty(X, \mathbb{R})$, the function*

$$\omega \mapsto \int_X f(x) d\nu_\omega$$

is measurable.
(iii) *The function $\omega \mapsto \nu_\omega$ of Ω into $\mathfrak{M}(X)$ is $(\mathscr{E}, \mathscr{B}(\mathfrak{M}(X)))$-measurable, where $\mathscr{B}(\mathfrak{M}(X))$ is the Borel σ-field on $\mathfrak{M}(X)$ generated by w^*-topology.*

Proof Since X is σ-compact, (i) and (ii) are equivalent by Theorem 2.1. It remains to show the equivalence of (ii) and (iii).

Assume that a family $\{\nu_\omega | \omega \in \Omega\}$ of finite measures on $(X, \mathscr{B}(X))$ satisfies $\sup_{\omega \in \Omega} \|\nu_\omega\| \leqq A < \infty$. S_A denotes the closed ball of center 0 and radius A in $\mathfrak{M}(X)$. Since $\mathfrak{C}_\infty(X, \mathbb{R})$ is separable,[24] S_A is metrizable *w.r.t.*[25] w^*-topology and compact (hence separable). It is also well-known that

$$\mathscr{B}(S_A) = \{B \cap S_A | B \in \mathscr{B}(\mathfrak{M}(X))\}.^{26}$$

($\mathscr{B}(S_A)$ and $\mathscr{B}(\mathfrak{M}(X))$ are Borel σ-fields generated by w^*-topology.)

Let $D_{\mathfrak{C}_\infty}$ be a countable dense subset of $\mathfrak{C}_\infty(X, \mathbb{R})$. Then the family

$$V(f_1, \cdots, f_p; \varepsilon) = \{\theta \in \mathfrak{M}(X) | \, |\int_X f_1 d\theta| < \varepsilon, \cdots, |\int_X f_p d\theta| < \varepsilon\},$$
$$f_j \in D_{\mathfrak{C}_\infty} (j = 1, 2, \cdots, p), \quad \varepsilon \in \mathbb{Q}, \quad \varepsilon > 0$$

[24] See footnote15.

[25] *w.r.t.* is an abbreviation of "with respect to" as usual.

[26] cf. Bourbaki [9, Part2, p. 200], Maruyama [36, pp. 391–392].

forms a neighborhood base of O of $\mathfrak{M}(X)$. Let $D_{S_A} = \{\theta_1, \theta_2, \cdots\}$ be a countable dense subset of S_A. Then the family of the intersections of S_A and

$$\theta_q + V(f_1, \cdots, f_p; \varepsilon), \qquad \theta_q \in D_{S_A}, \quad f_j \in D_{\mathfrak{C}_\infty}, \quad \varepsilon \in \mathbb{Q}, \quad \varepsilon > 0 \tag{2.6}$$

forms a countable base for w^*-topology on S_A. In order to prove (iii), that is the measurability of $\omega \mapsto \nu_\omega$, from (ii), we have only to show $\nu^{-1}(\theta_q + V(f_1, \cdots, f_p; \varepsilon)) \in \mathscr{E}$ for the set of the form (2.6).

$$\begin{aligned}
&\{\omega \in \Omega \mid \nu_\omega \in \theta_q + V(f_1, \cdots, f_p; \varepsilon)\} \\
&\quad = \{\omega \in \Omega \mid \nu_\omega - \theta_q \in V(f_1, \cdots, f_p; \varepsilon)\} \\
&\quad = \{\omega \in \Omega \mid |\int_X f_1 d(\nu_\omega - \theta_q)| < \varepsilon, \cdots, |\int_X f_p d(\nu_\omega - \theta_q)| < \varepsilon\} \\
&\quad = \bigcap_{j=1}^{p} \{\omega \in \Omega \mid |\int_X f_j d(\nu_\omega - \theta_q)| < \varepsilon\} \\
&\quad = \bigcap_{j=1}^{p} \{\omega \in \Omega \mid \int_X f_j d\theta_q - \varepsilon < \int_X f_j d\nu_\omega < \int_X f_j d\theta_q + \varepsilon\} \\
&\quad \in \mathscr{E}.
\end{aligned}$$

This proves (iii).

Conversely, assume (iii). The function

$$\omega \mapsto \int_X f d\nu_\omega \tag{2.7}$$

is a composition of $\omega \mapsto \nu_\omega$ and $\theta \mapsto \int_X f d\theta$. The former is measurable by assumption, and the latter is continuous *w.r.t.* w^*-topology. Hence the composite function (2.7) is measurable. □

Let $(\Omega, \mathscr{E}, \mu)$ be a finite measure space and X a Hausdorff topological space. We denote by $\mathfrak{P}(\Omega, \mu; X)$ (resp. $\mathfrak{P}_s(\Omega, \mu; X)$) the set of measurable family consisting of probability measures (resp. sub-probability measures).

Theorem 2.6 *Let $(\Omega, \mathscr{E}, \mu)$ be a finite measure space, and (X, ρ) a locally compact and separable metric space. (w^*-topology mentioned below is the one based upon the duality (2.5).)*

(i) $\mathfrak{P}_s(\Omega, \mu; X)$ is a w^*-closed set contained in the unit ball in $\mathfrak{L}^\infty(\Omega, \mathfrak{M}(X))$.
(ii) If μ is complete, any sequence in $\mathfrak{P}_s(\Omega, \mu; X)$ has a w^*-convergent subsequence.

(iii) If X is compact, $\mathfrak{P}(\Omega, \mu; X)$ is a w^*-closed set contained in the unit ball in $\mathfrak{L}^\infty(\Omega, \mathfrak{M}(X))$.

The concept of conditional expectation of a vector-valued function is required for the proof of the Theorem 2.6. We will briefly discuss about it for the sake of readers' convenience.

Let $(\Omega, \mathscr{E}, P)$ be a probability space, $\mathscr{E}'$ a sub-σ-field of $\mathscr{E}$, and X a real-valued random variable on Ω. Then there exists a real-valued integrable random variable Y on $(\Omega, \mathscr{E}', P)$ such that

$$\int_E X(\omega)dP = \int_E Y(\omega)dP \quad \text{for all } E \in \mathscr{E}'. \tag{2.8}$$

Such a function Y is unique (up to *a.e.* equivalent functions). Y is called the **conditional expectation** of X *w.r.t.* $\mathscr{E}'$ and is denoted by $\mathbb{E}^{\mathscr{E}'}(X)$. This result is an established fact in probability theory.[27]

This concept can be extended to Banach space-valued Bochner integrable functions. Let $(\Omega, \mathscr{E}, P)$ be a probability space as usual, $\mathscr{E}'$ a sub-σ-field of $\mathscr{E}$, and $\mathfrak{X}$ a separable Banach space. Then there exists a unique linear operator $\mathbb{E}_{\mathfrak{X}}^{\mathscr{E}'} : \mathfrak{L}^1(\Omega, \mathfrak{X}) \to \mathfrak{L}^1_{\mathscr{E}'}(\Omega, \mathfrak{X})$ which satisfies the following two conditions.[28]

1° For each $f \in \mathfrak{L}^1(\Omega, \mathfrak{X})$,

$$\int_E f dP = \int_E \mathbb{E}_{\mathfrak{X}}^{\mathscr{E}'}(f)dP \quad \text{for all} \quad E \in \mathscr{E}'. \tag{2.9}$$

2° $\|\mathbb{E}_{\mathfrak{X}}^{\mathscr{E}^1}\| = 1$ (operator norm).[29]

We now proceed to the proof of Theorem 2.6.[30]

Proof of Theorem 2.6 (i) Let $\mathfrak{M}_+^{\leqq 1}(X)$ be the set of all sub-probability measures. $\mathfrak{M}_+^{\leqq 1}(X)$ is w^*-compact and convex.

Let $\{g_1, g_2, \cdots\}$ be a countable dense set in $\mathfrak{C}_\infty(X, \mathbb{R})$, and $\delta^*(\cdot)$ the support function of $\mathfrak{M}_+^{\leqq 1}(X)$. We write

$$a_n = \delta^*(g_n) = \sup_{\zeta \in \mathfrak{M}_+^{\leqq 1}} \int_X g_n(x)d\zeta; \quad n = 1, 2, \cdots.$$

---

[27] cf. Malliavin [32, pp.183–190] and Dellacherie–Meyer [19, II-38]. Rigorously speaking, dP appearing on the right-hand side of (2.8) should be written as $dP|_{\mathscr{E}'}$ (the restriction of P to $\mathscr{E}'$).

[28] $\mathfrak{L}^1_{\mathscr{E}'}(\Omega, \mathfrak{X})$ is the space of Bochner integrable $\mathfrak{X}$-valued functions defined on $(\Omega, \mathscr{E}', P)$.

[29] Diestel–Uhl [20, pp. 121–125]. For Bochner integration, see Diestel–Uhl [20, Chap. II] and Maruyama [36, Chap.9].

[30] For the topological structures of the space of measures on a topological space, consult Billingsley [8], Choquet [18], Heyer [23], Schwartz [43] and Maruyama [36, Chap.8].

Then[31]

$$\zeta \in \mathfrak{M}_{+}^{\leqq 1}(X) \Leftrightarrow \int_X g_n(x)d\zeta \leqq a_n \quad \text{for all} \quad n.$$

Let S be the unit ball of $\mathfrak{L}^{\infty}(\Omega, \mathfrak{M}(X))$; i.e.

$$S = \{\nu \in \mathfrak{L}^{\infty}(\Omega, \mathfrak{M}(X)) | \ \|\nu\|_{\infty} \leqq 1\}.$$

S is w^*-compact by Alaoglu's theorem. For a measurable family $\nu = \{\nu_\omega | \omega \in \Omega\} \in S$, the following three statements are equivalent.

1° $\nu \in \mathfrak{P}_s(\Omega, \mu; X)$.

2° $\int_X g_n(x)d\nu_\omega \leqq a_n$ *a.e.* for all n.

3° For any nonnegative integrable function $\psi(\omega)$ on $(\Omega, \mathscr{E}, \mu)$,

$$\int_\Omega \psi(\omega)\{\int_X g_n(x)d\nu_\omega\}d\mu \leqq a_n \int_\Omega \psi(\omega)d\mu \quad \text{for all} \quad n.$$

We should note that the function $\omega \mapsto \psi(\omega)g_n(x)$ appearing in 3° is an element of $\mathfrak{L}^1(\Omega, \mathfrak{C}_\infty(X, \mathbb{R}))$.

Let $\nu^\alpha = \{\nu_\omega^\alpha | \omega \in \Omega\}$ be a net in $\mathfrak{P}_s(\Omega, \mu; X)$ which w^*-converges to some $\nu \in S$ ($\{\alpha\}$ is a directed set). The function ψ is specified in 3° above. Hence

$$\int_\Omega \{\int_X \psi(\omega)g_n(x)d\nu_\omega^\alpha\}d\mu \leqq a_n \int_\Omega \psi(\omega)d\mu \quad \text{for all} \quad n.$$

It follows that the same inequalities

$$\int_\Omega \{\int_X \psi(\omega)g_n(x)d\nu_\omega\}d\mu \leqq a_n \int_\Omega \psi(\omega)d\mu \quad \text{for all} \quad n$$

hold good for ν. Thus we obtain $\nu = \{\nu_\omega | \omega \in \Omega\} \in \mathfrak{P}_s(\Omega, \mu; X)$.

(iii) If X is compact, $\mathfrak{M}_{+}^{1}(X)$ is w^*-compact and convex. Let $\{g_1, g_2, \cdots\}$ be as in (i). $\delta^*(\cdot)$ is the support function of $\mathfrak{M}_{+}^{1}(X)$. Since

$$b_n = \delta^*(g_n) = \sup_{\eta \in \mathfrak{M}_{+}^{1}(X)} \int_X g_n(x)d\eta,$$

a similar argument as in (i) justifies (iii).

(ii) Let $\{\nu^n\}$ be a sequence in $\mathfrak{P}_s(\Omega, \mu; X) \subset \mathfrak{L}^1(\Omega, \mathfrak{M}(X))'$. We now prove that it has a convergent subsequence.

[31] See Castaing-Valadier [13, p. 48].

Case 1 Suppose that $\mathfrak{L}^1(\Omega, \mathfrak{C}_\infty(X, \mathbb{R}))$ is separable. In this case, S is metrizable and compact *w.r.t.* w^*-topology. So obviously, $\{\nu^n\}$ has a convergent subsequence.

Case 2 Suppose that $\mathfrak{L}^1(\Omega, \mathfrak{C}_\infty(X, \mathbb{R}))$ is not separable. The existence of a convergent subsequence can be proved by reducing the problem to Case 1. Let $\mathscr{E}_0$ be a σ-field on Ω which is generated by the sequence

$$\nu^n : \omega \mapsto \nu^n_\omega, \quad n = 1, 2, \cdots, \tag{2.10}$$

where the range $\mathfrak{M}^{\leqq 1}_+(X) \subset \mathfrak{M}(X)$ of ν^n is endowed with the Borel σ-field generated by w^*-topology. $\mathfrak{M}^{\leqq}_+(X)$ is separable and metrizable *w.r.t.* w^*-topology and so second countable. Consequently $\mathscr{E}_0$ is generated by some countable sets in $\mathfrak{M}^{\leqq}_+(X)$. Hence $\mathfrak{L}^1_{\mathscr{E}_0}(\Omega, \mathfrak{C}_\infty(X, \mathbb{R}))$ obtained by changing $\mathscr{E}$ by $\mathscr{E}_0$ is separable. Each of the functions (2.10) is in $\mathfrak{L}^\infty_{\mathscr{E}_0}(\Omega, \mathfrak{M}(X))$. By the argument in Case 1, there exists a subsequence $\{\nu^{n_k}\}$ which converges to some $\nu \in \mathfrak{L}^\infty_{\mathscr{E}_0}(\Omega, \mathfrak{M}(X))$.

We may assume, without loss of generality, that $\mu\Omega = 1$. Let $h \in \mathfrak{L}^1(\Omega, \mathfrak{C}_\infty(X, \mathbb{R}))$ and $\mathbb{E}^{\mathscr{E}_0}(h)$ its conditional expectation *w.r.t*, $\mathscr{E}_0$. Thus we obtain, by the measurability of ν^n and ν *w.r.t.* $\mathscr{E}_0$, that[32]

$$\int_\Omega \{\int_X h(\omega)x d\nu^{n_k}_\omega\} d\mu = \int_\Omega \{\int_X \mathbb{E}^{\mathscr{E}_0}(h)(\omega)x d\nu^{n_k}_\omega\} d\mu$$

---

[32]The first equality in (2.11) can be proved as follows. To start with, we specify $h(\omega)$ as a $\mathfrak{C}_\infty(X, \mathbb{R})$-valued simple function:

$$h(\omega) = \sum_{i=1}^n \chi_{E_i}(\omega) f_i, \quad f_i \in \mathfrak{C}_\infty(X, \mathbb{R}).$$

Then

$$\begin{aligned}
\int_\Omega \{\int_X h(\omega)x d\nu^{n_k}_\omega\} &= \int_\Omega \sum_{i=1}^n \chi_{E_i}(\omega) \underbrace{\int_X f_i(x) d\nu^{n_k}_\omega}_{\mathscr{E}_0-\text{measurable}} d\mu \\
&= \int_\Omega \sum_{i=1}^n \mathbb{E}^{\mathscr{E}_0}(\chi_{E_i})(\omega) \int_X f_i(x) d\nu^{n_k}_\omega d\mu \\
&= \int_\Omega \{\int_X \underbrace{\sum_{i=1}^n \mathbb{E}^{\mathscr{E}_0}(\chi_{E_i})(\omega) f_i(x)}_{=\mathbb{E}^{\mathscr{E}_0}(h)(\omega)} d\nu^{n_k}_\omega\} d\mu \\
&= \int_\Omega \{\int_X \mathbb{E}^{\mathscr{E}_0}(h)(\omega)x d\nu^{n_k}_\omega\} d\mu.
\end{aligned}$$

For a general $h \in \mathfrak{L}^1(\Omega, \mathfrak{C}_\infty(X, \mathbb{R}))$, approximate h by a sequence of simple functions.

$$\to \int_{\Omega}\{\int_{X} \mathbb{E}^{\mathscr{E}_0}(h)(\omega)x d\nu_{\omega}\}d\mu \tag{2.11}$$
$$= \int_{\Omega}\{\int_{X} h(\omega)x d\nu_{\omega}\}d\mu$$
$$\text{as} \quad k \to \infty.$$

This proves that $\{\nu^n\}$ has a convergent subsequence in $\mathfrak{L}^\infty(\Omega, \mathfrak{M}(X))$. □

2.3 Disintegration Theorem (A)

We assume that $(\Omega, \mathscr{E}, \mu)$ is a finite measure space, (X, ρ) is a metric space with certain properties and $\{\nu_\omega | \omega \in \Omega\}$ is a measurable family of finite measures. The set function γ defined by

$$\gamma(A) = \int_{\Omega}\{\int_{X} \chi_A(\omega, x) d\nu_{\omega}\}d\mu, \tag{2.12}$$
$$A \in \mathscr{E} \otimes \mathscr{B}(X)$$

is a measure on $(\Omega \times X, \mathscr{E} \otimes \mathscr{B}(X))$ as explained in Theorem 2.3.

If every $\nu_\omega (\omega \in \Omega)$ is a probability measure, in particular, γ is a Young measure since[33]

$$(\gamma \circ \pi_{\Omega}^{-1})(E) = \mu(E), \quad E \in \mathscr{E}.$$

We denote by $\mathfrak{M}(\Omega \times X)$ the set of all the finite signed measures on $(\Omega \times X, \mathscr{E} \otimes \mathscr{B}(X))$. Of course, $\mathfrak{Y}(\Omega, \mu; X) \subset \mathfrak{M}(\Omega \times X)$.

In this subsection, we show the converse result that any Young measure can be represented in the form (2.12) by means of some measurable family $\{\nu_\omega | \omega \in \Omega\}$ of probability measures, under certain conditions. We start by examining this problem in the case X is a metric space satisfying some strong conditions. We postpone a more general case to the next subsection.

[33] $\pi_{\Omega}^{-1}(E) = E \times X$. Hence

$$(\gamma \circ \pi_{\Omega}^{-1})(E) = \int_{\Omega}\{\int_{X} \chi_{E \times X}(\omega, x) d\nu_{\omega}\}d\mu$$
$$= \int_{\Omega} \chi_E(\omega) \int_{X} d\nu_{\omega} d\mu = \mu(E).$$

Theorem 2.7 (Disintegration: Metric Space) *Let $(\Omega, \mathscr{E}, \mu)$ be a complete finite measure, X a locally compact Polish space.*[34] *Then for any Young measure $\gamma \in \mathfrak{Y}(\Omega, \mu; X)$, there exists a unique measurable family $\nu = \{\nu_\omega | \omega \in \Omega\} \in \mathfrak{P}(\Omega, \mu; X)$ such that*

$$\gamma(A) = \int_\Omega \{\int_X \chi_A(\omega, x) d\nu_\omega\} d\mu, \quad A \in \mathscr{E} \otimes \mathscr{B}(X). \tag{2.13}$$

Proof We first prove the theorem in the case X is a compact metric space, and then proceed to the general case.

1° Suppose that X is a compact metric space. If we define a set function γ on $\mathscr{E} \otimes \mathscr{B}(X)$ by (2.13) by means of $\nu = \{\nu_\omega | \omega \in \Omega\} \in \mathfrak{P}(\Omega, \mu; X)$, then γ is a Young measure. We denote by Φ the mapping which associates ν with the corresponding γ. Φ is of the form:

$$\Phi : \mathfrak{P}(\Omega, \mu; X) \to \mathfrak{M}(\Omega \times X).$$

Since $\mathfrak{P}(\Omega, \mu; X) \subset \mathfrak{L}^\infty(\Omega, \mathfrak{M}(X))$, we can give the domain $\mathfrak{P}(\Omega, \mu; X)$ the relative topology induced from w^*-topology $\sigma(\mathfrak{L}^\infty(\Omega, \mathfrak{M}(X)), \mathfrak{L}^1(\Omega, \mathfrak{C}_\infty(X, \mathbb{R}))$ on $\mathfrak{L}^\infty(\Omega, \mathfrak{M}(X))$.

$\mathfrak{M}(\Omega \times X)$ is supposed to be endowed with the topology generated by the functions

$$\gamma \mapsto \int_{\Omega \times X} f(\omega, x) d\gamma, \quad f \in \mathfrak{G}_{\mathfrak{C}}(\Omega, \mu; X).$$

The topology on $\mathfrak{Y}(\Omega, \mu; X)$ is the relative topology induced from $\mathfrak{M}(\Omega \times X)$. The topology on $\mathfrak{M}(\Omega \times X)$ defined above is Hausdorff. Suppose that

$$\int_{\Omega \times X} f(\omega, x) d\gamma = 0 \quad \text{for all} \quad f \in \mathfrak{G}_{\mathfrak{C}}(\Omega, \mu; X). \tag{2.14}$$

For any compact set $K \subset X$, there exists a sequence $\{f_n\}$ in $\mathfrak{C}(X, \mathbb{R})$ such that $f_n \downarrow \chi_K$ as $n \to \infty$. (χ_K is a decreasing limit of f_n.) Then it follows that

$$\gamma(E \times K) = \lim_{n \to \infty} \int_{\Omega \times X} \underbrace{\chi_E(\omega) f_n(x)}_{\in \mathfrak{G}_{\mathfrak{C}}(\Omega, \mu; X)} d\gamma(\omega, x)$$

[34] See the footnote 21 for the reference about Polish spaces, Souslin spaces and Radon spaces.

for any $E \in \mathscr{E}$. However this must be equal to zero by (2.14). Hence $\gamma(E \times B) = 0$ for any $B \in \mathscr{B}(X)$.[35] It, in turn, implies that

$$\gamma(A) = 0 \quad \text{for all} \quad A \in \mathscr{E} \otimes \mathscr{B}(X).$$

Thus $\mathfrak{M}(\Omega \times X)$ is a Hausdorff space and $\mathfrak{Y}(\Omega, \mu; X)$ is its closed convex subset.

We show that Φ is injective. Let ν^1 and ν^2 be two distinct elements of $\mathfrak{P}(\Omega, \mu; X)$.

$$\mu\{\omega \in \Omega | \nu^1_\omega \neq \nu^2_\omega\} > 0.$$

Let $D \equiv \{g_1, g_2, \ldots\}$ be a countable dense subset of $\mathfrak{C}(X, \mathbb{R})$. There exists some $g_{n_0} \in D$ such that the measure of the set

$$E = \{\omega \in \Omega | \int_X g_{n_0} d\nu^1_\omega \neq \int_X g_{n_0} d\nu^2_\omega\}$$

is not 0; i.e. $\mu E > 0$. Without loss of generality, we may assume that the measure of

$$E' = \{\omega \in \Omega | \int_X g_{n_0} d\nu^1_\omega > \int_X g_{n_0} d\nu^2_\omega\}$$

is not 0; i.e. $\mu E' > 0$. It follows that

$$\int_\Omega \{\int_X \chi_{E'}(\omega) g_{n_0}(x) d\nu^1_\omega\} d\mu > \int_\Omega \{\int_X \chi_{E'}(\omega) g_{n_0}(x) d\nu^2_\omega\} d\mu.$$

That is, $\Phi(\nu^1) \neq \Phi(\nu^2)$.

The continuity of Φ can be proved easily from the definition of the topology. $\mathfrak{P}(\Omega, \mu; X)$ is metrizable and compact. So we can safely use a sequence-argument in order to confirm the continuity of Φ. Let $\{\nu^n\}$ be a sequence in $\mathfrak{P}(\Omega, \mu; X)$ which w^*-converges to $\nu^* \in \mathfrak{P}(\Omega, \mu; X)$. γ_n and γ_* are Young measures defined by ν^n and ν^*, respectively.

$$\gamma_n = \int_\Omega \delta_\omega \otimes \nu^n_\omega d\mu, \quad \gamma_* = \int_\Omega \delta_\omega \otimes \nu_\omega d\mu.$$

Since $\nu^n \to \nu^*$(w^*-convergence), it holds for any $f \in \mathfrak{G}_{\mathfrak{C}}(\Omega, \mu; X) \cong \mathfrak{L}^1(\Omega, \mathfrak{C}(X, \mathbb{R}))$ that

$$\lim_{n\to\infty} \int_\Omega \{\int_X f(\omega, x) d\nu^n_\omega\} d\mu = \int_\Omega \{\int_X f(\omega, x) d\nu^*_\omega\} d\mu,$$

---

[35]The set function $B \mapsto \gamma(E \times B)$ is a Radon measure on $(X, \mathscr{B}(X))$. Hence $\gamma(E \times B)$ can be approximated by $\gamma(E \times K_n)(= 0)$ for a sequence of some compact sets $K_n \subset B$.

that is

$$\lim_{n\to\infty}\int_{\Omega\times X} f(\omega,x)d\gamma_n=\int_{\Omega\times X} f(\omega,x)d\gamma_*.$$

We obtain $\Phi(\nu^n)=\gamma_n\to\Phi(\nu^*)=\gamma_*$. This proves the continuity of Φ.

Consequently, $\Phi(\mathfrak{P}(\Omega,\mu;X))$ is a compact set in $\mathfrak{Y}(\Omega,\mu;X)$. Its convexity is also clear.

We now go over to the surjectivity of Φ. It is enough to show that their support functions, $\delta^*_{\Phi(\mathfrak{P})}$ and $\delta^*_{\mathfrak{Y}}$ are identical, keeping in mind that $\Phi(\mathfrak{P}(\Omega,\mu;X))$ and $\mathfrak{Y}(\Omega,\mu;X)$ are closed and convex sets in $\mathfrak{M}(\Omega\times X)$ (Hausdorff locally convex).[36]

For each $f\in\mathfrak{S}_{\mathfrak{C}}(\Omega,\mu;X)$, we obtain

$$\begin{aligned}\delta^*_{\mathfrak{Y}}(f)&=\sup\{\int_{\Omega\times X} f d\gamma|\gamma\in\mathfrak{Y}(\Omega,\mu;X)\}\\&\leqq\sup\{\int_{\Omega\times X}\sup_{x\in X} f(\omega,x)d\gamma|\gamma\in\mathfrak{Y}(\Omega,\mu;X)\}\\&=\int_{\Omega}\sup_{x\in X} f(\omega,x)d\mu\\&\leqq\int_{\Omega}\|f(\omega,\cdot)\|_\infty d\mu\\&<\infty.\end{aligned}\tag{2.15}$$

Thanks to the measurable version of the maximum theorem due to C. Berge[37] there exists a measurable function $u:\Omega\to X$ such that

$$f(\omega,u(\omega))=\sup_{x\in X} f(\omega,x)\quad\text{for all}\quad\omega\in\Omega.$$

[36]There exists a one-to-one correspondence between the family of closed convex sets in a locally convex Hausdorff vector space and the family of their support functions. cf. Castaing–Valadier [13, II. 16, p. 48]. It is a Valadier's idea to make use of this result to show the surjectivity of Φ. (Valadier [48, pp. 180–181].)

[37]Let $(\Omega,\mathscr{E})$ be a measurable space and X a Polish space. Assume that a function $f:\Omega\times X\to\mathbb{R}$ is $\mathscr{E}\otimes\mathscr{B}(X)$-measurable and $x\mapsto f(\omega,x)$ is upper semi-continuous for all $\omega\in\Omega$. Furthermore a compact-valued, $(\mathscr{E},\mathscr{B}(X))$-measurable correspondence (multi-valued function) $\Gamma:\Omega\twoheadrightarrow X$ is also assumed to be given. Then the function $v(\omega)=\mathrm{Max}\{f(\omega,x)|x\in\Gamma(\omega)\}$ is $(\hat{\mathscr{E}},\mathscr{B}(X))$-measurable, and the correspondence $\Delta(\omega)=\{x\in\Gamma(\omega)|f(\omega,x)=v(\omega)\}$ has a $(\hat{\mathscr{E}},\mathscr{B}(X))$-measurable selection (cf. Maruyama [36, p. 428.]).

$\hat{\mathscr{E}}$ is the universal completion of $\mathscr{E}$ and $\twoheadrightarrow$ denotes the domain and the range of a correspondence. We used here the completeness of Ω in the text.

In general, suppose that $(\Omega,\mathscr{E})$ is a measurable space and U a topological space. A correspondence $\Theta:\Omega\twoheadrightarrow U$is said to be measurable if $\{\omega\in\Omega|\Theta(\omega)\cap G\neq\emptyset\}\in\mathscr{E}$ for any open set G in U.

If we define a function $\omega \mapsto \nu_\omega \in \mathfrak{P}(\Omega, \mu; X)$ by

$$\nu_\omega = \delta_{u(\omega)} \quad \text{(Dirac measure)},$$

we obtain

$$\begin{aligned} \delta^*_{\Phi(\mathfrak{P})}(f) &\geqq \int_\Omega \{\int_X f(\omega, x) d\nu_\omega\} d\mu \\ &= \int_\Omega f(\omega, u(\omega)) d\mu \\ &= \int_\Omega \sup_{x \in X} f(\omega, x) d\mu \\ &\geqq \delta^*_{\mathfrak{Y}}(f) \quad \text{(by (2.15))}. \end{aligned} \tag{2.16}$$

The inverse inequality is obvious since $\Phi(\mathfrak{P}(\Omega, \mu; X)) \subset \mathfrak{Y}(\Omega, \mu; X)$. We conclude that $\delta^*_{\Phi(\mathfrak{P})} = \delta^*_{\mathfrak{Y}}$, and so $\Phi(\mathfrak{P}(\Omega, \mu; X) = \mathfrak{Y}(\Omega, \mu; X)$.

Summing up, Φ is a continuous bijection of $\mathfrak{P}(\Omega, \mu; X)$ onto $\mathfrak{Y}(\Omega, \mu; X)$. Since the domain $\mathfrak{P}(\Omega, \mu; X)$ is compact, Φ is actually a homeomorphism.

We shall now go over to a general case.

2° General case. X is a Radon space because it is Polish, by assumption. Consequently $\gamma \circ \pi_X^{-1}$ is a Radon measure on X, and there exists a sequence $\{X_n\}$ of compact sets in X which satisfies[38]

$$X_n \bigcap X_m = \emptyset \quad \text{if} \quad n \neq m, \quad \text{and}$$

$$\gamma \circ \pi_X^{-1}(X \setminus \bigcup_{n=1}^{\infty} X_n) = 0.$$

We denote by γ_n the restriction of γ to $\Omega \times X_n$. Then γ_n is a Young measure on $\Omega \times X_n$. According to 1°, γ_n is represented in the form

$$\gamma_n = \int_\Omega \delta_\omega \otimes \nu^n_\omega d\mu_n$$

If $(\Omega, \mathscr{E})$ is universally complete and U is a Polish space, then a closed-valued correspondence $\Theta : \Omega \twoheadrightarrow U$ is measurable if and only if the graph of $\Theta = \{(\omega, x) \in \Omega \times U | x \in \Theta(\omega)\} \in \mathscr{E} \otimes \mathscr{B}(U)$.

A measurable function $\theta : \Omega \to G$ which satisfies $\theta(\omega) \in \Theta(\omega)$ for all $\omega \in \Omega$ is called a measurable selection of $\Theta(\omega)$. If either 1° or 2° below is satisfied, then there exists a measurable selection of Θ.

1° (Kuratowski–Nardzewski [29] U is a Polish space and Θ is closed-valued and measurable.

2° (Saint–Beuve [42]) $(\Omega, \mathscr{E})$ is complete *w.r.t.* a finite measure μ. U is a Souslin space, and the graph of Θ is an element of $\mathscr{E} \otimes \mathscr{B}(U)$. cf. Maruyama [36, chap.11].

[38] For any $\varepsilon_n > 0$ with $\varepsilon_n \downarrow 0$, there is a sequence $\{X_n\}$ of disjoint compact sets in X such that $\nu(X \setminus \bigcup_{i=1}^{n} X_i) < \varepsilon_n$. This sequence satisfies the required property.

by means of $\nu^n \in \mathfrak{P}(\Omega, \mu; X_n)$, where $\mu_n = \gamma_n \circ \pi_\Omega^{-1}$. We can interprete ν^n as an element of $\mathfrak{P}(\Omega, \mu; X)$, the support of which is contained in X_n. It can be observed that μ_n is absolutely continuous *w.r.t.* μ and so it has the Radon–Nikodým derivative $\xi_n = d\mu_n/d\mu$. In fact, $\mu_n \ll \mu$ follows from

$$\mu_n = \gamma_n \circ \pi_\Omega^{-1} = \gamma|_{\Omega\times X_n} \circ \pi_\Omega^{-1} \leqq \gamma \circ \pi_\Omega^{-1} = \mu$$

Then

$$\mu(E) = \int_E 1 d\mu = \sum_{n=1}^{\infty} \mu_n(E) = \int_E \sum_{n=1}^{\infty} \xi_n(\omega) d\mu.$$

By the uniqueness of the Radon–Nikodým derivative, we obtain

$$\sum_{n=1}^{\infty} \xi_n(\omega) = 1 \quad a.e.\, (\mu).$$

We define ν_ω by

$$\nu_\omega(B) = \sum_{n=1}^{\infty} \xi_n(\omega)\nu_\omega^n(B \cap X_n), \quad B \in \mathscr{B}(X).$$

Each term of the right-hand side is measurable in ω. Hence $\{\nu_\omega | \omega \in \Omega\}$ is a measurable family.

Since $\gamma \circ \pi_X^{-1}(X \setminus \cup_{n=1}^{\infty} X_n) = 0$, it follows that

$$\begin{aligned} \gamma(A) &= \sum_{n=1}^{\infty} \gamma(A \cap (\Omega \times X_n)) = \sum_{n=1}^{\infty} \gamma_n(A) \\ &= \sum_{n=1}^{\infty} \int_A \delta_\omega \otimes \nu_\omega^n d\mu_n = \int_A \delta_\omega \otimes \underbrace{\sum_{n=1}^{\infty} \nu_\omega^n \xi_n(\omega)}_{\nu_\omega} d\mu \qquad (2.17) \\ &= \int_A \delta_\omega \otimes \nu_\omega d\mu, \quad \text{for} \quad A \in \mathscr{E} \otimes \mathscr{B}(X). \end{aligned}$$

This proves the possibility of disintegration for a general case. It remains to show the uniqueness.

3° Uniqueness. Finally we show the uniqueness of representation. Assume that there are two measurable families $\nu^1 = \{\nu_\omega^1 | \omega \in \Omega\}$, $\nu^2 = \{\nu_\omega^2 | \omega \in \Omega\} \in \mathfrak{P}(\Omega, \mu; X)$ which satisfy

$$\gamma = \int_\Omega \delta_\omega \otimes \nu_\omega^1 d\mu = \int_\Omega \delta_\omega \otimes \nu_\omega^2 d\mu \qquad (2.18)$$

for a $\gamma \in \mathfrak{Y}(\Omega, \mu; X)$. We would like to show that $\nu_\omega^1 = \nu_\omega^2\, a.e.$ (2.18) means, of course, that

$$\int_\Omega \{\int_X f(\omega, x) d\nu_\omega^1\} d\mu = \int_\Omega \{\int_X f(\omega, x) d\nu_\omega^2\} d\mu \quad \text{for any} \quad f \in \mathfrak{G}_{\mathfrak{C}}(\Omega, \mu; X). \tag{2.18'}$$

Suppose that $\nu^1 \neq \nu^2$. If we write $S = \{\omega \in \Omega | \nu_\omega^1 \neq \nu_\omega^2\}$, we have $\mu S > 0$. Then $\nu = \{\nu_\omega = \nu_\omega^1 - \nu_\omega^2 | \omega \in \Omega\}$ forms a measurable family of signed measures on X.

Since $\nu_\omega \neq 0$ for $\omega \in S$, there exists some $f_\omega \in \mathfrak{C}_\infty(X, \mathbb{R})$ such that

$$\int_X f_\omega(x) d\nu_\omega \neq 0. \tag{2.19}$$

Without loss of generality, we may assume that

$$\int_X f_\omega(x) d\nu_\omega \geqq \varepsilon > 0 \tag{2.20}$$

for $\omega \in S$. We define $f_\omega(x) \equiv 0$ for $\omega \notin S$.

We should note here that the function $(\omega, x) \mapsto f_\omega(x)$ is not necessarily a Carathéodory function because the measurability of $\omega \mapsto f_\omega(x)$ has not been ascertained yet.

Define a closed-valued correspondence $A : \mathfrak{M}(X) \twoheadrightarrow \mathfrak{C}_\infty(X, \mathbb{R})$ by

$$A : \theta \mapsto \begin{cases} \{g \in \mathfrak{C}_\infty(X, \mathbb{R}) | \int_X g(x) d\theta \geqq \varepsilon\} & \text{for} \quad \theta \neq 0, \\ \{0\} \quad \text{for} \quad \theta = 0. \end{cases} \tag{2.21}$$

It is easy to see that $A\theta \neq \emptyset$ for any $\theta \in \mathfrak{M}(X)$ and graph of A is measurable. Note that $\mathfrak{M}(X)$ (resp. $\mathfrak{C}_\infty(X, \mathbb{R})$) is endowed with the Borel σ-field generated by w^*-topology (resp. uniform convergence topology).

A function $B : \Omega \to \mathfrak{M}_+^1(X)$ is defined by

$$B : \omega \mapsto \nu_\omega,$$

which is measurable by Theorem 2.5. Hence the correspondence $A \circ B : \Omega \twoheadrightarrow \mathfrak{C}_\infty(X, \mathbb{R})$ is closed-valued and measurable. Hence it has a measurable selection $\eta(\omega)$ thanks to Kuratowski–Nardzewski's theorem.[39] We then define a function $f : \Omega \times X \to \mathbb{R}$ by

$$f(\omega, x) = \eta(\omega)(x).$$

[39] cf. The footnote 37.

The function $f_x : \omega \mapsto f(\omega, x)$ is measurable for any fixed $x \in X$. For any closed set $F \subset \mathbb{R}$, $f_x^{-1}(F)$ is equal to $\eta^{-1}\{g \in \mathfrak{C}_\infty(X, \mathbb{R}) | g(x) \in F\}$. $\{\cdots\}$ is clearly closed in $\mathfrak{C}_\infty(X, \mathbb{R})$. Hence $f_x^{-1}(F) = \eta^{-1}\{\cdots\} \in \mathscr{E}$. Thus $f \in \mathfrak{G}_{\mathfrak{C}}(\Omega, \mu; X)$ and it satisfies

$$\int_\Omega \{\int_X f(\omega, x) d\nu_\omega\} d\mu \geqq \varepsilon > 0.$$

This contradicts (2.18'). □

Equation (2.13) is symbolically written in the form

$$\gamma = \int_\Omega \delta_\omega \otimes \nu_\omega d\mu. \tag{2.22}$$

This representation is called the **disintegration** of a Young measure γ by means of a measurable family $\{\nu_\omega | \omega \in \Omega\}$.

Remark 2 Valadier [48, (pp. 179–181)] also gives a detailed proof under the assumption that X is a compact metric space. For the proof in the case of non-compact space (2°), we owe to Valadier [45, pp. 10.20] and [46]. If X is σ-compact and locally compact, it is not automatically a Radon space. Hence it is required to assume additionally that $\gamma \circ \pi_X^{-1}$ is a Radon measure in this case.

The completeness of the measure space Ω should also be assumed in order to apply the Berge theorem and verify the measurability of the correspondence A defined by (2.21).

Valadier does not seem to discuss the uniqueness in non-compact case.

2.4 Auxiliary Preparations

A more general version of the disintegration theorem will be stated and proved in the next subsection. We would like to prepare a couple of important results in advance. The first result is the theorem due to Dellacherie–Meyer [19], which evaluates the size of a class of bounded functions. The second one is the lifting theorem due to von Neumann [50] and Maharam [31].

Lemma 2.3 (Dellacherie–Meyer) *Assume that $\mathfrak{B}$ is a family of bounded real-valued functions, and $\mathfrak{H}$ is a vector space contained in $\mathfrak{B}$ which satisfies the following conditions.*

(a) $\mathfrak{H}$ contains all the constants.
(b) $\mathfrak{H}$ is closed for the uniform convergence.
(c) Let $\{f_n\}$ be a sequence of nonnegative functions in $\mathfrak{H}$ which is monotonically increasing and uniformly bounded.. Then $f = \lim_{n\to\infty} f_n$ (pointwise limit) is contained in $\mathfrak{H}$.

Furthermore, suppose that $\mathfrak{F}$ is a sub-family of $\mathfrak{H}$, which is closed for multiplication (i.e. $f, g \in \mathfrak{F} \Rightarrow f \cdot g \in \mathfrak{F}$).

Under these assumptions, $\mathfrak{H}$ contains all the bounded functions which are measurable w.r.t. the σ-field $\sigma(\mathfrak{F})$ on Ω generated by $\mathfrak{F}$.

Proof Let $\mathfrak{A}_*$ be an algebra generated by $\mathfrak{F}$ and 1. Since $\mathfrak{F}$ is closed for multiplication and $\mathfrak{H}$ is a vector space satisfying (a), it is clear that $\mathfrak{A}_* \subset \mathfrak{H}$. If we define a class A of algebras by

$$A = \{\mathfrak{A} | \mathfrak{A} \text{ is an algebra such that } \mathfrak{A}_* \subset \mathfrak{A} \subset \mathfrak{H}\},$$

then there is a maximal element $\mathfrak{A}^*$ in A *w.r.t.* the partial ordering by inclusions (Zorn's lemma).

It is not hard to confirm that $\mathfrak{A}^*$ has several properties as listed below.

1° $\mathfrak{A}_*$ is closed for uniform convergence. this is due to (b) and the maximality of $\mathfrak{A}^*$.

2° $\mathfrak{A}^*$ contains all the constants.

3° $f \in \mathfrak{A}^* \Rightarrow |f| \in \mathfrak{A}^*$.

Assume that $f \in \mathfrak{A}^*$ with $\|f\|_\infty \leqq C$. There exists, for each $\varepsilon > 0$, a polynomial P_ε such that

$$||\lambda| - P_\varepsilon(\lambda)| \leqq \varepsilon \quad \text{for all} \quad \lambda \in [-C, C],$$

by Weierstrass' approximation theorem. Hence

$$||f(\omega)| - P_\varepsilon(f(\omega))| \leqq \varepsilon \quad \text{for all} \quad \omega \in \Omega.$$

Thus $|f| \in \mathfrak{A}^*$ since $\mathfrak{A}^*$ is an algebra satisfying 1°.

4° $\mathfrak{A}^*$ is closed for the operations $\bigvee$ and $\bigwedge$, (i.e. $f, g \in \mathfrak{A}^* \Rightarrow f \bigvee g, f \bigwedge g \in \mathfrak{A}^*$), where $(f \bigvee g)(x) = \mathrm{Max}\{f(x), g(x)\}$, $(f \bigwedge g)(x) = \mathrm{Min}\{f(x), g(x)\}$.

This follows from **3°** and

$$f \bigvee g = (f+g)/2 + |f-g|/2,$$

$$f \bigwedge g = (f+g)/2 - |f-g|/2.$$

5° Let $\{f_n\}$ be a sequence of nonnegative functions in $\mathfrak{A}^*$ which is monotonically increasing (or decreasing) and uniformly bounded. Then $f = \lim_{n\to\infty} f_n$ (pointwise convergence) is contained in $\mathfrak{A}^*$.

Suppose first that $\{f_n\}$ is monotonically increasing. Since $alg(\mathfrak{A}^*, f)$, the algebra generated by $\mathfrak{A}^*$ and f, is contained in $\mathfrak{H}$ and $\mathfrak{A}^*$ is a maximal element in A, we have $alg(\mathfrak{A}^*, f) = \mathfrak{A}^*$. Hence $f \in \mathfrak{A}^*$. On the other hand, suppose that $\{f_n\}$ is monotonically decreasing. By the uniform boundedness of $\{f_n\}$. there is a large $C > 0$ such that $-f_n + C \geqq 0$ for all n. $-f_n + C \in \mathfrak{A}^*$ by 2°. The monotonically increasing sequence $\{-f_n + C\}$ of nonnegative functions in $\mathfrak{A}^*$

converges to $-f + C$. The above arguments justifies $-f + C \in \mathfrak{A}^*$. Hence $f \in \mathfrak{A}^*$ again by 2°.

We define a family $\mathscr{E}$ of sets in Ω by

$$\mathscr{E} = \{E \subset \Omega | \chi_E \in \mathfrak{A}^*\}.$$

$\mathscr{E}$ is a field and a monotone family.[40] So $\mathscr{E}$ is a σ-field.

The conclusion of the lemma follows by combining the two propositions below.

(i) The algebra $\mathfrak{A}^*$ contains all the bounded functions which are measurable *w.r.t.* $\mathscr{E}$.

Let f be a $\mathscr{E}$-measurable bounded function. Without loss of generality, we may assume that $f \geqq 0$. Then there exists a monotonically increasing sequence of nonnegative simple functions

$$\varphi_n(\omega) = \sum_{j=1}^{p} c_j^n \chi_{E_j}(\omega), \quad E_j \in \mathscr{E}, \quad n = 1, 2, \cdots$$

which converges to $f(\omega)$ (as $n \to \infty$). Since $\varphi_n \in \mathfrak{A}^*$, we obtain $f \in \mathfrak{A}^*$ by 5°.

(ii) $\sigma(\mathfrak{F}) \subset \mathscr{E}$.

In order to show this, it is sufficient to prove

$$E \equiv \{\omega \in \Omega | f(\omega) \geqq 1\} \in \mathscr{E}$$

for any $f \in \mathfrak{A}^*$. If we define[41]

$$g = (f \bigwedge 1) \bigvee 0,$$

$g \in \mathfrak{A}^*$ by 2° and 4° (Fig. 2).

Since $g \in \mathfrak{A}^*$, $g^n (n = 1, 2, \cdots)$ is a monotonically decreasing sequence of $\mathfrak{A}^*$, and converges to χ_E. We obtain $\chi_E \in \mathfrak{A}^*$ again by 5°. So $E \in \mathscr{E}$. □

We now proceed to the second preparation, the theory of liftings.

Let X be a set endowed with a partial ordering $\leq$. A is supposed to be a non-empty subset of X. When A has the least upper bound (*l.u.b.*) in X, it is denoted by $\bigvee(A)$. When there is its greatest lower bound (*g.l.b.*) in X, it is denoted by $\bigwedge(A)$.

---

[40] $\chi_E, \chi_{E'} \in \mathfrak{A}^*$ for $E, E' \in \mathscr{E}$ by definition. Since $\chi_{E \cup E'} = \chi_E \bigvee \chi_{E'}$, $\chi_{E \cup E'} \in \mathfrak{A}^*$ by 4°. Hence $E \cup E' \in \mathscr{E}$. Furthermore $\chi_{E^c} = 1 - \chi_E \in \mathfrak{A}^*$ by 2° for each $E \in \mathscr{E}$; i.e. $E^c \in \mathscr{E}$. It is obvious that $\Omega, \emptyset \in \mathfrak{A}^*$. Therefore $\mathscr{E}$ is a field. If $E_n \in \mathscr{E} (n = 1, 2, \cdots)$ is a monotonically increasing sequence, we have $\chi_{\bigcup_{n=1}^\infty E_n} = \lim_{n\to\infty} \chi_{E_n}$. Since $\{\chi_{E_n}\}$ is a monotonically increasing and uniformly bounded sequence in $\mathfrak{A}^*$. 5° justifies $\chi_{\bigcup_{n=1}^\infty E_n} \in \mathfrak{A}^*$; i.e. $\bigcup_{n=1}^\infty E_n \in \mathscr{E}$.

[41] In other words, $g(\omega) = \text{Max}\|\text{Min}\{f(\omega), 1\}, 0\}$.

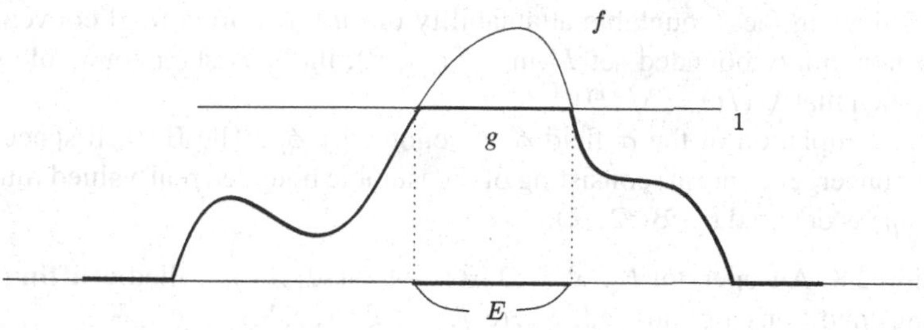

Fig. 2 Graph of g

When any two-point set in X has *l.u.b.* and *g.l.b.*, X is called a **lattice**. If any non-empty set $A \subset X$ which is bounded from above has *l.u.b.*, then X is said to be **complete** *w.r.t.* $\leq$.[42]

When a real vector space $\mathfrak{X}$ endowed with a partial ordering is called a **vector lattice** if it is a lattice and the following conditions (i), (ii) are satisfied.

(i) For $x, y, z \in \mathfrak{X}$,

$$x \leq y \Rightarrow x + z \leq y + z,$$

(ii) For $x, y \in \mathfrak{X}$,[43]

$$x \leq y \Rightarrow \alpha x \leq \alpha y \quad \text{for all} \quad \alpha \geqq 0.$$

Let $(\Omega, \mathscr{E}, \mu)$ be a finite measure space. $L^\infty(\Omega, \mathbb{R})$, the space of essentially bounded measurable real-valued functions on Ω, is a complete vector lattice *w.r.t.* the partial ordering $\leq$ defined by

$$f \leq g \Leftrightarrow f(\omega) \leqq g(\omega) \quad \text{for all} \quad \omega \in \Omega.$$

The quotient space $\mathfrak{L}^\infty(\Omega, \mathbb{R})$ of $L^\infty(\Omega, \mathbb{R})$ modulo $\mathfrak{M} = \{f \in L^\infty(\Omega, \mathbb{R}) | f(\omega) = 0 \;\; a.e.\}$ is also a complete vector lattice *w.r.t.* the partial ordering induced naturally from $L^\infty(\Omega, \mathbb{R})$.

42 The following two statements are equivalent for a set $A \neq \emptyset$ in X.

(i) If A is bounded from above, it has the *l.u.b.*
(ii) If A is bounded from below, it has the *g.l.b.*

The completeness defined here is called **Dedekind-complete** in Zaanen [53].

43 An equivalent form is:

$$x \leq y \Rightarrow \alpha y \leq \alpha x \quad \text{for all} \quad \alpha \leqq 0.$$

The following fact (countable attainability of *l.u.b.*) is often used conveniently. For any non-empty bounded set H in $L^\infty(\Omega, \mathbb{R})$, there exists a countable subset $D \subset H$ such that $\bigvee(H) = \bigvee(D)$.[44]

The μ-completion of the σ-field $\mathscr{E}$ is denoted by $\mathscr{E}_\mu$. The Banach space (with uniform convergence norm) consisting of measurable bounded real-valued functions on $(\Omega, \mathscr{E}_\mu)$ is denoted by $\mathfrak{B}(\Omega, \mathbb{R})$.

Definition 2.8 An operator $l : \mathfrak{L}^\infty(\Omega, \mathbb{R}) \to \mathfrak{B}(\Omega, \mathbb{R})$ is called a **lifting** if the following conditions are satisfied, where $f, g \in \mathfrak{L}^\infty(\Omega, \mathbb{R}), \alpha, \beta \in \mathbb{R}$.

(i) $f(\omega) \leqq g(\omega) a.e. \Rightarrow l(f) \leq l(g)$.[45]
(ii) $l(\alpha f + \beta g) = \alpha l(f) + \beta l(g)$.
(iii) $l(f \cdot g) = l(f) \cdot l(g)$.
(iv) $l(1) = 1$.
(v) $l(f) \in f$.

If we denote the image $l(\mathfrak{L}^\infty(\Omega, \mathbb{R}))$ by L, L is a subset of $\mathfrak{B}(\Omega, \mathbb{R})$ and it is a complete vector lattice. A lifting is an isometric algebra-isomorphism of $\mathfrak{L}^\infty(\Omega, \mathbb{R})$ onto L which preserves the partial ordering.

It is known that there exists a lifting provided that $(\Omega, \mathscr{E}, \mu)$ is a finite measure space.[46]

The following delicate lemma concerning the family of bounded measurable functions is due to Hoffmann–Jørgensen [24].

Lemma 2.4 *A subset L_0 of L is assumed to be bounded. $sup(L_0)$ is defined as*

$$\sup(L_0)(\omega) = \sup\{f(\omega) | f \in L_0\}, \omega \in \Omega.$$

(i) $\sup(L_0)(\omega) \leqq \bigvee(L_0)(\omega)$ *for all* $\omega \in \Omega$.
(ii) $\sup(L_0)$ *is measurable and*

$$\sup(L_0) = \bigvee(L_0) \ a.e.$$

Proof
(i) By the definition of $\bigvee(L_0)$, we obtain

$$f(\omega) \leqq \bigvee(L_0)(\omega) \quad \text{for all} \quad \omega \in \Omega$$

---

44 See Dunford–Schwartz [21, p. 336], Schwartz [44, pp. 36–37].

45 $l(f) \leq l(g)$ means that "$l(f)(\omega) \leqq l(g)(\omega)$ for all $\omega \in \Omega$".

46 See Ionescu Tulcea [26] for details about the theory of liftings. Although the origin of this problem can be found in von Neumann [50], the complete solution was given by Maharam [31]. Schwartz [44] pp.33–39 is quite useful as a brief exposition.

for any $f \in L_0$. Consequently

$$\sup_{f \in L_0} f(\omega) = \sup(L_0)(\omega) \leqq \bigvee(L_0)(\omega) \quad \text{for all} \quad \omega \in \Omega.$$

(ii) By the countable attainability of *l.u.b.*, there exists a countable set $D = \{f_1, f_2, \cdots\}$ in L_0 which satisfies

$$\bigvee(L_0)(\omega) = \bigvee(D)(\omega) \quad \text{for all} \quad \omega \in \Omega. \tag{2.23}$$

$\sup(D)(\omega) = \sup\{f_n(\omega)|n = 1, 2, \cdots\}$ is measurable and

$$\sup(D)(\omega) \leqq \sup(L_0)(\omega) \leqq \bigvee(L_0)(\omega) \quad \text{for all} \quad \omega \in \Omega. \tag{2.24}$$

Since $f_n \leq \sup(D)$, we obtain

$$l(f_n)(\omega) = f_n(\omega) \leqq l(\sup(D))(\omega) \quad \text{for all} \quad \omega \in \Omega.$$

That is, $l(\sup(D))$ is an upper bound for D. Hence

$$\bigvee(L_0)(\omega) \underset{(2.23)}{=} \bigvee(D)(\omega) \leqq l(\sup(D)).$$

By the property (v) of liftings,

$$\bigvee(L_0)(\omega) \leqq \sup(D)(\omega) \quad a.e. \tag{2.25}$$

(ii) follows from (2.24) and(2.25). □

Lemma 2.5 *Assume that $L_0 \subset L$ is bounded and the following condition* (F) *is satisfied.*[47]

$$(\mathrm{F}) \begin{cases} \text{For any} \quad f, g \in L_0, \quad \text{there exists some} \quad h \in L_0 \quad \text{such that} \\ \sup(f, g) \leq h. \end{cases}$$

Then

$$\int_\Omega \sup(L_0) d\mu = \sup\{\int_\Omega f d\mu | f \in L_0\}. \tag{2.26}$$

[47]The condition (F) is nothing other than what Hoffmann–Jørgensen [24] called "filtering to the right".

Proof By the countable attainability of *l.u.b.*, there exists a countable set $D = \{f_1, f_2, \cdots\}$ in L_0 such that

$$\bigvee(L_0)(\omega) = \bigvee(D)(\omega) \quad \text{for all} \quad \omega \in \Omega.$$

We may assume, without loss of generality, that $f_1 \leq f_2 \leq \cdots$ by the condition (F). Thanks to Lemma 2.4,

$$\sup(L_0) = \bigvee(L_0) = \sup_n f_n = \lim_{n\to\infty} f_n \;\; a.e.$$

It follows that

$$\int_\Omega \sup(L_0)d\mu = \lim_{n\to\infty}\int_\Omega f_n d\mu \leqq \sup\{\int_\Omega f d\mu | f \in L_0\}$$

by the dominated convergence theorem.

The converse inequality is obvious. Thus we obtain (2.26). □

Counter-Example The relation (2.26) does not necessarily hold good without the condition (F). We try to illuminate it by an example. Suppose that $\Omega = [0, 1]$ with Lebesgue measure m. Let L_0 be a family of Rademacher functions on [0, 1]. This family L_0 does not satisfy (F). Taking account of $\sup(L_0)(\omega) = 1$, we obtain

$$\int_\Omega \sup(L_0)d\mu = 1, \quad \int_\Omega f d\mu = 0 \quad \text{for any} \quad f \in L_0.$$

Hence (2.26) does not hold good.

2.5 Disintegration Theorem (B)

We have already examined, in Sect. 2.3, the possibility of the disintegration representation of Young measures on $\Omega \times X$ in case X is a Polish space. In this subsection, we study the same problem in a more general case.[48]

Theorem 2.8 (Disintegration: General Topological Space) *Let $(\Omega, \mathscr{E}, \mu)$ be a finite measure space, X a Hausdorff topological space, and γ a Young measure on $(\Omega \times X, \mathscr{E} \otimes \mathscr{B}(X))$ such that $\gamma \circ \pi_X^{-1}$ is a Radon measure on X. Then there exists a unique measurable family $\nu = \{\nu_\omega | \omega \in \Omega\} \in \mathfrak{P}(\Omega, \mu; X)$ such that*

$$\gamma(A) = \int_\Omega \{\int_X \chi_A(\omega, x)d\nu_\omega\}d\mu, \tag{2.27}$$
$$A \in \mathscr{E} \otimes \mathscr{B}(X).$$

[48] We owe the basic ideas of Theorem 2.8 to Valadier [48, 49] and Hoffmann–Jørgensen [24].

Proof The proof of uniqueness can be done in a quite similar manner as in Theorem 2.7. So we have only to prove the existence.

It is enough to show that (2.27) holds good in a particular case where A is a rectangular set $E \times B$, $E \in \mathscr{E}$, $B \in \mathscr{B}(X)$.[49]

Furthermore since $\gamma \circ \pi_X^{-1}$ is a Radon measure by assumption, we may restrict our attention to the case of compact X, The general case can be reduced to this simple case similarly as in 2° in the proof of Theorem 2.7.[50] So we will **assume that X is compact**.

Fix a function $\psi \in \mathfrak{C}(X, \mathbb{R})$. We now define a functional $\Lambda_\psi : \mathfrak{L}^1(\Omega, \mathbb{R}) \to \mathbb{R}$ by

$$\Lambda_\psi : f \mapsto \int_{\Omega\times X} f(\omega)\psi(x)d\gamma, \quad f \in \mathfrak{L}^1(\Omega, \mathbb{R}).$$

Since Λ_ψ is a bounded linear functional on $\mathfrak{L}^1(\Omega, \mathbb{R})$ ($\|\Lambda_\psi\| \leqq \|\psi\|_\infty$), there exists some $h_\psi \in \mathfrak{L}^\infty(\Omega, \mathbb{R})$ such that

$$\int_{\Omega\times X} f(\omega)\psi(x)d\gamma = \int_\Omega f(\omega)h_\psi(\omega)d\mu, \qquad f \in \mathfrak{L}^1(\Omega, \mathbb{R}). \tag{2.28}$$

The operator $\psi \mapsto h_\psi (\mathfrak{C}(X, \mathbb{R}) \to \mathfrak{L}^\infty(\Omega, \mathbb{R}))$ is a positive linear operator of norm 1.[51]

Let $l : \mathfrak{L}^\infty(\Omega, \mathbb{R}) \to \mathfrak{B}(\Omega, \mathbb{R})$ be a lifting. Then the function

$$\nu_\omega : \psi \mapsto (l(h_\psi))(\omega), \tag{2.29}$$

$\omega \in \Omega$ being fixed, is a bounded linear functional $\nu_\omega : \mathfrak{C}(X, \mathbb{R}) \to \mathbb{R}(\|\nu_\omega\| = 1)$; i.e. $\nu_\omega \in \mathfrak{C}(X, \mathbb{R})'$. Since ν_ω is a positive linear functional on $\mathfrak{C}(X, \mathbb{R})$, ν_ω is a Radon probability measure on $(X, \mathscr{B}(X))$; i.e. $\nu_\omega \in \mathfrak{M}_+^1(X)$.

Substituting

$$l(h_\psi)(\omega) = \nu_\omega(\psi) = \int_X \psi(x)d\nu_\omega \tag{2.30}$$

into $h_\psi(\omega)$ appearing on the right-hand side of (2.28), we obtain

$$\int_{\Omega\times X} f(\omega)\psi(x)d\gamma = \int_\Omega f(\omega)\{\int_X \psi(x)d\nu_\omega\}d\mu. \tag{2.31}$$

---

[49]Assume that it is done. Then, the family of all sets in $\mathscr{E} \otimes \mathscr{B}(X)$ which satisfy (2.27) is a σ-field containing all the rectangular sets $E \times B$, $E \in \mathscr{E}$, $B \in \mathscr{B}(X)$. Hence (2.27) holds good on $\mathscr{E} \otimes \mathscr{B}(X)$.

[50]In theorem 2.7, we assume that X is Polish. In this case, $\gamma \circ \pi_X^{-1}$ is automatically a Radon measure.

[51]Since $\|\Lambda_\psi\| = \|h_\psi\|_\infty \leqq \|\psi\|_\infty$, the norm of the operator in question $\leqq 1$. If $\psi = 1$, then the corresponding $h_1 = 1$. So the norm of the operator$= 1$.

Fixing $E \in \mathscr{E}$ arbitrarily, we define a set function $\nu_E : B \mapsto \gamma(E \times B)$ $(B \in \mathscr{B}(X))$. Then ν_E is a Radon measure on X since $\gamma \circ \pi_X^{-1}$ is a Radon measure.[52] For any open set U in X,[53]

$$\begin{aligned}\gamma(E \times U) &= \sup\{\int_{\Omega \times X} \chi_E(\omega)\psi(x)d\gamma \mid \psi \in \mathfrak{C}(X, \mathbb{R}), \quad 0 \leqq \psi \leqq \chi_U\}, \\ &\underset{(2.31)}{=} \sup\{\int_E \{\int_X \psi(x)d\nu_\omega\}d\mu \mid \psi \in \mathfrak{C}(X, \mathbb{R}), \quad 0 \leqq \psi \leqq \chi_U\} \qquad (2.32) \\ &= \sup\{\int_E \nu_\omega(\psi)d\mu \mid \psi \in \mathfrak{C}(X, \mathbb{R}), \quad 0 \leqq \psi \leqq \chi_U\},\end{aligned}$$

where $E \in \mathscr{E}$. When $\psi \in \mathfrak{C}(X, \mathbb{R})$ is fixed, $\omega \mapsto \nu_\omega(\psi)$ is $\mathscr{E}_\mu$-measurable by (2.30).[54] So

$$\omega \mapsto \int_E \psi(x)d\nu_\omega$$

is also $\mathscr{E}_\mu$-measurable by (2.29). This justifies the integration appearing in (2.32).

Looking at (2.29) as a function of ω, we observe that $\omega \mapsto \nu_\omega(\psi)$ is contained in the image L of the lifting l and the set

$$L_0 = \{\nu_\omega(\psi) \mid \psi \in \mathfrak{C}(X, \mathbb{R}),\ 0 \leqq \psi \leqq \chi_U\}$$

is bounded in L. Hence

$$\omega \mapsto \nu_\omega(U) = \sup(L_0) \tag{2.33}$$

is $\mathscr{E}_\mu$-measurable.

[52]For any sequence $\varepsilon_n \downarrow 0$, and $B \in \mathscr{B}(X)$, there is a sequence $\{K_n\}$ of compact sets in B such that $\gamma \circ \pi_X^{-1}(B \setminus K_n) < \varepsilon_n$. Consequently we obtain

$$\nu_E(B \setminus K_n) = \gamma(E \times B \setminus E \times K_n) \leqq \mu(A)(\gamma \circ \pi_X^{-1})(B \setminus K_n) \leqq \mu(A)\varepsilon_n.$$

Hence ν_E is a Radon measure.

[53]The first equality of (2.32) is justified by the following reasoning. For any sequence $\varepsilon_n \downarrow 0$, there exists a sequence $\{K_n\}$ of compact sets in U such that $\gamma \circ \pi_X^{-1}(U \setminus K_n) < \varepsilon_n$. X is now assumed to be Hausdorff and compact. Hence their exists a continuous function $\psi_n : X \to [0, 1]$ such that $\psi_n(K_n) = \{1\}$ and $\psi_n(U^c) = \{0\}$. Then

$$\int_{\Omega \times X} \chi_E(\omega)\psi_n(x)d\gamma \to \gamma(E \times U) \quad \text{as } n \to \infty.$$

[54]$l(h_\psi)$ is measurable *w.r.t.*, the σ-field $\mathscr{E}_\mu$ (the completion of $\mathscr{E}$ *w.r.t.* μ), by the definition of the lifting. Here we have to use $\mathscr{E}_\mu$ instead of $\mathscr{E}$.

L_0 satisfies the condition (F) in Lemma 2.5. It follows that

$$\gamma(E \times V) = \sup\{\int_E \nu_\omega(\psi)d\mu | \nu_\omega(\psi) \in L_0\}$$
$$= \int_E \sup(L_0)d\mu$$
$$\underset{(2.33)}{=} \int_E \nu_\omega(U)d\mu.$$

Let $\mathfrak{H}$ be the set of bounded Borel functions $h : X \to \mathbb{R}$ which satisfy the following two conditions.

1° $\omega \mapsto \int_X h d\nu_\omega$ is $\mathscr{E}_\mu$-measurable.

2° $\int_{\Omega \times X} \chi_E(\omega)h(x)d\gamma = \int_E \{\int_X h d\nu_\omega\}d\mu \quad$ for any $E \in \mathscr{E}$.

Then $\mathfrak{H}$ satisfies (a), (b) and (c) in Lemma 2.3 and closed *w.r.t.* multiplication. Hence, by Lemma 2.3, $\mathfrak{H}$ contains all the bounded Borel functions.[55] It follows that

$$\chi_B \in \mathfrak{H} \quad \text{for all} \quad B \in \mathscr{B}(X);$$

that is $\omega \mapsto \nu_\omega(B)$ $(B \in \mathscr{B}(X))$ is $\mathscr{E}_\mu$-measurable, and

$$\gamma(E \times B) = \int_E \nu_\omega(B)d\mu.$$

□

2.6 *Conditional Probabilities*

In this subsection, the relationship between the concept of disintegrations of Young measures and that of regular conditional probabilities is clarified. Although the similarity between these two has been noticed in the long history of Young measures, some vagueness seems to have been left. Our target is to prove the existence of regular conditional probabilities through the disintegrability theorem.[56]

Let $(\Omega, \mathscr{E}, P)$ be a probability space and $\mathscr{F}$ a sub-σ-field of $\mathscr{E}$. As anyone knows, the existence of the conditional probability $P(A|\mathscr{F})(\omega)(A \in \mathscr{E})$ is assured by means of Radon–Nikodým theorem. $P(A|\mathscr{F})(\omega)$ is said to be **regular** if (i) the mapping $\omega \mapsto P(A|\mathscr{F})(\omega)$ is $\mathscr{F}$-measurable for any $A \in \mathscr{E}$ and (ii) $P(\cdot|\mathscr{F})(\omega)$ is a probability measure on $(\Omega, \mathscr{E})$ for *a.e.* ω. A similarity between the concept of regular conditional probabilities and that of measurable families is quite clear.

---

[55] Both $\mathfrak{H}$ and $\mathfrak{F}$ in Lemma 2.3 should be regarded as equal to $\mathfrak{H}$ here defined. It is obvious that $\sigma(\mathfrak{F}) \subset \mathscr{B}(X)$.

[56] This subsection is based upon Maruyama [38]. See Loève [30, pp. 337–370] for conditional probabilities.

We now proceed to the existence problem of regular conditional probabilities. The following theorem is basically obtained by Hoffmann–Jørgensen [24] and Chatterji [17]. However we now give an alternative proof based upon Theorem 2.8. (See also Jirina [28] and Schwartz [44, pp. 55–56].)

Theorem 2.9 *Let $(\Omega, \mathscr{E})$ be a measurable space and X a Hausdorff topological space with a Radon probability measure θ. A function $p : X \to \Omega$ is assumed to be $(\mathscr{B}(X), \mathscr{E})$-measurable. A measure μ on $(\Omega, \mathscr{E})$ is defined by $\mu = \theta \circ p^{-1}$. Then there exists a $\mathscr{E}_\mu$-measurable family $\{\nu_\omega | \omega \in \Omega\}$ of Radon probabilities on X which satisfies*

$$\int_\Omega \nu_\omega(B \cap p^{-1}(A)) d\mu = \theta(B \cap p^{-1}(A)) \quad \textit{for} \quad A \in \mathscr{E}, B \in \mathscr{B}(X).$$

(The extension of μ is also denoted by the same notation.)

Proof To start with, we define a measure γ on $(\Omega \times X, \mathscr{E} \otimes \mathscr{B}(X))$ by

$$\gamma(E) = \int_\Omega \{\int_X \chi_E(\omega, x) d\theta\} d\mu \quad \text{for} \quad E \in \mathscr{E}_\mu \otimes \mathscr{B}(X).$$

We can easily confirm that γ is a Young measure, taking account of the relation

$$\begin{aligned} \gamma \circ \pi_\Omega^{-1}(A) &= \int_\Omega \{\int_X \chi_{A \times X}(\omega, x) d\theta\} d\mu \\ &= \int_\Omega \chi_A(\omega) d\mu \\ &= \mu(A) \quad \text{for} \quad A \in \mathscr{E}. \end{aligned}$$

We also note that

$$\begin{aligned} \gamma \circ \pi_X^{-1}(B) &= \int_\Omega \{\int_X \chi_{\Omega \times B}(\omega, x) d\theta\} d\mu \\ &= \int_\Omega \theta(B) d\mu \qquad (2.34) \\ &= \theta(B) \quad \text{for} \quad B \in \mathscr{B}(X). \end{aligned}$$

Hence $\gamma \circ \pi_X^{-1}$ is a Radon measure on X.

By Theorem 2.8, there exists a $\mathscr{E}_\mu$-measurable family $\{\nu_\omega | \omega \in \Omega\}$ of Radon probabilities on X which satisfies

$$\gamma(E) = \int_\Omega \{\int_X \chi_E(\omega, x) d\nu_\omega\} d\mu \quad \text{for} \quad E \in \mathscr{E}_\mu \otimes \mathscr{B}(X).$$

If $E = A \times B \in \mathscr{E} \otimes \mathscr{B}(X)$ in particular, we obtain

$$\begin{aligned}\gamma(A \times B) &= \int_{\Omega} \{\chi_{A\times B}(\omega, x) d\nu_{\omega}\} d\mu \\ &= \int_{\Omega} \chi_A(\omega)\nu_{\omega}(B) d\mu \\ &= \int_A \nu_{\omega}(B) d\mu.\end{aligned} \tag{2.35}$$

Since $\theta = \gamma \circ \pi_X^{-1}$ by (2.34), we find that

$$\begin{aligned}\theta(B \cap p^{-1}(A)) &= \gamma \circ \pi_X^{-1}(B \cap p^{-1}(A)) \\ &= \gamma(\Omega \times (B \cap p^{-1}(A)) \\ &= \int_{\Omega} \nu_{\omega}(B \cap p^{-1}(A)) d\mu\end{aligned}$$

by (2.35). This completes the proof. □

The following result is actually a corollary of Theorem 2.9.

Theorem 2.10 *Let X be a Hausdorff topological space endowed with the Borel σ-field $\mathscr{B}(X)$. μ is a Radon probability measure on X. For any sub-σ-field $\mathscr{E}$ of $\mathscr{B}(X)$, there exists a $\mathscr{E}_{\mu}$ -measurable family $\{\nu_{\omega} | \omega \in X\}$ which satisfies*

$$\mu(A \cap B) = \int_A \nu_{\omega}(B) d\mu \quad \text{for} \quad A \in \mathscr{E}, B \in \mathscr{B}(X).$$

(The extension of μ is also denoted by the same notation.)

Proof Let Ω be a copy of X (i.e. $\Omega = X$) endowed with $\mathscr{E}$. We specify a mapping $p : X \to \Omega$ as the identity $p(\omega) = \omega$, which is $(\mathscr{B}(X), \mathscr{E})$-measurable.

Since θ in Theorem 2.9 corresponds to the Radon measure μ on $(X, \mathscr{B}(X))$ here, $\mu = \theta \circ p^{-1}$ in Theorem 2.9 should be interpreted as $\mu|_{\mathscr{E}}$. the restriction of μ to $\mathscr{E}$.

Applying Theorem 2.9 to this simple setting, there exists a $\mathscr{E}_{\mu}$-measurable family $\{\nu_{\omega} | \omega \in \Omega\}$ of Radon probability measures on X which satisfies

$$\int_{\Omega} \chi_A(\omega)\nu_{\omega}(B) d\mu = \mu(B \cap p^{-1}(A)) = \mu(A \cap B)$$
$$\text{for} \quad A \in \mathscr{E},\ B \in \mathscr{B}(X).$$

This proves Theorem 2.10. □

3 Convergence

3.1 Narrow Topology

Let $(\Omega, \mathscr{E}, \mu)$ be a complete finite measure space and X a locally compact Polish space.

According to Theorem 2.7, any Young measure on $(\Omega \times X, \mathscr{E} \otimes \mathscr{B}(X))$ can be represented in the form of disintegration. That is, for any Young measure γ, there exists uniquely a measurable family $\{\nu_\omega | \omega \in \Omega\}$ of probability measures which satisfies

$$\gamma(A) = \int_\Omega \{\int_X \chi_A(\omega, x) d\nu_\omega\} d\mu, \quad A \in \mathscr{E} \otimes \mathscr{B}(X). \tag{3.1}$$

Conversely any $\nu = \{\nu_\omega | \omega \in \Omega\} \in \mathfrak{P}(\Omega, \mu; X)$ defines a Young measure γ by (3.1). We denote, by Φ, the mapping which corresponds to each $\nu \in \mathfrak{P}(\Omega, \mu; X)$ the Young measure γ defined by (3.1). Then Φ is a bijection of the form Φ : $\mathfrak{P}(\Omega, \mu; X) \to \mathfrak{Y}(\Omega, \mu; X)$.

In the framework stated above, the following two duality relations hold good.

$$\mathfrak{C}_\infty(X, \mathbb{R})' \cong \mathfrak{M}(X), \tag{3.2}$$

$$\mathfrak{L}^1(\Omega, \mathfrak{C}_\infty(X, \mathbb{R}))' \cong \mathfrak{L}^\infty(\Omega, \mathfrak{M}(X)). \tag{3.3}$$

We have to note that $\mathfrak{M}(X)$ appearing in (3.3) is endowed with the Borel σ-field generated by the weak*-topology $\sigma(\mathfrak{M}(X), \mathfrak{C}_\infty(X, \mathbb{R}))$ based upon (3.2).

Since $\mathfrak{P}(\Omega, \mu; X)$ is a subset of $\mathfrak{L}^\infty(\Omega, \mathfrak{M}(X))$, a natural topology on $\mathfrak{P}(\Omega, \mu; X)$ is the relative topology induced from the w^*-topology $\sigma(\mathfrak{L}^\infty(\Omega, \mathfrak{M}(X)), \mathfrak{L}^1(\Omega, \mathfrak{C}_\infty(X, \mathbb{R})))$ on $\mathfrak{L}^\infty(\Omega, \mathfrak{M}(X))$.

On the other hand,[57] $\mathfrak{M}(\Omega \times X)$ is endowed with a topology which is generated by

$$\gamma \mapsto \int_{\Omega \times X} f(\omega, x) d\gamma, \tag{3.4}$$
$$f \in \mathfrak{G}_{\mathfrak{C}_\infty}(\Omega, \mu; X) \cong \mathfrak{L}^1(\Omega, \mathfrak{C}_\infty(X)).$$

We can define a natural topology on $\mathfrak{Y}(\Omega, \mu; X)$ as the relative topology induced from $\mathfrak{M}(\Omega \times X)$.

Then Φ is a homeomorphism *w.r.t.* this combination of topologies on the domain and the range of Φ. In fact, it follows from the relation.

$$\int_{\Omega \times X} f(\omega, x) d\gamma = \int_\Omega \{\int_X f(\omega, x) d\nu_\omega\} d\mu$$

[57] $\mathfrak{M}(\Omega \times X)$ is the space of all the signed measures with finite total variations on $(\Omega \times X, \mathscr{E} \otimes \mathscr{B}(X))$.

for any Young measure γ which is represented in the form (3.1) by means of $\{\nu_\omega | \omega \in \Omega\}$. Hence $\mathfrak{Y}(\Omega, \mu; X)$ with the topology generated by (3.4) and $\mathfrak{P}(\Omega, \mu; X)$ with w^*-topology can be identified as a topological space.

If X is a compact metric space, in particular, the argument as above holds good in a simpler form, by substituting $\mathfrak{C}_\infty(X, \mathbb{R})$ and $\mathfrak{G}_{\mathfrak{C}_\infty}(\Omega, \mu; X)$ by $\mathfrak{C}(X, \mathbb{R})$ and $\mathfrak{G}_{\mathfrak{C}}(\Omega, \mu; X)$, respectively.

When X is not locally compact, the concept of "vanishing at infinity" does not make sense. However we define a topology on $\mathfrak{Y}(\Omega, \mu; X)$ similarly in this case, too.

Definition 3.1 Let $(\Omega, \mathscr{E}, \mu)$ be a finite measure space and X a separable metric space. The topology on $\mathfrak{Y}(\Omega, \mu; X)$ generated by

$$\gamma \mapsto \int_{\Omega \times X} f(\omega, x) d\gamma, \quad f \in \mathfrak{G}_{\mathfrak{C}}(\Omega, \mu; X) \tag{3.5}$$

is called the **narrow topology**.

The narrow topology is Hausdorff as confirmed in the course of the proof of Theorem 2.7. The integral in (3.5) makes sense because any $f \in \mathfrak{G}_{\mathfrak{C}}(\Omega, \mu; X)$ is measurable *w.r.t*, $\mathscr{E} \otimes \mathscr{B}(X)$ when X is a separable metric space. The definition of the narrow topology also applies to the case where X is a metrizable Souslin space or a Polish space.

Definition 3.2 Let $(\Omega, \mathscr{E})$ be a measurable space and X a topological space. A function $f : \Omega \times X \to \overline{\mathbb{R}}$ is called a **normal integrand** if the following two conditions are satisfied.

(i) f is $(\mathscr{E} \otimes \mathscr{B}(X), \mathscr{B}(\overline{\mathbb{R}}))$-measurable.
(ii) The function $x \mapsto f(\omega, x)$ is lower semi-continuous (*l.s.c.*) for every fixed $\omega \in \Omega$.

The set of all the normal integrands on $\Omega \times X$ is denoted by $\mathfrak{G}(\Omega, \mathscr{E}; X)$. The set of all the nonnegative-valued normal integrands is denoted by $\mathfrak{G}_+(\Omega, \mathscr{E}; X)$. The functions of this kind, together with Carathéodory functions, play crucial roles in basic studies in calculus of variations.

Any element of $\mathfrak{G}_+(\Omega, \mathscr{E}; X)$ can be represented as a limit of some monotonically increasing sequence of Carathéodory functions. We start by showing this important fact.

Lemma 3.1 *Let (X, ρ) be a separable metric space. There exists a sequence $\varphi_n : X \to \mathbb{R} (n = 1, 2, \cdots)$ of nonnegative continuous functions such that*

$$f(x) = \sup\{\varphi_n(x) | \varphi_n \leqq f\}, \quad x \in X$$

for any nonnegative l.s.c. function $f : X \to \overline{\mathbb{R}}$.

Proof Since X is second countable, there exists a countable base $\mathscr{B} = \{U_1, U_2, \cdots\}$. Define countable open sets $G_n^{(m)}(m, n = 1, 2, \cdots)$ by

$$G_n^{(m)} = \{x \in X | \rho(x, \overline{U}_m) < \frac{1}{n}\}.$$

Then there is, for each positive rational r, a continuous function $\varphi_{nr}^{(m)} : X \to [0, r]$ such that

$$\varphi_{nr}^{(m)}(x) = \begin{cases} r & \text{on} \quad \overline{U_m}, \\ 0 & \text{on} \quad (G_n^{(m)})^c. \end{cases}$$

The existence of such a function is guaranteed by Urysohn's theorem.

By renumbering the countable functions $\{\varphi_{nr}^{(m)}\}$ thus constructed, we obtain a sequence $\{\varphi_n\}$ with required properties.[58] □

Theorem 3.1 *Let $(\Omega, \mathscr{E}, \mu)$ be a complete finite measure space, and X a metrizable Souslin space. Then for any $f \in \mathfrak{S}_+(\Omega, \mathscr{E}; X)$, there exists a sequence $\{\psi_n\}$ of nonnegative integrable Carathéodory functions (i.e. in $\mathfrak{S}_{\mathfrak{C}}(\Omega, \mu; X)$) which satisfies*

$$f(\omega, x) = \sup_n \psi_n(\omega, x). \tag{3.6}$$

[58] I will give a little bit detailed explanation. For any $x_0 \in X$ and $\varepsilon > 0$, there exists an open neighborhood V of x_0 such that

$$f(x) \geqq f(x_0) - \varepsilon \quad \text{for all} \quad x \in V.$$

There exists some $U_m \in \mathscr{B}$ which satisfies $x_0 \in U_m$ and $G_n^{(m)} \subset V$. Choose $r \in \mathbb{Q}$ so that $f(x_0) - \varepsilon < r \leqq f(x_0)$. Then $|\varphi_{nr}^{(m)}(x_0) - f(x_0)| < \varepsilon$.

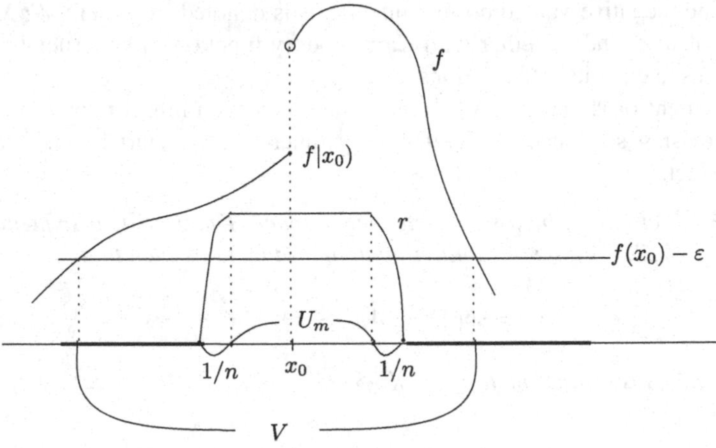

Proof By Lemma 3.1, there exists a sequence $\{\varphi_n : X \to \mathbb{R}\}$ of nonnegative continuous functions such that

$$h(x) = \sup\{\varphi_n(x) | \varphi_n \leqq h\}, \ x \in X$$

for all nonnegative *l.s.c.* functions $h : X \to \bar{\mathbb{R}}$. If we define

$$E_n = \{\omega \in \Omega | f(\omega, x) \geqq \varphi_n(x) \quad \text{for all} \quad x \in X\}, \quad n = 1, 2, \cdots,$$

then E_n^c is the projection

$$G_n = \{(\omega, x) \in \Omega \times X | f(\omega, x) < \varphi_n(x)\}$$

into Ω; i.e. $E_n^c = \pi_\Omega(G_n)$. Since $G_n \in \mathscr{E} \otimes \mathscr{B}(X)$, we obtain

$$E_n^c = \pi_\Omega(G_n) \in \mathscr{E}, \quad n = 1, 2, \cdots, \tag{3.7}$$

by the projection theorem.[59] Hence $E_n \in \mathscr{E}$ $(n = 1, 2, \cdots)$.

Define a function $\psi_n : \Omega \times X \to \mathbb{R}$ by

$$\psi_n : (\omega, x) \mapsto \chi_{E_n}(\omega)\varphi_n(x); \ n = 1, 2, \cdots. \tag{3.8}$$

Then ψ_n is an integrable Carathéodory function (i.e., $\psi_n \in \mathfrak{G}_\mathfrak{C}(\Omega, \mu; X)$ and it holds good that

$$f(\omega, x) = \sup_n \psi_n(\omega, x).$$

□

We define another sequence $\{\theta_n\}$ of functions by making use of $\{\psi_n\}$ obtained above:

$$\theta_1 = \psi_1, \theta_2 = \text{Max}\{\psi_1, \psi_2\}, \cdots, \theta_n = \text{Max}\{\psi_1, \psi_2, \cdots, \psi_n\}, \cdots.$$

Then $\{\theta_n\}$ is a monotonically increasing sequence of nonnegative Carathéodory functions, which converges to $f(\omega, x)$ pointwise.

---

[59]Let $(X, \mathscr{E})$ be a measurable space and Y a Souslin space. Then (π_X = projection of $\Omega \times X$ into X)

$$G \in \mathscr{E} \otimes \mathscr{B}(Y) \Rightarrow \pi_X G \in \hat{\mathscr{E}},$$

where $\hat{\mathscr{E}}$ is the universal completion of $\mathscr{E}$. In Theorem 3.1, we do not need the universal completion since we are assuming that $(\Omega, \mathscr{E}, \mu)$ is complete. cf. Castaing–Valadier [13, pp.75–80], Maruyama [36, pp. 411–426].

Corollary 3.1 *Under the same assumptions as in Theorem 3.1, there exists a monotonically increasing sequence* $\{\psi_n\}$ *of nonnegative Carathéodory functions which satisfies*

$$f(\omega, x) = \lim_{n\to\infty} \psi_n(\omega, x). \tag{3.9}$$

Remark 3 Restricting Ω and X to more specific spaces, we obtain further results.

1° If $(\Omega, \mathscr{E}, \mu)$ is a complete finite measure space and X a locally compact Polish space, the following two statements are equivalent for a function $f : \Omega \times X \to \overline{\mathbb{R}}$.

(i) $f \in \mathfrak{S}(\Omega, \mathscr{E}; X)$.
(ii) The correspondence

$$\omega \mapsto\!\!\!\rightarrow \text{Epi} f(\omega) = \{(x, \alpha) \in X \times \mathbb{R} | f(\omega, x) \leqq \alpha\}$$

is closed-valued and measurable.

2° Suppose that Ω is a locally compact Polish space, μ a (positive) Radon measure on Ω, and X a Polish space. Then the following two statements are equivalent for a function $f : \Omega \times X \to \overline{\mathbb{R}}$.

(i) $f \in \mathfrak{S}(\Omega, \mathscr{B}(\Omega)_\mu; X)$.
(ii) For any compact set $K \subset \Omega$ and arbitrary $\varepsilon > 0$, there exists some compact set $H \subset K$ such that

$$\mu(K \setminus H) < \varepsilon \quad \text{and} \quad f|_{H\times X} \text{ is } l.s.c.$$

3° Under the same assumptions as in 2°, the following two statements are equivalent for a function $f : \Omega \times X \to \overline{\mathbb{R}}$.

(i) $f \in \mathfrak{S}_+(\Omega, \mathscr{E}; X)$.
(ii) There exists a sequence $\{\psi_n\}$ of integrable nonnegative Carathéodory functions such that

$$f(\omega, x) = \sup_n \psi_n(\omega, x).$$

That is, the converse assertion of Theorem 3.1 holds good in this case.

Remark 4 Valadier [48] provided an alternative proof of the same result as our Corollary 3.1. His proof is based upon Urysohn's embedding theorem.

Urysohn's Embedding Theorem *Any separable and metrizable topological space is homeomorphic to a subset of certain metrizable and compact topological space* $\hat{X}$.

A sequence $\{\tilde{\psi}_n : \Omega \times \hat{X} \to \mathbb{R}\}$ of functions is defined by

$$\tilde{\psi}_n(\omega, z) = \text{Min}\{n, f_n(\omega, z)\}$$

where

$$f_n(\omega, z) = \inf_{x \in X}\{f(\omega, x) + n\rho(x, z)\}, \quad n = 1, 2, \cdots.$$

Of course, ρ is a metric on $\hat{X}$ which is compatible with its topology. Then the restriction $\tilde{\psi_n}|_{\Omega \times X}$ of $\tilde{\psi_n}$ to $\Omega \times X$ enjoys the desired properties.

Theorem 3.2 *Let $(\Omega, \mathscr{E}, \mu)$ be a complete finite measure space and X a metrizable Souslin space.*

(i) For any $f \in \mathfrak{G}_+(\Omega, \mathscr{E}; X)$, the function

$$\gamma \mapsto \int_{\Omega \times X} f(\omega, x) d\gamma$$

is l.s.c. on $\mathfrak{Y}(\Omega, \mu; X)$ w.r.t. the narrow topology.

(ii) The narrow topology on $\mathfrak{Y}(\Omega, \mu; X)$ coincides with the topology generated by the family of functions

$$\gamma \mapsto \int_{\Omega \times X} g|_{\Omega \times X} d\gamma; \quad g \in \mathfrak{G}_{\mathfrak{C}}(\Omega, \mu; \hat{X}).$$[60]

Proof

(i) Let f be a nonnegative normal integrand. By Corollary 3.1 there exists a monotonically increasing sequence $\{\psi_n : \Omega \times X \to \mathbb{R}\}$ of nonnegative Carathéodory functions which converges to f pointwise. The function

$$\gamma \mapsto \int_{\Omega \times X} \psi_n d\gamma$$

is continuous on $\mathfrak{Y}(\Omega, \mu; X)$ by the definition of the narrow topology. Since f is a monotonically increasing limit of $\{\psi_n\}$,

$$\int_{\Omega \times X} f d\gamma = \sup_n \int_{\Omega \times X} \psi_n d\gamma,$$

which proves (i).

(ii) Let $\mathscr{T}_1$ be the narrow topology on $\mathfrak{Y}(\Omega, \mu; X)$, and $\mathscr{T}_2$ the topology generated by

$$\gamma \mapsto \int_{\Omega \times X} g|_{\Omega \times X} d\gamma \;\; ; \;\; g \in \mathfrak{G}_{\mathfrak{C}}(\Omega, \mu; \hat{X}).$$

[60] $\hat{X}$ is a metrizable compact topological space in which X is embedded homeomorphically.

We have to prove that $\mathscr{T}_1 = \mathscr{T}_2$. It is clear that $\mathscr{T}_2 \subset \mathscr{T}_1$ since

$$\{h : \Omega \times X \to \mathbb{R} | h = g|_{\Omega \times X}, g \in \mathfrak{G}_{\mathfrak{C}}(\Omega, \mu; \hat{X})\} \subset \mathfrak{G}_{\mathfrak{C}}(\Omega, \mu; X).$$

Let $f \in \mathfrak{G}_{\mathfrak{C}}(\Omega, \mu; X)$. If we define

$$\alpha(\omega) = \|f(\omega, \cdot)\|_\infty,$$

it is obvious that $f(\omega, x) + \alpha(\omega) \geqq 0$.

In the same manner as in Theorem 3.1 and Corollary 3.1, we can construct a sequence $\{\psi_n : \Omega \times \hat{X} \to \mathbb{R}\}$ in $\mathfrak{G}_{\mathfrak{C}}(\Omega, \mu; \hat{X})$ such that (a)$\psi_n \geqq 0$, (b)$\{\psi_n\}$ is monotonically increasing, and (c) $\{\psi_n|_{\Omega \times X}\}$ converges to $f(\omega, x) + \alpha(\omega)$. If $\mathfrak{Y}(\Omega, \mu; X)$ is endowed with the topology $\mathscr{T}_2$, then

$$\gamma \mapsto \int_{\Omega \times X} \{f(\omega, x) + \alpha(\omega)\} d\gamma$$

is *l.s.c.* (*w.r.t.* $\mathscr{T}_2$) on $\mathfrak{Y}(\Omega, \mu; X)$ since

$$\gamma \mapsto \int_{\Omega \times X} \psi_n|_{\Omega \times X} d\gamma$$

is continuous, by assumption. Clearly we obtain

$$\int_{\Omega \times X} f(\omega, x) d\gamma = \int_{\Omega \times X} \{f(\omega, x) + \alpha(\omega)\} d\gamma - \int_{\Omega} \alpha(\omega) d\mu.$$

The second term of the right-hand side is independent of γ. Consequently,

$$\gamma \mapsto \int_{\Omega \times X} f(\omega, x) \alpha\gamma \tag{3.10}$$

is *l.s.c.* (*w.r.t.* $\mathscr{T}_2$). If $f \in \mathfrak{G}_{\mathfrak{C}}(\Omega, \mu; X)$, then $-f \in \mathfrak{G}_{\mathfrak{C}}(\Omega, \mu; X)$ since $\mathfrak{G}_{\mathfrak{C}}(\Omega, \mu; X)$ is a vector space. Therefore

$$\gamma \mapsto \int_{\Omega \times X} -f(\omega, x) d\gamma$$

is also *l.s.c.* (*w.r.t.*, $\mathscr{T}_2$). That is, (3.10) is *u.s.c.*. This proves the continuity of (3.10) (*w.r.t.*, $\mathscr{T}_2$) and so $\mathscr{T}_1 \subset \mathscr{T}_2$. □

The second statement of the above theorem has a remarkable implication. Since $\mathfrak{C}(\hat{X}, \mathbb{R})$ is separable,[61] $\mathfrak{G}_{\mathfrak{C}}(\Omega, \mu; \hat{X}) \cong \mathfrak{L}^1(\Omega, \mathfrak{C}(\hat{X}, \mathbb{R}))$ is also separable provided

[61] Dunford–Schwartz [21, p. 340], Maruyama [34, pp. 155–157].

that $\mathscr{E}$ has a countable base.[62] Hence the unit ball of $\mathfrak{L}^\infty(\Omega, \mathfrak{M}(\hat{X}))$ is metrizable *w.r.t.* w^*-topology. Consequently, $\mathfrak{P}(\Omega, \mu; X)$ viewed as a subset of $\mathfrak{P}(\Omega, \mu; \hat{X})$ is also metrizable. Thus **arguments in terms of "sequences" are justified when we talk about the narrow topology provided that $\mathscr{E}$ has a countable base.**

3.2 *Convergence of Young Measures Associated with Measurable Functions*

We now examine some general relations between the convergence of measurable functions and that of Young measures associated with them.[63]

Theorem 3.3 *Let $(\Omega, \mathscr{E}, \mu)$ be a finite measure space, (X, ρ) a separable metric space, and $\{u_n : \Omega \to X\}$ a sequence of $(\mathscr{E}, \mathscr{B}(X))$-measurable functions. $\{\gamma_n\}$ is a sequence of Young measures on $(\Omega \times X, \mathscr{E} \otimes \mathscr{B}(X))$ associated with $\{u_n\}$; i.e.*

$$\gamma_n = \int_\Omega \delta_\omega \otimes \delta_{u_n(\omega)} d\mu; \quad n = 1, 2, \cdots. \tag{3.11}$$

Then the following two statements are equivalent.

(i) $\{u_n\}$ converges to some $(\mathscr{E}, \mathscr{B}(X))$-measurable function $u_* : \Omega \to X$ in measure.
(ii) $\{\gamma_n\}$ converges to

$$\gamma_* = \int_\Omega \delta_\omega \otimes \delta_{u_*(\omega)} d\mu$$

in the narrow topology.[64]

Proof We first note that $\{\delta_{u(\omega)} | \omega \in \Omega\}$is a measurable family since

$$\omega \mapsto \int_X f(x) d\delta_{u_n(\omega)} = f(u_n(\omega))$$

is measurable for any $f \in \mathfrak{C}^b(X, \mathbb{R})$. (cf. Theorem 2.1)

[62] It can be proved in a similar way as in Maruyama [36, pp. 230–231].

[63] I am much indebted to Valadier [48, pp. 160–165] for the expositions of the basic results included in subsections 3.2–3.4 except Theorem 3.4 and Theorem 3.7.

[64] The convergence in the narrow topology is sometimes expressed as the "narrow convergence". We also use casual expressions like "narrowly converges". I hope no confusion occurs by such informal terminologies.

(i)⇒ (ii): Suppose that $u_n \to u_*$ as $n \to \infty$ (in measure). Then

$$\int_{\Omega \times X} f d\gamma_n = \int_{\Omega} f(\omega, u_n(\omega)) d\mu \tag{3.12}$$

$$\to \int_{\Omega \times X} f d\gamma_* = \int_{\Omega} f(\omega, u_*(\omega)) d\mu \quad \text{as } n \to \infty$$

for any $f \in \mathfrak{S}_{\mathfrak{C}}(\Omega, \mu; X)$. If this is not true, there exist some $f \in \mathfrak{S}_{\mathfrak{C}}(\Omega, \mu; X)$, $\varepsilon > 0$ and a subsequence $\{u_{n'}\}$ of $\{u_n\}$ which satisfy

$$\left| \int_{\Omega} f(\omega, u_{n'}(\omega)) d\mu - \int_{\Omega} f(\omega, u_*(\omega)) d\mu \right| \geqq \varepsilon. \tag{3.13}$$

Since $\mu\Omega < \infty$, $\{u_{n'}\}$ has a further subsequence $\{u_{n''}\}$ which converges to $u_*(\omega)$ *a.e.*; i.e.

$$u_{n''}(\omega) \to u_*(\omega) \ \ a.e.$$

It follows that

$$f(\omega, u_{n''}(\omega)) \to f(\omega, u_*(\omega)) \quad a.e. \quad as \ n'' \to \infty$$

since f is a Carathéodory function. Clearly $|f(\omega, u_{n''}(\omega))| \leqq \|f(\omega, \cdot)\|_\infty$ and the right-hand side is integrable. So we obtain

$$\int_{\Omega} f(\omega, u_{n''}(\omega)) d\mu \to \int_{\Omega} f(\omega, u_*(\omega)) d\mu \quad \text{as } n'' \to \infty, \tag{3.14}$$

by the dominated convergence theorem. However (3.13) and (3.14) contradict.

(ii)⇒(i): Suppose conversely that $\gamma_n \to \gamma_*$ (narrowly). If we define a function $f : \Omega \times X \to \mathbb{R}$ by

$$f(\omega, x) = \text{Min}\{1, \rho(x, u_*(\omega))\},$$

$0 \leqq f \in \mathfrak{S}_{\mathfrak{C}}(\Omega, \mu; X)$. Hence we obtain, by the definition of the narrow topology, that

$$\int_{\Omega \times X} f d\gamma_n = \int_{\Omega} \text{Min}\{1, \rho(u_n(\omega), u_*(\omega))\} d\mu$$

$$\to \int_{\Omega \times X} f d\gamma_* = 0 \quad \text{as } n \to \infty.$$

It follows that

$$\varepsilon \cdot \mu\{\omega \in \Omega | \rho(u_n(\omega), u_*(\omega)) \geqq \varepsilon\} \leqq \int_{\Omega \times X} f d\gamma_n \to 0 \quad \text{as } n \to \infty. \tag{3.15}$$

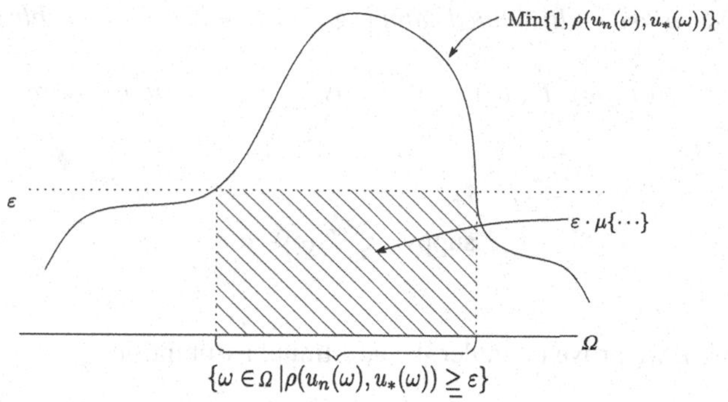

Fig. 3 The situation of (3.15)

(In the case $\varepsilon > 1$, the left-hand side of (3.15) is zero.) By (3.15), we obtain $u_n \to u_*$ (in measure). □

Suppose that a sequence $\{\gamma_n\}$ of Young measures of the form (3.11) narrowly converges to some Young measure γ_* which is not necessarily associated with a measurable function, what can we know about the limit measure γ_*? I now give a few answers to this question (Fig. 3).

Theorem 3.4 *Let $(\Omega, \mathscr{E}, \mu)$ be a complete finite measure space and X a metrizable Souslin space. Suppose that a sequence $\{\gamma_n\}$ of Young measures of the form (3.11) narrowly converges to some*

$$\gamma = \int_{\Omega} \delta_\omega \otimes \nu_\omega d\mu.^{65}$$

Then the following propositions hold good.

(i) $\mathrm{supp}\nu_\omega \subset L_s(u_n(\omega))$ *a.e.*[66]

---

[65]In this framework, any Young measure can be expressed in the form of disintegration (cf. Theorem 2.8). However note that the limit measure γ is not necessarily associated with a function.

[66]Let $\{M_n\}$ be a sequence of sets in a metric space X. The **topological superior limit** $L_s(M_n)$ and the **topological inferior limit** $L_i(M_n)$ of $\{M_n\}$ are defined as follows, respectively,

$$x \in L_s(M_n) \Leftrightarrow \text{For any neighborhood } V \text{ of } x,$$
$$V \cap M_n \neq \emptyset \text{ for infinitely many } n.$$
$$x \in L_i(M_n) \Leftrightarrow \text{For any neighborhood } V \text{ of } x,$$
$$\text{there exists some } n_0 \in \mathbb{N} \text{ such that}$$
$$V \cap M_n \neq \emptyset \text{ for all } n \geqq n_0.$$

(ii) If a closed-valued multi-valued mapping $\Gamma : \Omega \twoheadrightarrow X$ is measurable and satisfies

$$\rho(u_n(\omega), \Gamma(\omega)) \to 0^{67} \quad \textit{as } n \to \infty \quad \textit{(in measure)},$$

then

$$\mathrm{supp}\nu_\omega \subset \Gamma(\omega) \ \ a.e.$$

Proof

1° To start with, we prove (i) under the additional assumption

$$u_n(\omega) \in \Gamma(\omega) \ \ a.e. \quad \text{for all } n. \tag{3.16}$$

If we define a function $f : \Omega \times X \to \overline{\mathbb{R}}$ by

$$f(\omega, x) = \begin{cases} 0 & \text{if } \ x \in \Gamma(\omega), \\ \infty & \text{if } \ x \notin \Gamma(\omega), \end{cases}$$

it follows from Theorem 3.2 that

$$\begin{aligned} \int_{\Omega \times X} f(\omega, x) d\gamma &\leqq \liminf_n \int_{\Omega \times X} f(\omega, x) d\gamma_n \\ &= \liminf_n \int_\Omega f(\omega, u_n(\omega)) d\mu \\ &= 0. \quad \text{(by (3.16))} \end{aligned} \tag{3.17}$$

By the definition of f and (3.17), we obtain

$$\int_\Omega \{\int_X f(\omega, x) d\nu_\omega\} d\mu = 0,$$

which implies

$$\int_X f(\omega, x) d\nu_\omega = 0 \quad a.e.$$

Hence $\nu_\omega(X \setminus \Gamma(\omega)) = 0 \ \ a.e.,$ i.e.

$$\mathrm{supp}\nu_\omega \subset \Gamma(\omega) \quad a.e.$$

[67] $\rho(u_n(\omega), \Gamma(\omega))$ is the distance between the point $u_n(\omega)$ and the set $\Gamma(\omega)$.

2° We now prove (i) without the additional assumption (3.16). Define a sequence $\{\Gamma_p : \Omega \twoheadrightarrow X\}$ of correspondences by

$$\Gamma_p(\omega) = cl.\{u_n(\omega) | n \geqq p\}, \quad p = 1, 2, \cdots .$$

Then Γ_p is a closed-valued and measurable correspondence,[68] which satisfies

$$u_n(\omega) \in \Gamma_p(\omega) \quad \text{for all} \quad n \geqq p.$$

It follows from 1° that

$$\begin{aligned} &\nu_\omega(X \setminus \Gamma_p(\omega)) = 0 \quad a.e. \quad \text{for all} \quad p; \\ &\text{i.e.} \quad \text{supp}\nu_\omega \subset \Gamma_p(\omega) \quad a.e. \quad \text{for all} \quad p. \end{aligned} \tag{3.18}$$

Since $L_s(u_n(\omega)) = \bigcap_{p=1}^{\infty} \Gamma_p(\omega)$,

$$\text{supp}\nu_\omega \subset L_s(u_n(\omega)) \quad a.e.$$

by (3.18).

3° It remains to show (ii). There exists a subsequence $\{u_{n'}\}$ of $\{u_n\}$ such that

$$\rho(u_{n'}(\omega), \Gamma(\omega)) \to 0 \quad a.e. \quad \text{as} \quad n' \to \infty. \tag{3.19}$$

Since any element x of $L_s(u_{n'}(\omega))$ is a limit of some subsequence of $\{u_{n'}(\omega)\}$, (3.19) implies

$$L_s(u_{n'}(\omega)) \subset \Gamma(\omega) \ a.e.$$

We know that $\text{supp}\nu_\omega \subset L_s(u_{n'}(\omega))$ $a.e.$ by (i). Thus we conclude that

$$\text{supp}\nu_\omega \subset \Gamma(\omega) \ a.e. \qquad \square$$

3.3 *Tightness*

The relation between the uniform tightness and the relative compactness of a set of Borel probability measures on a topological space is quite well-known.[69] In this subsection, we are going to examine similar problems in $\mathfrak{Y}(\Omega, \mu; X)$.

[68] See the footnote 37.

[69] See Billingsley [8, pp. 35–41] and Maruyama [36, pp. 334–340].

Definition 3.3 Let $(\Omega, \mathscr{E}, \mu)$ be a finite measure space and X a Hausdorff topological space. A subset H of $\mathfrak{Y}(\Omega, \mu; X)$ is said to be **uniformly tight** if there exists a compact subset K_ε of X, for each $\varepsilon > 0$, such that

$$\sup_{\gamma \in H} \gamma(\Omega \times (X \setminus K_\varepsilon)) \leqq \varepsilon. \tag{3.20}$$

Remark 5 If we assume, in addition, that μ is complete and X is a metrizable Souslin space, then the uniform tightness of H and that of $\bar{H}$ (the closure of H *w.r.t*, the narrow topology) are equivalent.

It is enough to show that the uniform tightness of H implies that of $\bar{H}$. (The converse is obvious.) If H is uniformly tight, there exists a compact subset K_ε of H, for each $\varepsilon > 0$, such that (3.20) is satisfied. If we define a function $f : \Omega \times X \to \mathbb{R}$ by

$$f(\omega, x) = \chi_{\Omega \times (X \setminus K_\varepsilon)}(\omega, x),$$

then $f \in \mathfrak{G}_+(\Omega, \mathscr{E}; X)$. It is easy to see that, for $\gamma \in \mathfrak{Y}(\Omega, \mu; X)$,

$$\int_{\Omega \times X} f(\omega, x) d\gamma \leqq \varepsilon \Leftrightarrow \gamma(\Omega \times (X \setminus K_\varepsilon)) \leqq \varepsilon.$$

Thanks to Theorem 3.2 (i),

$$\gamma(\Omega \times (X \setminus K_\varepsilon)) \leqq \varepsilon \quad \text{for all} \quad \gamma \in \bar{H}.$$

Theorem 3.5 *Let $(\Omega, \mathscr{E}, \mu)$ be a complete finite measure space, and X a metrizable Souslin space. If $H \subset \mathfrak{Y}(\Omega, \mu; X)$ is uniformly tight, then the following statements hold good.*

(i) H is sequentially compact.
(ii) H is relatively compact.
(iii) If X is a metrizable, separable and locally compact topological space, then the narrow topology on H coincides with the topology generated by

$$\gamma \mapsto \int_{\Omega \times X} f d\gamma; \ f \in \mathfrak{G}_{\mathfrak{C}_\infty}(\Omega, \mu; X).$$

Proof

(i) X can be embedded in a metrizable compact space $\hat{X}$ (Urysohn's embedding theorem). Let

$$\gamma_n = \int_\Omega \delta_\omega \otimes \nu_\omega^n d\mu \ ; \ n = 1, 2, \cdots$$

be a sequence in H. If we define

$$\widehat{\nu_\omega^n}(\hat{B}) = \nu_\omega^n(\hat{B} \cap X); \ \ \hat{B} \in \mathscr{B}(\hat{X}),$$

then $\widehat{\nu^n} = \{\widehat{\nu_\omega^n} | \omega \in \Omega\} \in \mathfrak{P}(\Omega, \mu; \hat{X})$. By Theorem 2.7, $\mathfrak{P}(\Omega, \mu; \hat{H})$ is a w^*-compact set contained in the unit ball in $\mathfrak{L}^\infty(\Omega, \mathfrak{M}(\hat{X}))$.[70] There is an homeomorphism Φ of $\mathfrak{P}(\Omega, \mu; \hat{X})$ onto $\mathfrak{Y}(\Omega, \mu; \hat{X})$ as discussed in Theorem 2.7. Hence $\mathfrak{Y}(\Omega, \mu; \hat{X})$ is narrowly compact.

Let $\{\widehat{\nu^{n'}}\}$ be a w^*-convergent subsequence of $\{\widehat{\nu^n}\}$ and $\hat{\nu}$ its limit. By the continuity of Φ,

$$\widehat{\gamma_{n'}} = \int_\Omega \delta_\omega \otimes \widehat{\nu_\omega^{n'}} d\mu; \ \ n = 1, 2, \cdots$$

narrowly converges to

$$\hat{\gamma} = \int_\Omega \delta_\omega \otimes \widehat{\nu_\omega} d\mu.$$

There exists a compact set $K_p \subset X$, for each $p \geqq 1$, such that

$$\gamma_n(\Omega \times (X \setminus K_p)) \leqq \frac{1}{p} \ \ \text{for all } n.$$[71]

The function

$$x \mapsto \chi_{\Omega \times (\hat{X} \setminus K_p)}(\omega, x)$$

is *l.s.c.* Hence, by Theorem 3.2 (i),

$$\hat{\gamma}(\Omega \times (\hat{X} \setminus K_p)) \leqq \frac{1}{p}.$$

It follows that

$$\hat{\gamma}([\Omega \times \bigcup_{p=1}^{\infty} K_p]^c) = 0.$$

---

[70]The w^*-topology is, of course, based upon the duality relation

$$\mathfrak{L}^1(\Omega, \mathfrak{C}(\hat{X}, \mathbb{R}))' \cong \mathfrak{L}^\infty(\Omega, \mathfrak{M}(\hat{X})).$$

[71]Since $\{\gamma_1, \gamma_2, \cdots\}$ is uniformly tight, we can choose a common K_p for all γ_n.

We now define $\nu = \{\nu_\omega | \omega \in \Omega\} \in \mathfrak{P}(\Omega, \mu; X)$ by[72]

$$\nu_\omega = \widehat{\nu_\omega}|_{\mathscr{B}(X)}$$

and

$$\gamma = \int_{\Omega \times X} \delta_\omega \otimes \nu_\omega d\mu.$$

Then we obtain, for any $f \in \mathfrak{G}_{\mathfrak{C}}(\Omega, \mu; \hat{X})$,

$$\int_{\Omega \times \hat{X}} f d\widehat{\gamma_{n'}} = \int_{\Omega \times X} f|_{\Omega \times X} d\gamma_{n'}, \tag{3.21}$$

$$\int_{\Omega \times \hat{X}} f d\hat{\gamma} = \int_{\Omega \times X} f|_{\Omega \times X} d\gamma. \tag{3.22}$$

Since $\widehat{\gamma_{n'}} \to \hat{\gamma}$ (narrowly),

$$\int_{\Omega \times \hat{X}} f d\widehat{\gamma_{n'}} \to \int_{\Omega \times \hat{X}} f d\hat{\gamma} \quad \text{as} \quad n' \to \infty. \tag{3.23}$$

By (3.21), (3.22) and (3.23),

$$\int_{\Omega \times X} f|_{\Omega \times X} d\gamma_{n'} \to \int_{\Omega \times X} f|_{\Omega \times X} d\gamma \quad \text{as} \quad n' \to \infty. \tag{3.24}$$

(3.24) holds good for all $f \in \mathfrak{G}_{\mathfrak{C}}(\Omega, \mu; \hat{X})$. Therefore $\gamma_{n'} \to \gamma$ (narrowly) by Theorem 3.2 (ii).
(ii) A similar argument as in (i) applies if we make use of a net $\{\gamma_\alpha\}$ instead of a sequence $\{\gamma_n\}$.
(iii) In this case,

$$\begin{aligned} \mathfrak{G}_{\mathfrak{C}_\infty}(\Omega, \mu; X) &\cong \mathfrak{L}^1(\Omega, \mathfrak{C}_\infty(X, \mathbb{R})) \\ &\subset \mathfrak{G}_{\mathfrak{C}}(\Omega, \mu; X) \end{aligned}$$

by Theorem 2.4. Hence the topology on $\mathfrak{Y}(\Omega, \mu; X)$ generated by

$$\gamma \mapsto \int_{\Omega \times X} f d\gamma; \quad f \in \mathfrak{G}_{\mathfrak{C}_\infty}(\Omega, \mu; X) \tag{3.25}$$

[72] $\widehat{\nu_\omega}|_{\mathscr{B}(X)}$ is the restriction of $\hat{\nu}_\omega$ to $\mathscr{B}(X)$.

is weaker than the narrow topology. Keeping the compactness of $\bar{H}$ in mind (by (ii)), we consider the identity mapping $I : \bar{H} \to \bar{H}$, where the domain is endowed with the narrow topology and the range with the topology generated by (3.25). I is a continuous bijection on the compact domain $\bar{H}$. Therefore I is a homeomorphism. □

Remark 6 When X is a Polish space, a partial converse of Theorem 3.5 holds good:

$$H \text{ is relatively compact w.r.t, the narrow topology} \Rightarrow H \text{ is uniformly tight.}$$

We can verify this fact as follows. Define a function $S : \mathfrak{Y}(\Omega, \mu; X) \to \mathfrak{M}(X)$ by

$$S : \gamma \mapsto \gamma \circ \pi_X^{-1}.$$

The domain is endowed with the narrow topology and the range with w^*-topology. Then S is continuous.[73] $S(H)$ is uniformly tight since it is relatively compact in $\mathfrak{M}_+(X)$.[74] That is, there exists some compact set $K_\varepsilon \subset X$, for each $\varepsilon > 0$, such that

$$S(\gamma)(X \setminus K_\varepsilon) = \gamma(\Omega \times (X \setminus K_\varepsilon)) \leqq \varepsilon \quad \text{for all} \quad \gamma \in H.$$

Thus H is uniformly tight.

Let $\mathscr{U}_H$ be a family of measurable functions of Ω into X, and H a set of Young measures determined by members of $\mathscr{U}_H$:

$$H = \{\int_\Omega \delta_\omega \otimes \delta_{u(\omega)} d\mu | u \in \mathscr{U}_H\}.$$

The family $\mathscr{U}_H$ of functions is said to be **uniformly tight** if H is uniformly tight.

Theorem 3.6 *Let $(\Omega, \mathscr{E}, \mu)$ be a finite measure space, and X a metric space. The following three statements are equivalent for a set $\mathscr{U} = \{u : \Omega \to X\}$ of measurable functions.*

(i) $\mathscr{U}$ is uniformly tight.

---

[73] Assume that a net $\{\gamma_\alpha\}$ in $\mathfrak{Y}(\Omega, \mu; X)$ converges to γ_*. If $f \in \mathfrak{C}^b(X, \mathbb{R})$ and $1(\omega)$ is the constant function on Ω which is identically 1 for all $\omega \in \Omega$, then $1(\omega) f(x) \in \mathfrak{G}_{\mathfrak{C}}(\Omega, \mu; X)$

$$\int_X f(x) dS(\gamma_\alpha) = \int_{\Omega \times X} 1(\omega) f(x) d\gamma_\alpha$$
$$\to \int_{\Omega \times X} 1(\omega) f(x) d\gamma_* = \int_X f(x) dS(\gamma_*).$$

This proves the continuity of S.

[74] Billingsley [8, pp. 10 12], Maruyama [36, pp. 334–340]. See also Piccinini–Valadier [41].

(ii) There exists a compact set $K_\varepsilon \subset X$, for each $\varepsilon > 0$, such that

$$\sup_{u \in \mathscr{U}} \mu\{\omega \in \Omega | u(\omega) \notin K_\varepsilon\} \leqq \varepsilon.$$

(iii) There exists a function $\psi : X \to [0, \infty]$ which satisfies a and b.

a. $\{x \in X | \psi(x) \leqq \alpha\}$ is compact for any $\alpha \in \mathbb{R}$.

b. $\sup_{u \in \mathscr{U}} \int_\Omega \psi(u(\omega)) d\mu < \infty$.

Remark 7 A function ψ is said to be **inf-compact** if it satisfies the condition a. In this case, ψ is automatically measurable.

Proof of Theorem 3.6 (i)$\Rightarrow$ (ii): We denote

$$\mathfrak{Y}_{\mathscr{U}} = \{\int_\Omega \delta_\omega \otimes \delta_{u(\omega)} d\mu | u \in \mathscr{U}\}.$$

By (i), there exists a compact set $K_\varepsilon \subset X$, for each $\varepsilon > 0$, such that

$$\sup_{\gamma \in \mathfrak{Y}_{\mathscr{U}}} \gamma(\Omega \times (X \setminus K_\varepsilon)) \leqq \varepsilon. \tag{3.26}$$

Consequently, it follows that

$$\begin{aligned}
\gamma(\Omega \times (X \setminus K_\varepsilon)) &= \int_\Omega \{\int_X \chi_{X \setminus K_\varepsilon}(x)\} d(\delta_\omega \otimes \delta_{u(\omega)}\} d\mu \\
&= \int_\Omega \chi_{X \setminus K_\varepsilon}(u(\omega)) d\mu \\
&= \mu\{\omega \in \Omega | u(\omega) \notin K_\varepsilon\} \\
&\underset{(3.26)}{\leqq} \varepsilon \quad \text{for any} \quad \gamma = \int_\Omega \delta_\omega \otimes \delta_{u(\omega)} d\mu \in \mathfrak{Y}_{\mathscr{U}}.
\end{aligned}$$

This proves (ii).

(ii)$\Rightarrow$(iii): Assume (ii). Then there exists a compact set $K_n \subset X$, for each $n \in \mathbb{N}$, such that

$$\sup_{u \in \mathscr{U}} \mu\{\omega \in \Omega | u(\omega) \notin K_n\} \leqq \frac{1}{2^n}.$$

We may assume, without loss of generality, that $K_n \subset K_{n+1}$. Define a function $\psi : X \to [0, \infty]$ by

$$\psi(x) = \sum_{n=1}^{\infty} n \cdot \chi_{X \setminus K_n}(x).$$

Then $\psi(x) \in \{0, 1, 2, \cdots\}$. This function ψ is inf-compact, and so satisfies the condition a since

$$K_{n+1} = \{x \in X | \psi(x) \leqq 1 + 2 + \cdots + n\}.$$

ψ also satisfies the condition b because

$$\int_X \psi(u(\omega)) d\mu \leqq \sum_{n=1}^{\infty} n \cdot \frac{1}{2^n} < \infty$$

for any $u \in \mathcal{U}$.

(iii)⇒(i): Suppose that there exists a function ψ which satisfies the condition (iii). Then

$$\alpha\gamma(\Omega \times (X \setminus K_\alpha)) \leqq \int_\Omega \psi(u(\omega)) d\mu \leqq M < \infty$$

for any $\alpha > 0$ and $\gamma \in \mathfrak{Y}_{\mathcal{U}}$. Hence

$$\sup_{\gamma \in \mathfrak{Y}_{\mathcal{U}}} \gamma(\Omega \times (X \setminus K_\varepsilon)) \to 0 \quad \text{as} \quad \alpha \to \infty.$$

□

The following result due to Evans [22, pp.16–17] is an easy consequence of Theorem 3.5 and Theorem 3.6 (Fig. 4).

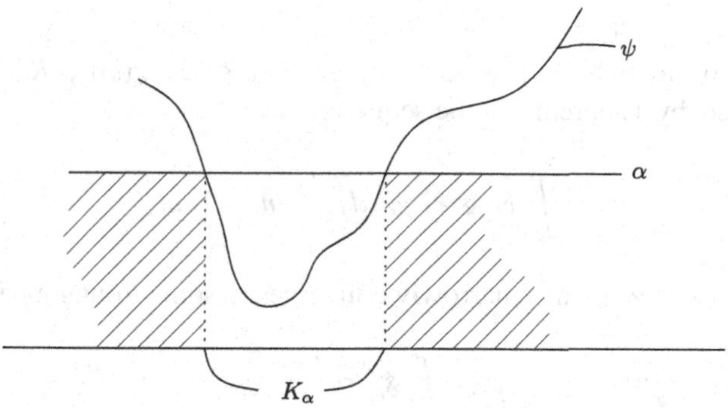

Fig. 4 $\alpha\gamma(\Omega \times (X \setminus K_\alpha))$

Theorem 3.7 (Evans) *Let $(\Omega, \mathscr{E}, \mu)$ be a complete finite measure space, and $\{u_n\}$ a bounded*[75] *sequence in $\mathfrak{L}^\infty(\Omega, \mathbb{R}^l)$. Then for any $F \in \mathfrak{C}(\mathbb{R}^l, \mathbb{R})$, there exist a subsequence $\{u_{n'}\}$ of $\{u_n\}$ and a measurable family $\{\nu_\omega | \omega \in \Omega\} \in \mathfrak{P}(\Omega, \mu; \mathbb{R}^l)$ which satisfy*

$$w^*\text{-} \lim_{n' \to \infty} F(u_{n'}) = \int_{\mathbb{R}^l} F(x) d\nu_\omega \quad a.e. \quad \omega \in \Omega \tag{3.27}$$

in $\mathfrak{L}^\infty(\Omega, \mathbb{R})$.[76] □

Proof Let g be any element of $\mathfrak{L}^1(\Omega, \mathbb{R})$. If we define a function $f : \Omega \times \mathbb{R}^l \to \mathbb{R}$ by

$$f(\omega, x) = g(\omega) F(x); \quad (\omega, x) \in \Omega \times \mathbb{R}^l,$$

then f is a Carathéodory function. Since $\{u_n\}$ is bounded,

$$u_n(\omega) \in \overline{B_r(0)} \equiv K \ a.e. \text{ for all } n$$

for a sufficiently large $r > 0$.[77]

There exists a function $\tilde{F} \in \mathfrak{C}(\mathbb{R}^l, \mathbb{R})$ which satisfies

a. $\tilde{F}(x) = F(x)$ on K,
b. $\text{supp}\tilde{F}$ is compact.

Furthermore we define a function $\tilde{f} : \Omega \times \mathbb{R}^l \to \mathbb{R}$ by

$$\tilde{f}(\omega, x) = g(\omega) \tilde{F}(x).$$

Then $\tilde{f} \in \mathfrak{G}_{\mathfrak{C}}(\Omega, \mu; \mathbb{R}^l)$

$\{u_n\}$ is uniformly tight by Theorem 3.6, since $\{\omega \in \Omega | u_n(\omega) \notin K\}$ is measure 0 for all n. Hence, by Theorem 3.5, the sequence

$$\gamma_n = \int_\Omega \delta_\omega \otimes \delta_{u_n(\omega)} d\mu; \quad n = 1, 2, \cdots$$

has a subsequence $\{\gamma_{n'}\}$ which narrowly converges to some Young measure

$$\gamma = \int_\Omega \delta_\omega \otimes \nu_\omega d\mu,$$

[75] That is, $\sup_n \|u_n\|_\infty < \infty$.

[76] $F(u_n) \in \mathfrak{L}^\infty(\Omega, \mathbb{R})$.

[77] $B_r(0)$ is the open ball with center 0 and radius r. $\overline{B_r(0)}$ is its closure.

where $\{\nu_\omega|\omega \in \Omega\} \in \mathfrak{P}(\Omega, \mu; \mathbb{R}^l)$. We have

$$\mathrm{supp}\nu_\omega \subset K \quad a.e.$$

by Theorem 3.4.

We now prove that this $\{\nu_\omega|\omega \in \Omega\}$ satisfies (3.27).

$$\begin{aligned}
\int_\Omega g(\omega)F(u_{n'}(\omega))d\mu &= \int_{\Omega\times\mathbb{R}^l} g(\omega)F(x)d\gamma_{n'} \\
&= \int_{\Omega\times\mathbb{R}^l} g(\omega)\tilde{F}(x)d\gamma_{n'} \quad (\mathrm{supp}\delta_{u_{n'}(\omega)} \subset K \;\; a.e.) \\
&= \int_{\Omega\times\mathbb{R}^l} \tilde{f}(\omega, x)d\gamma_{n'} \\
&\to \int_{\Omega\times\mathbb{R}^l} \tilde{f}(\omega, x)d\gamma \quad (\tilde{f} \in \mathfrak{G}_{\mathfrak{C}}(\Omega, \mu; \mathbb{R}^l)) \\
&= \int_{\Omega\times\mathbb{R}^l} g(\omega)\tilde{F}(x)d\gamma \\
&= \int_{\Omega\times\mathbb{R}^l} g(\omega)F(x)d\gamma \quad (\mathrm{supp}\nu_\omega \subset K).
\end{aligned}$$

This relation holds good for any $g \in \mathfrak{L}^1(\Omega, \mathbb{R})$. □

3.4 Multiple Young Measures

This chapter is devoted to a brief exposition of topological properties of "compounded" Young measures.[78] The first result concerns the "product" of two Young measures. The notations are a little bit complicated.

Theorem 3.8 *Let $(\Omega_i, \mathscr{E}_i, \mu_i)(i = 1, 2)$ be complete finite measure spaces, and X_i $(i = 1, 2)$ metrizable Souslin spaces. Furthermore γ_i $(i = 1, 2)$ is an element of $\mathfrak{Y}(\Omega_i, \mu_i; X_i)$ the disintegration of which is given by*

$$\gamma_i = \int_{\Omega_i} \delta_{\omega_i} \otimes \nu^i_{\omega_i} d\mu_i$$

[78]The exposition is basically due to Valadier [48, pp. 163–164].

by means of $\{\nu^i_{\omega_i} | \omega_i \in \Omega_i\} \in \mathfrak{P}(\Omega_i, \mu_i; X_i)$. *We define a Young measure* $\gamma_1 \otimes \gamma_2 \in \mathfrak{Y}(\Omega_1 \times \Omega_2, \mu_1 \otimes \mu_2; X_1 \times X_2)$ *as the one disintegrated by means of*

$$[\nu^1 \otimes \nu^2]_\omega = \nu^1_{\omega_1} \otimes \nu^2_{\omega_2} \in \mathfrak{P}(\Omega_1 \times \Omega_2, \mu_1 \otimes \mu_2, X_1 \times X_2).$$

If a sequence $\{\gamma_i{}^n\}$ $(i = 1, 2)$ *narrowly converges to* $\gamma_i{}^*$ *in* $\mathfrak{Y}(\Omega_i, \mu_i; X_i)$, *then*

$$\gamma_1{}^n \otimes \gamma_2{}^n \to \gamma_1{}^* \otimes \gamma_2{}^* \quad (\textit{narrowly}) \;\; \textit{as} \quad n \to \infty.$$

Proof We denote by $\hat{X}_i$ $(i = 1, 2)$ metrizable and compact spaces, in which X_i's are embedded in the sense of Urysohn. For the sake of simplicity of notations, we write

$$\Omega = \Omega_1 \times \Omega_2, \quad X = X_1 \times X_2, \quad \hat{X} = \hat{X}_1 \times \hat{X}_2.$$

We have to show that

$$\int_{\Omega \times X} f|_{\Omega \times X} d(\gamma_1{}^n \otimes \gamma_2{}^n) \qquad (3.28)$$
$$\to \int_{\Omega \times X} f|_{\Omega \times X} d(\gamma_1{}^* \otimes \gamma_2{}^*) \quad \text{as} \quad n \to \infty,$$

for any $f \in \mathfrak{G}_{\mathfrak{C}}(\Omega, \mu \otimes \mu_2; \hat{X})$ by Theorem 3.2 (ii).

We regard $\gamma_1 \otimes \gamma_2 \in \mathfrak{Y}(\Omega, \mu_1 \otimes \mu_2; X)$ as an operator on

$$\{h|_{\Omega \times X} | h \in \mathfrak{L}^1_{\mu_1 \otimes \mu_2}(\Omega, \mathfrak{C}(\hat{X}, \mathbb{R}))\}.$$

keeping in mind that $\mathfrak{G}_{\mathfrak{C}}(\Omega\, \mu_1 \otimes \mu_2; \hat{X})$ can be identified with $\mathfrak{L}^1_{\mu_1 \otimes \mu_2}(\Omega, \mathfrak{C}(\hat{X}, \mathbb{R}))$. We use a notation

$$\langle \gamma_1 \otimes \gamma_2, h|_{\Omega \times X} \rangle = \int_{\Omega \times X} h|_{\Omega \times X} d(\gamma_1 \otimes \gamma_2),$$

again for the sake of simplicity.

$$\begin{aligned} |\langle \gamma_1 \otimes \gamma_2, h|_{\Omega \times X} \rangle| &\leqq \int_\Omega \sup_{(\hat{x}_1, \hat{x}_2) \in X} |h(\omega_1, \omega_2)(\hat{x}_1, \hat{x}_2)| d(\mu_1 \otimes \mu_2) \\ &= \int_\Omega \|h(\omega_1, \omega_2)\|_\infty d(\mu_1 \otimes \mu_2) \qquad (3.29) \\ &= \|h\|_1. \end{aligned}$$

This evaluation will be used afterwards.

In order to prove (3.28), it is sufficient to consider only simple functions of the form

$$\varphi(\omega_1, \omega_2) = \sum_{j=1}^{m} \chi_{A_j}(\omega_1, \omega_2) f_j, \tag{3.30}$$

$$f_j \in \mathfrak{C}(\hat{X}, \mathbb{R}); \quad j = 1, 2, \cdots, m,$$
$$A_j \in \mathscr{E}_1 \otimes \mathscr{E}_2, \quad A_j \cap A_k = \emptyset \text{ if } j \neq k.$$

The following is the reason why this is so. Suppose that

$$\langle {\gamma_1}^n \otimes {\gamma_2}^n, \varphi|_{\Omega\times X}\rangle \to \langle {\gamma_1}^* \otimes {\gamma_2}^*, \varphi|_{\Omega\times X}\rangle \quad \text{as } n \to \infty \tag{3.31}$$

for any simple function of the form (3.30).

For any $h \in \mathfrak{L}^1_{\mu_1\otimes\mu_2}(\Omega, \mathfrak{C}(\hat{X}, \mathbb{R}))$, there exists a sequence $\{\varphi_p\}$ of simple functions such that $\|\varphi_p - h\|_1 \to 0$ (as $p \to \infty$). Since (3.31) holds good for every φ_p $(p = 1, 2, \cdots)$,

$$\begin{aligned}
&|\langle {\gamma_1}^n \otimes {\gamma_2}^n, h|_{\Omega\times X}\rangle - \langle {\gamma_1}^* \otimes {\gamma_2}^*, h|_{\Omega\times X}\rangle| \\
&\qquad \leqq |\langle {\gamma_1}^n \otimes {\gamma_2}^n - {\gamma_1}^* \otimes {\gamma_2}^*, h|_{\Omega\times X} - \varphi_p|_{\Omega\times X}\rangle| \\
&\qquad\qquad + |\langle {\gamma_1}^n \otimes {\gamma_2}^n - {\gamma_1}^* \times {\gamma_2}^*, \varphi_p|_{\Omega\times X}\rangle| \qquad (3.32)\\
&\qquad \underset{(3.29)}{\leqq} 2\|h|_{\Omega\times X} - \varphi_p|_{\Omega\times X}\|_1 + |\langle {\gamma_1}^n \otimes {\gamma_2}^n - {\gamma_1}^* \otimes {\gamma_2}^*, \varphi_p|_{\Omega\times X}\rangle|.
\end{aligned}$$

We used the evaluation (3.29) to obtain the last inequality. If we choose p sufficiently large for any $\varepsilon > 0$, the first term of (3.32) $< \varepsilon/2$. Fixing such a large p, we obtain that the second term of (3.32) $< \varepsilon/2$ for n large enough. Hence

$$\begin{aligned}
&|\langle {\gamma_1}^n \otimes {\gamma_2}^n, h|_{\Omega\times X}\rangle - \langle {\gamma_1}^* \otimes {\gamma_2}^*, h|_{\Omega\times X}\rangle| \\
&\qquad < \varepsilon/2 + \varepsilon/2 = \varepsilon \quad \text{for large } n.
\end{aligned}$$

Furthermore we can restrict our attention only to the following special cases[79]:

(a) each A_j is of the form

$$A_j = A_{j1} \times A_{j2}; \ A_{ji} \in \mathscr{B}(\hat{X}_i),$$
$$i = 1, 2; \ j = 1, 2, \cdots, m.$$

[79]This argument is sustained by the following two observations. The first is that the monotone family generated by the field consisting of finite unions of rectangles in $\hat{X}$ is equal to $\mathscr{B}(\hat{X})$. (cf. Maruyama [36, pp. 10–13]. The second is the well-known Stone-Weierstrass theorem. (cf. Maruyama [34, pp. 162–163].)

(b) each f_j is of "separable" type:

$$f_j(\hat{x}_1, \hat{x}_2) = \sum_{q=1}^{r_j} g_{j1}^q(\hat{x}_1) g_{j2}^q(\hat{x}_2),$$

$$g_{ji}^q \in \mathfrak{C}(\hat{X}_i, \mathbb{R}),$$

$$i = 1, 2; \ j = 1, 2, \cdots, m,$$

$$q = 1, 2, \cdots, r_j.$$

So it is sufficient to consider a special case of the form

$$\varphi(\omega_1, \omega_2) = \sum_{j=1}^{m} \chi_{A_{j1} \times A_{j2}}(\omega_1, \omega_2) \sum_{q=1}^{r_j} g_{j1}^q(\hat{x}_1) g_{j2}^q(\hat{x}_2) \tag{3.33}$$

as φ in (3.30).

$$\begin{aligned}
&\langle \gamma_1 \otimes \gamma_2, \chi_{A_{j1} \times A_{j2}} g_{j1}^q g_{j2}^q |_{\Omega \times X} \rangle \\
&= \int_\Omega \{ \int_X \chi_{A_{j1} \times A_{j2}} g_{j1}^q g_{j2}^q |_{\Omega \times X} d(\nu_{\omega_1}^{1,n} \otimes \nu_{\omega_2}^{2,n}) \} d(\mu_1 \otimes \mu_2) \\
&= \int_\Omega \{ \int_{X_1} \chi_{A_{j1}} g_{j1}^q |_{\Omega_1 \times X_1} d\nu_{\omega_1}^{1,n} \} d\mu_1 \times \int_\Omega \{ \int_{X_2} \chi_{A_{j2}} g_{j2}^q |_{\Omega_2 \times X_2} d\nu_{\omega_2}^{2,n} \} d\mu_2 \\
&= \langle {\gamma_1}^n, \chi_{A_{j1}} g_{j1}^q \rangle \times \langle {\gamma_2}^n, \chi_{A_{j2}} g_{j2}^q \rangle \\
&\to \langle {\gamma_1}^*, \chi_{A_{j1}} g_{j1}^q \rangle \times \langle {\gamma_2}^*, \chi_{A_{j2}} g_{j2}^q \rangle \quad \text{(by assumption)} \\
&= \langle {\gamma_1}^* \otimes {\gamma_2}^*, \chi_{A_{j1} \times A_{j2}} g_{j1}^q g_{j2}^q |_{\Omega \times X} \rangle \quad \text{as } n \to \infty.
\end{aligned}$$

Adding up this result for all q and j, we obtain (3.31) for any function of the form (3.33). □

The next result deals with the convergence of Young measures associated with a measurable family consisting of product measures of the form $\delta_{u(\omega)} \otimes \nu_\omega$.

Theorem 3.9 *Let $(\Omega, \mathscr{E}, \mu)$ be a complete finite measure space, and X_i $(i = 1, 2)$ metrizable Souslin spaces. Suppose that $u_n, u_* : \Omega \to X_1$ $(n = 1, 2, \cdots)$ are $(\mathscr{E}, \mathscr{B}(X_1))$-measurable functions, and $\nu^n = \{\nu_\omega^n | \omega \in \Omega\}$ $(n = 1, 2, \cdots)$ as well as $\nu^* = \{\nu_\omega^* | \omega \in \Omega\}$ are elements of $\mathfrak{P}(\Omega, \mu; X_2)$. θ_n $(n = 1, 2, \cdots)$ and θ_* are Young measures associated with ν^n's and ν^*; i.e,*

$$\theta_n = \int_\Omega \delta_\omega \otimes \nu_\omega^n d\mu, \ \theta_* = \int_\Omega \delta_\omega \otimes \nu_\omega^* d\mu.$$

Assume that the following a and b.

a. $\{u_n\}$ converges to u_* in measure.
b. $\{\theta_n\}$ narrowly converges to θ_*.

Define

$$\gamma_n = \int_\Omega \delta_\omega \otimes (\delta_{u_n(\omega)} \otimes \nu_\omega^n) d\mu,$$

$$\gamma_* = \int_\Omega \delta_\omega \otimes (\delta_{u_*(\omega)} \otimes \nu_\omega^*) d\mu.$$

Then γ_n $(n = 1, 2, \cdots)$ and γ_* are Young measures on $\Omega \times X_1 \times X_2$, and

$$\gamma_n \to \gamma_* \quad (narrowly) \quad as\ n \to \infty.$$

Proof Assume, on the contrary, that

$$\int_{\Omega \times X_1 \times X_2} f d\gamma_n \nrightarrow \int_{\Omega \times X_1 \times X_2} f d\gamma_*$$

for some $f \in \mathfrak{S}_{\mathfrak{C}}(\Omega, \mu; X_1 \times X_2)$. We may assume, without loss of generality, that

$$\left| \int_{\Omega \times X_1 \times X_2} f d\gamma_n - \int_{\Omega \times X_1 \times X_2} f d\gamma_* \right| \geqq \varepsilon \tag{3.34}$$

for some $\varepsilon > 0$.

$\{u_n\}$ has a subsequence $\{u_{n'}\}$ which converges to u_* *a.e.* since $\{u_n\}$ converges to u_* in measure and $\mu\Omega < \infty$. We assume that $\{u_n\}$ itself converges to u_* *a.e.* for the sake of simplicity of notations.

Let $\hat{\mathbb{N}} = \mathbb{N} \cup \{\infty\}$ be the Alexandrov compactificaion of $\mathbb{N}$. Define a function $F : \Omega \times \hat{\mathbb{N}} \times X_2 \to \mathbb{R}$ by

$$F(\omega, n, z) = f(\omega, u_n(\omega), z).$$

1° F is a Carathéodory function.

In fact, it is clear that F is measurable in ω.

Assume next that $(n_p, z_p) \to (n_0, z_0)$. In case $n_0 \in \mathbb{N}, n_p = n_0$ for all sufficiently large p. Hence, trivially, $F(\omega, n_p, z_p) \to F(\omega, n_0, z_0)$. If $n_0 = \infty$, then we have

$$F(\omega, n_p, z_p) = f(\omega, u_{n_p}(\omega), z_p) \to f(\omega, u_*(\omega), z_0) = F(\omega, \infty, z_0).$$

2° $\omega \mapsto \|F(\omega, \cdot, \cdot)\|_\infty$ is integrable.

In fact, this fact is easily verified by the evaluation

$$\|F(\omega, \cdot, \cdot)\|_\infty \leqq \|f(\omega, \cdot, \cdot)\|_\infty$$

and the assumption $f \in \mathfrak{C}_{\mathfrak{C}}(\Omega, \mu; X_1 \times X_2)$.

Formally, we introduce another measure space $\Omega_1 = \{\omega_1\}$, $\mu_1\Omega_1 = 1$. Applying Theorem 3.8, by substituting $\Omega_1, \Omega_2, X_1, X_2, \nu^1, \nu^2$ there by $\Omega_1, \Omega, \hat{\mathbb{N}}, X_2, \delta_n, \nu^n$, we obtain

$$\begin{aligned}
\int_{\Omega \times X_1 \times X_2} f d\gamma_n &= \int_\Omega \{\int_{X_2} f(\omega, u_n(\omega), z) d\nu_\omega^n\} d\mu \\
&= \int_\Omega \{\int_{X_2} F(\omega, n, z) d\nu_\omega^n\} d\mu \\
&= \int_{\Omega_1 \times \Omega} \{\int_{\hat{\mathbb{N}} \times X_2} F(\omega, n, z) d(\delta_n \otimes \nu_\omega^n)\} d(\mu_1 \otimes \mu) \\
&\to \int_{\Omega_1 \times \Omega} \{\int_{\hat{\mathbb{N}} \times X_2} F(\omega, n, z) d(\delta_\infty \otimes \nu_\omega^*) d(\mu_1 \otimes \mu) \\
&= \int_\Omega \{\int_{X_2} F(\omega, \infty, z) d\nu_\omega^*\} d\mu \\
&= \int_{\Omega \times X_1 \times X_2} f d\gamma^*.
\end{aligned}$$

This contradicts (3.34). □

3.5 Simple Variational Problems via Theory of Young Measures

Let $x_1, x_2, \cdots, x_p$ be any finite elements of $\mathbb{R}^l$. We denote by X the convex hull of $\{x_1, x_2, \cdots, x_p\}$. X is, of course, a convex compact set. X is endowed with the Borel σ-field $\mathscr{B}(X)$. $(\Omega, \mathscr{E}, \mu)$ is a complete finite measure space and function $f : \Omega \times X \to \mathbb{R}$ is a nonnegative normal integrand.

The set of all measurable functions of Ω into $\{x_1, x_2, \cdots, x_p\}$ (resp. X) is denoted by $\mathscr{P}$ (resp. $\mathscr{M}$). We also denote by $\mathfrak{Y}(\Omega, \mu; X)$ the set of all the Young measures on $(\Omega \times X, \mathscr{E} \otimes \mathscr{B}(X))$. The definition and basic properties of Young measures are discussed in the preceding sections.[80]

[80]This section is based upon Maruyama [37].

Compare a triple of minimization problems:

(I) $\underset{x(\cdot)\in\mathscr{P}}{Minimize} \int_{\Omega} f(\omega, x(\omega))d\mu,$

(II) $\underset{x(\cdot)\in\mathscr{M}}{Minimize} \int_{\Omega} f(\omega, x(\omega))d\mu,$

(III) $\underset{\gamma\in\mathfrak{Y}(\Omega,\mu;X)}{Minimize} \int_{\Omega\times X} f(\omega, x)d\gamma.$

Our targets are twofold. The first question is the existence of solutions for each problem of the three. The second puzzle concerns their relations to each other. Assume that there exists a solution $y^*(\cdot) \in \mathscr{M}$ for Problem(II). Is there any equivalent solution $y^*(\cdot) \in \mathscr{P}$ for Problem(I) in the sense that

$$\int_{\Omega} f(\omega, x^*(\omega))d\mu = \int_{\Omega} f(\omega, y^*(\omega))d\mu?$$

A similar question concerning the relation between Problems(II) and (III) should also be answered.

We have to remind of the fact that for any positive normal integrand $f : \Omega\times X \to \mathbb{R}$, the functional

$$J : \gamma \mapsto \int_{\Omega\times X} f(\omega, x)d\gamma$$

is narrowly *l.s.c.* on $\mathfrak{Y}(\Omega, \mu; X)$ by Theorem 3.2.

We start with examining Problems(II) and (III).

$(\Omega, \mathscr{E}, \mu)$ is a finite complete measure space and $X = \text{co}\{x_1, x_2, \cdots, x_p\}$. However it is required to impose some additional assumptions concerning $(\Omega, \mathscr{E}, \mu)$ and $f(\omega, x)$.

It is obvious that Problem(III) has a solution

$$\gamma^* = \int_{\Omega} \delta_\omega \otimes \nu_\omega^* d\mu$$

in $\mathfrak{Y}(\Omega, \mu; X)$ in view of the narrow compactness of $\mathfrak{Y}(\Omega, \mu; X)$ (Theorem 2.7, 1°) and *l.s.c.* of J (Theorem 3.2).

Theorem 3.10 *There exists a solution for Problem(III).*

How about the existence of a solution for Problem(II)? The answer is positive again. However we need some additional assumptions as well as a little bit sophisticated reasonings for its proof.

Assumption 1 $(\Omega, \mathscr{E}, \mu)$ is non-atomic.

Assumption 2 $x \mapsto f(\omega, x)$ is convex for each fixed $\omega \in \Omega$.

A normal integrand which satisfies Assumption 2 is called a convex normal integrand.

Definition 3.4 Let $\mathfrak{X}$ be a real vector space and C a nonempty convex set of $\mathfrak{X}$. Suppose that x is any point of C. Then $v \in \mathfrak{X}$ is called a **facial direction** of C at x if $x \pm tv \in C$ for sufficiently small $t > 0$. The set of all the facial directions of C at x is called the **facial space** of C at x, and is denoted by $L(x|C)$.[81]

$L(x|C)$ is a vector subspace of $\mathfrak{X}$. It is easy to see that $x \in C$ is an extreme point of C if and only if $\dim L(x|C) = 0$.

Theorem 3.11 *Under Assumption 1, there exists a solution for Problem (II).*

Proof By Theorem 3.10, we know that Problem(III) has a solution γ^* of the form

$$\gamma^* = \int_{\Omega} \delta_{\omega} \otimes \nu^*_{\omega} d\mu.$$

Let me recapitulate the reasoning of Theorem 2.7 briefly. The space of measurable families $\{\nu_{\omega}|\omega \in \Omega\}$ with $\sup_{\omega\in\Omega} \|\nu_{\omega}\| < \infty$[82] can be identified with $\mathfrak{L}^{\infty}(\Omega, \mathfrak{M}(X))$. Let $\Phi : \mathfrak{L}^{\infty}(\Omega, \mathfrak{M}(X)) \to \mathfrak{M}(\Omega \times X)$ (the set of all the signed measures on $(\Omega \times X, \mathscr{E} \otimes \mathscr{B}(X))$with finite total variations) be the operator which associates with each $\{\nu_{\omega}\}$ the corresponding measure

$$\gamma = \int_{\Omega} \delta_{\omega} \otimes \nu_{\omega} d\mu.$$

Since $\mathfrak{Y}(\Omega, \mu; X) = \Phi(\mathfrak{P}(\Omega, \mu; X))$, it is narrowly compact and convex.

If we denote by K the set of Young measures which are equivalent to γ^* in the following sense:

$$\int_{\Omega\times X} f(\omega, x) d\gamma - \int_{\Omega\times X} f(\omega, x) d\gamma^* = 0. \tag{3.35}$$

Since K is nonempty, convex and narrowly compact, K has an extreme point, say $\hat{\gamma}$ (i.e. $\hat{\gamma} \in \ddot{K}$: the set of extreme points of K). We now show that $\hat{\gamma}$ can be expressed in the form

$$\hat{\gamma} = \int_{\Omega} \delta_{\omega} \otimes \delta_{x^*(\omega)} d\mu$$

for some measurable mapping $x^* : \Omega \to X$.

---

[81] For the concept of facial spaces, consult Arrow and Hahn [1, pp. 389–390]. See also Artstein [2].

[82] $\|\nu_{\omega}\|$ is the norm defined by the total variation of ν_{ω}.

If we define a linear operator $T : \Phi(\mathfrak{L}^\infty(\Omega, \mathfrak{M}(X)) \to \mathbb{R}$ by

$$T\gamma = \int_{\Omega \times X} f(\omega, x) d\gamma$$

and denote by F^*_{III} the optimized value of Problem(III); i.e.

$$F^*_{\text{III}} = \int_{\Omega \times X} f(\omega, x) d\gamma^*.$$

It is clear that $K = T^{-1}(F^*_{\text{III}}) \cap \mathfrak{Y}(\Omega, \mu; X)$.

We observe that the restriction of T to $L(\hat{\gamma} \mid \mathfrak{Y}(\Omega, \mu; X))$ is an injection.[83] Consequently,

$$\dim L(\gamma \mid \mathfrak{Y}(\Omega, \mu; X)) \leqq 1.$$

γ is an element of the compact convex set

$$H = [\hat{\gamma} + L(\hat{\gamma} \mid \mathfrak{Y}(\Omega, \mu; X))] \cap \mathfrak{Y}(\Omega, \mu; X).$$

By Carathéodory's theorem, γ can be expressed as a convex combination of at most two extreme points of H. Any extreme point of H is an extreme point of $\mathfrak{Y}(\Omega, \mu; X)$. So γ can be expressed as a convex combination of at most (or exactly) two extreme points of $\mathfrak{Y}(\Omega, \mu; X)$.

By Karlin–Castaing's theorem,[84] these extreme points are of the form

$$\int_\Omega \delta_\omega \otimes \delta_{y(\omega)} d\mu, \quad \int_\Omega \delta_\omega \otimes \delta_{z(\omega)} d\mu$$

[83] If T is not injective on $L(\hat{\gamma} \mid \mathfrak{Y}(\Omega, \mu; X))$, there must exist some $0 \neq \theta \in L(\hat{\gamma} \mid \mathfrak{Y}(\Omega, \mu; X))$ such that $T\theta = 0$. Since

$$\eta_+ = \hat{\gamma} + t\theta, \quad \eta_- = \hat{\gamma} - t\theta \in \mathfrak{Y}(\Omega, \mu; X)$$

for sufficiently small $t > 0$, it is clear that $T\eta_+ = T\eta_- = F^*_{\text{III}}$. That is, $\eta_+, \eta_- \in T^{-1}(F^*_{\text{III}}) \cap \mathfrak{Y}(\Omega, \mu; X) = K$. $\hat{\gamma} = 1/2(\eta_+ + \eta_-)$ which contradicts the fact $\hat{\gamma} \in \ddot{K}$.

[84] Let $\mathfrak{X}$ be a locally convex topological vector space. Assume that a correspondence $\Gamma : \Omega \twoheadrightarrow \mathfrak{X}$ is compact, convex-valued, measurable and $\mathfrak{L}^1$-integrably bounded. A correspondence $\ddot{\Gamma} : \Omega \twoheadrightarrow \mathfrak{X}$ is defined by $\ddot{\Gamma} : \omega \twoheadrightarrow (\Gamma(\omega))^{\cdot\cdot}$ (the set of extreme points of $\Gamma(\omega)$). $\mathscr{F}_\Gamma$(resp. $\mathscr{F}_{\ddot{\Gamma}}$) denotes the set of all measurable selections of Γ (resp. $\ddot{\Gamma}$). Then

$$\text{(i)} \ddot{\mathscr{F}}_\Gamma \neq \emptyset, \quad \text{(ii)} \mathscr{F}_{\ddot{\Gamma}} \neq \emptyset, \quad \text{(iii)} \ddot{\mathscr{F}}_\Gamma = \mathscr{F}_{\ddot{\Gamma}}.$$

See Castaing–Valadier [13, Theorem IV. 15, p.109], Maruyama [36, Chap.12,§2].

for some measurable mappings $y(\cdot), z(\cdot) : \Omega \to X$. Consequently $\hat{\gamma}$ can be expressed as

$$\hat{\gamma} = (1-t)\int_\Omega \delta_\omega \otimes \delta_{y(\omega)} d\mu + t\int_\Omega \delta_\omega \otimes \delta_{z(\omega)} d\mu.$$

for some measurable mapping $y(\cdot), z(\cdot) : \Omega \to X$ and $t \in [0, 1]$.[85] Hence

$$\int_{\Omega\times X} f(\omega, x) d\hat{\gamma} = (1-t)\int_\Omega f(\omega, y(\omega)) d\mu + t\int_\Omega f(\omega, z(\omega)) d\mu.$$

By Ljapunov's convexity theorem, there exists a decomposition $E_1, E_2 \in \mathscr{E}$ of Ω such that[86]

$$\int_{\Omega\times X} f(\omega, x) d\hat{\gamma} = \int_{E_1} f(\omega, y(\omega)) d\mu + \int_{E_2} f(\omega, z(\omega)) d\mu.$$

Defining

$$x^*(\omega) = \chi_{E_1}(\omega) y(\omega) + \chi_{E_2}(\omega) z(\omega)$$

(χ_{E_i} is the characteristic function of E_i), we obtain

$$\int_\Omega f(\omega, x^*(\omega)) d\mu = F^*_{\mathrm{II}}.$$

Thus $x^*(\cdot)$ is clearly a solution for Problem(II). □

Remark 8 Theorem 3.11 can be proved without the theory of Young measures. First of all, the set $\mathscr{M}$ of all the measurable mappings of Ω into X is weakly compact

---

[85] We had recourse to a geometric theory of the facial space due to Arrow–Hahn [1]. Berliocchi–Lasry [7, pp.145–146] established the following result sharing a common idea.

Let $\mathfrak{X}$ be a locally convex Hausdorff real topological vector space. Assume that $K \subset \mathfrak{X}$ is a nonempty compact convex set and $\varphi_i : \mathfrak{X} \to (-\infty, \infty](i = 1, 2, \cdots, n)$ are affine mappings. Then any extreme point of the set $G = \{x \in K | \varphi_i(x) \leqq 0 : i = 1, 2, \cdots, n\}$ can be expressed as a convex combination of $(n+1)$ extreme points of K.

[86] Let $(\Omega, \mathscr{E}, \mu)$ be a finite complete non-atomic measure space. Assume that $f_1, f_2, \cdots, f_m$ are any elements of $\mathfrak{L}^1(\Omega, \mathbb{R}^l)$ and a mapping $\lambda : \Omega \to \Lambda_m$ is measurable, where Λ_m is the fundamental simplex in $\mathbb{R}^m$: i.e.$\Lambda_m = \{\lambda \in \mathbb{R}^m | \lambda_i \geqq 0 (i = 1, 2, \cdots, m), \sum_{i=1}^m \lambda_i = 1\}$. Then there exists a decomposition $E_1, E_2, \cdots, E_m$ of $\Omega (E_i \in \mathscr{E}$ for all $i)$ such that

$$\int_\Omega \sum_{i=1}^m \lambda_i(\omega) f_i(\omega) d\mu = \sum_{i=1}^m \int_{E_i} f_i(\omega) d\mu.$$

cf. Castaing–Valadier [13, Theorem IV. 17, pp. 112–117]. Maruyama [36, Chap.12. §2].

in $\mathfrak{L}^1(\Omega, \mathbb{R}^l)$ by the Dunford-Pettis-Nagumo theorem. Furthermore we know that, under Assumptions 1 and 2, the integral functional $J' : \mathfrak{L}^1(\Omega, \mathbb{R}^l) \to \mathbb{R}$ defined by

$$J' : u \mapsto \int_{\Omega} f(\omega, u(\omega)) d\mu$$

is sequentially *l.s.c.* with respect to the weak topology of $\mathfrak{L}^1(\Omega, \mathbb{R}^l)$. This comes from Ioffe's [25] fundamental theorem. Combining these observations, we can assure the existence of a solution for Problem(II) (cf. Maruyama [37, pp. 286–295]).

The proof of Theorem 3.11 guarantees the equivalence of Problems(II) and (III) in the sense that the optimized values of the two problems are equal.

We turn next to Problem(I).

Theorem 3.12 *Under Assumption 1, Problem(I) has a solution.*

Proof The finite set $K = \{x_1, x_2, \cdots, x_p\}$ is compact. It is easy to check that the set of all the measurable families $\mathfrak{P}(\Omega, \mu; K)$ consisting of probability measures on K is compact in $\mathfrak{L}^\infty(\Omega, \mathfrak{M}(X))$. Hence, by Theorem 2.7, the set $\mathfrak{Y}(\Omega, \mu; K)$ of Young measures on $(\Omega \times K, \mathscr{E} \otimes \mathscr{B}(K))$ is narrowly compact. Since the integral functional J defined by

$$J : \gamma \mapsto \int_{\Omega \times X} f(\omega, x) d\gamma$$

is narrowly *l.s.c.* on $\mathfrak{Y}(\Omega, \mu; X)$, so is on $\mathfrak{Y}(\Omega, \mu; K)$. Thus there exists a solution $\gamma^* \in \mathfrak{Y}(\Omega, \mu; K)$ of the problem:

$$\underset{\gamma \in \mathfrak{Y}(\Omega, \mu; K)}{Minimize} \int_{\Omega \times K} f(\omega, x) d\gamma.$$

The measurable family ν_ω^* which determines γ^* is of the form:

$$\nu_\omega^* = \sum_{i=1}^{p} \lambda_i(\omega) \delta_{x_i}$$

where $\lambda(\omega) = (\lambda_1(\omega), \lambda_2(\omega), \cdots, \lambda_p(\omega)) : \Omega \to \Lambda_p$ (the fundamental simplex in $\mathbb{R}^p$) is measurable. Hence the optimized value F_I^* of Problem(I) can be calculated as

$$F_I^* = \int_{\Omega \times K} f(\omega, x) d\gamma^* = \int_{\Omega} \sum_{i=1}^{p} \lambda_i(\omega) f(\omega, x_i) d\mu.$$

Again by Ljapunov's convexity theorem, there exists a decomposition $E_1, E_2, \cdots, E_p$ of $\Omega(E_i \in \mathscr{E}$ for all i) such that

$$F_I^* = \sum_{i=1}^{p} \int_{E_i} f(\omega, x_i) d\mu.$$

Defining

$$x^*(\omega) = \sum_{i=1}^{p} \chi_{E_i}(\omega) x_i,$$

we obtain

$$F_I^* = \int_{\Omega} f(\omega, x^*(\omega)) d\mu.$$

$x^*(\cdot)$ is clearly a solution for Problem(I). □

As we already saw, Problems(II) and (III) are equivalent. Then it is natural to ask if Problems(I) and (II) are equivalent. Is there any solution for Problem(I) which attains the optimized value F_{II}^* of Problem(II)?

Game theorists interprete each element of the set $\mathscr{P}$ (resp. $\mathscr{M}$) as a pure strategy (resp. mixed strategy). If the optimized value F_{II}^* of Problem(II) is attained by some pure strategy, game theorists say that Problem(II) can be **purified**. So the problem stated above is expressed as "Can Problem(II) be purified?"

The answer is negative as the following counter-example illuminates.

Counter-Example Let Ω be the unit interval [0,1] with Lebesgue measure m. X is also specified as $X = [0, 1] = \text{co}\{0, 1\}$. We define an integrand $f : [0, 1] \times [0, 1] \to \mathbb{R}$ by

$$f(\omega, x) = \begin{cases} -2x + 1 & \text{on} \quad [0, 1/2], \\ 2x - 1 & \text{on} \quad [1/2, 1] \end{cases}$$

for any $\omega \in [0, 1]$. Then the only solution $x^*(\omega)$ for Problem(II)

$$\underset{x(\cdot):[0,1]\to[0,1]}{Minimize} \int_0^1 f(\omega, x(\omega)) dm(\omega)$$

is given by $x^*(\omega) = 1/2 \; a.e.$ The optimized value $V^* = 0$. However it is impossible to find a measurable function $y^*(\omega)$ of [0, 1] into $\{0, 1\}$ which is equivalent to $x^*(\omega)$: i.e.

$$\int_0^1 f(\omega, y^*(\omega)) dm = \int_0^1 f(\omega, x^*(\omega)) dm = 0.$$

Thus $x^*(\omega)$ can not be purified.

However the purification is proved to be possible if we impose an additional condition on $f(\omega, x)$.

Assumption 3 $f(\omega, (1-\lambda)x_1 + \lambda x_2) = (1-\lambda)f(\omega, x_1) + \lambda f(\omega, x_2)$ for any $\omega \in \Omega, x_1, x_2 \in X$ and $\lambda \in [0, 1]$.

Assumption 3 requires the graph of the function $x \mapsto f(\omega, x)$ to be flat for each fixed $\omega \in \Omega$.

Let $x^*(\cdot)$ be a solution for Problem(II). Fillipov's measurable implicit function theorem[87] assures the existence of measurable function $\lambda^* : \Omega \to \Lambda_p$ such that

$$x^*(\omega) = \sum_{i=1}^{p} \lambda_i^*(\omega)x_i.$$

Since the optimized value of Problem(II) is attained by $x^*(\cdot)$, it follows that

$$F_{\mathrm{II}}^* = \int_\Omega f(\omega, x^*(\omega))d\mu = \int_\Omega f(\omega, \sum_{i=1}^{p} \lambda_i^*(\omega)x_i)d\mu = \int_\Omega \sum_{i=1}^{p} \lambda_i^*(\omega) f(\omega, x_i)d\mu$$

by Assumption 3. We now apply Ljapunov's convexity theorem again to get a function $y^* : \Omega \to K$ which satisfies

$$F_{\mathrm{II}}^* = \int_\Omega f(\omega, y^*(\omega))d\mu.$$

We conclude the possibility of the purification of Problem(II).

Theorem 3.13 *Under Assumptions 1 and 3, Problem(II) can be purified.*

Since Problem(II) and Problem(III) are equivalent, Problem(III) can also be purified, that is, there exists a solution $z^*(\cdot)$ for Problem(I) which realizes the optimized value F_{III}^* of Problem(III).

Although the full purification is difficult without very strict conditions, what can we say about an approximate purification? Let $x^*(\cdot)$ be a solution for Problem(II). If there exists a solution $y^*(\cdot)$ for Problem(I) such that

$$|\int_\Omega f(\omega, x^*(\omega))d\mu - \int_\Omega f(\omega, y^*(\omega))d\mu| < \varepsilon$$

[87] Suppose that $(\Omega, \mathscr{E})$ and $(\Omega', \mathscr{E}')$ are two measurable spaces, and X is a Souslin space. A correspondence $\Gamma : \Omega \twoheadrightarrow X$ and a couple of functions, $f : \Omega \to \Omega'$ and $g : \Omega \times X \to \Omega'$, are assumed to satisfy the followings: a. f is $(\mathscr{E}, \mathscr{E}')$-measurable, b. g is $(\mathscr{E} \otimes \mathscr{B}(X), \mathscr{E}')$-measurable, c. the graph of Γ belongs to $\mathscr{E} \otimes \mathscr{B}(X)$, d. $f(\omega) \in g(\omega, \Gamma(\omega))$ for all ω. Then there exists a $(\hat{\mathscr{E}}, \mathscr{B}(X))$-measurable selection γ of Γ which satisfies $f(\omega) = g(\omega, \gamma(\omega))$ for all ω. cf. Castaing–Valadier [13, pp. 85–86], Maruyama [36, pp. 426–427].

for some $\varepsilon > 0$, we say that Problem(II) can be ε-**purified**. Aumann et al. [4] examined a similar problem in the context of game theory. However we leave it to another occasion.[88]

4 Continuity of Nonlinear Integral functionals

4.1 Roles of Uniform Integrability

The goals of this chapter is to examine the continuity (or semi-continuity) of nonlinear integral functionals via theories of Young measures. This gives an indispensable foundation for the existence problem in calculus of variations.

We start by preparing a few results concerning the concept of uniform integrability.[89]

Definition 4.1 Let $(\Omega, \mathscr{E}, \mu)$ be a measure space. A subset $\mathscr{F}$ of $\mathfrak{L}^1(\Omega, \mathbb{R})$ is said to be **uniformly integrable** if there exists some $\delta > 0$, for each $\varepsilon > 0$, such that

$$\mu(E) \leqq \delta,\ E \in \mathscr{E} \Rightarrow \sup_{f\in\mathscr{F}} \int_E |f(\omega)| d\mu \leqq \varepsilon.$$

The following well-known proposition clarifies the relation between the uniform integrability and the relative compactness of $\mathscr{F} \subset \mathfrak{L}^1(\Omega, \mathbb{R})$ in the weak topology. (cf. Maruyama [36, pp. 275–284] for detailed discussions.)

Proposition 4.1 *Let $(\Omega, \mathscr{E}, \mu)$ be a finite measure space, and $\mathscr{F}$ a subset of $\mathfrak{L}^1(\Omega, \mathbb{R})$. The following statements are all equivalent.*

(i) $\mathscr{F}$ is weakly relatively compact.
(ii) $\mathscr{F}$ is weakly sequentially compact.
(iii) $\mathscr{F}$ is strongly bounded, and there exists a function $\Phi : [0, \infty) \to [0, \infty)$ which is continuous a.e., nondecreasing and satisfies the following a and b.

a. $\displaystyle\lim_{u\to\infty} \frac{\Phi(u)}{u} = \infty.$

b. $\displaystyle\sup_{f\in\mathscr{F}} \int_\Omega \Phi(|f(\omega)|) d\mu \leqq B$ *for some* $B \in (0, \infty)$.

(iv) $\mathscr{F}$ is strongly bounded and uniformly integrable.
(v) If we define a set E_α^f for $f \in \mathscr{F}$ and $\alpha > 0$ by

$$E_\alpha^f = \{\omega \in \Omega \,|\, |f(\omega)| \geqq \alpha\},$$

[88] cf. Aumann et al. [4].

[89] The contents of this subsection are due to Valadier [48, pp. 166–168].

then

$$\lim_{\alpha\to\infty}\sup_{f\in\mathscr{F}}\int_{E_\alpha^f}|f(\omega)|d\mu=0.$$

The equivalence of (i) and (ii) is just a special case of Eberlein–Šmulian's theorem.[90] The criterion (iii) is due to M. Nagumo[91] and (iv) to N. Dunford.[92]

Theorem 4.1 *Let $(\Omega,\mathscr{E},\mu)$ be a complete finite measure space, and X a metrizable Souslin space. Suppose that $\{u_n:\Omega\to X\}$ is a sequence of $(\mathscr{E},\mathscr{B}(X))$-measurable functions which satisfies either a or b.*

a. $\{u_n\}$ is uniformly tight.
b. The sequence

$$\gamma_n=\int_\Omega \delta_\omega\otimes\delta_{u_n(\omega)}d\mu;\ n=1,2,\cdots$$

of Young measures is narrowly convergent.

Furthermore a function $f:\Omega\times X\to\bar{\mathbb{R}}$ is a normal integrand such that $\{\omega\mapsto f(\omega,u_n(\omega))^-\}$ is $\mathfrak{L}^1$-bounded and uniformly integrable.[93]

Then there exists a $\gamma\in\mathfrak{Y}(\Omega,\mu;X)$ which satisfies the following (i)–(iii).

(i) γ *is a limiting point of $\{\gamma_n\}$ (w.r.t, the narrow topology).*

(ii) $\liminf_n\int_\Omega f(\omega,u_n(\omega))d\mu<\infty\Rightarrow\int_{\Omega\times X}f(\omega,x)^+d\gamma<\infty.$

(iii) $\int_{\Omega\times X}f(\omega,x)d\gamma\leqq\liminf_n\int_\Omega f(\omega,u_n(\omega))d\mu.$

Proof By the uniform tightness of $\{u_n\}$ and Theorem 3.5, there is a subsequence $\{u_{n_k}\}$ of $\{u_n\}$ which satisfies

$$\int_\Omega f(\omega,u_{n_k}(\omega))d\mu\to\liminf_n\int_\Omega f(\omega,u_n(\omega))d\mu\quad\text{as }k\to\infty,\tag{4.1}$$

$$\gamma_{n_k}\to\gamma\quad\text{(narrowly)}\quad\text{as }k\to\infty.\tag{4.2}$$

If we define a function f_α by

$$f_\alpha=\sup\{-\alpha,f\}$$

[90] Bourbaki [10, IV. 35–36], Dunford–Schwartz [21, pp. 430–433].

[91] Nagumo [39].

[92] Dunford–Schwartz [21, pp. 294–295], Diestel–Uhl [20, pp. 101–102].

[93] $f(\omega,u_n(\omega))^-$ denotes the negative part of $f(\omega,u_n(\omega))$; i.e. $f(\omega,u_n(\omega))=-\mathrm{Min}\{f(\omega,u_n(\omega)),0\}$.

for $\alpha \in [0, \infty)$, then $f_\alpha + \alpha \in \mathfrak{G}_+(\Omega, \mathscr{E}; X)$. So we have, by Theorem 3.2.

$$\int_{\Omega \times X} (f_\alpha(\omega, x) + \alpha) d\gamma \leqq \liminf_k \int_\Omega \{f_\alpha(\omega, u_{n_k}(\omega)) + \alpha\} d\mu. \tag{4.3}$$

Hence

$$\int_{\Omega \times X} f_\alpha(\omega, x) d\gamma \leqq \liminf_k \int_\Omega f_\alpha(\omega, u_{n_k}(\omega)) d\mu. \tag{4.4}$$

Defining

$$E_\alpha^n = \{\omega \in \Omega \,|\, f(\omega, u_n(\omega))^- \geqq \alpha\}$$

for each $n \in \mathbb{N}$ and $\alpha > 0$, we obtain

$$\lim_{\alpha \to \infty} \sup_n \int_{E_\alpha^n} f(\omega, u_n(\omega))^- d\mu = 0 \tag{4.5}$$

by Proposition 4.1 and the assumption that $\{\omega \mapsto f(\omega, u_n(\omega))^-\}$ is $\mathfrak{L}^1$-bounded and uniformly integrable. We also note that

$$f_\alpha(\omega, u_n(\omega)) = \begin{cases} f(\omega, u_n(\omega)) & \text{on} \quad \Omega \setminus E_\alpha^n, \\ -\alpha & \text{on} \quad E_\alpha^n. \end{cases}$$

Choosing $\alpha > 0$ sufficiently large for any $\varepsilon > 0$, we obtain

$$\sup_n \int_{E_\alpha^n} f(\omega, u_n(\omega))^- d\mu \leqq \varepsilon, \tag{4.6}$$

by (4.5). Consequently,

$$\begin{aligned}
&\int_\Omega f(\omega, u_n(\omega)) d\mu \\
&\quad = \int_{E_\alpha^n} f(\omega, u_n(\omega)) d\mu + \int_{\Omega \setminus E_\alpha^n} f_\alpha(\omega, u_n(\omega)) d\mu \\
&\quad = \int_\Omega f_\alpha(\omega, u_n(\omega)) d\mu + \int_{E_\alpha^n} f(\omega, u_n(\omega)) d\mu - \int_{E_\alpha^n} \underbrace{f_\alpha(\omega, u_n(\omega))}_{=-\alpha} d\mu \\
&\quad \geqq \int_\Omega f_\alpha(\omega, u_n(\omega)) d\mu + \int_{E_\alpha^n} f(\omega, u_n(\omega)) d\mu \\
&\quad \geqq \int_\Omega f_\alpha(\omega, u_n(\omega)) d\mu - \varepsilon \quad \text{(by (4.6))}.
\end{aligned} \tag{4.7}$$

Therefore

$$\int_{\Omega\times X} f d\gamma \leqq \int_{\Omega\times X} f_\alpha d\gamma \underset{(4.4)}{\leqq} \liminf_k \int_\Omega f_\alpha(\omega, u_{n_k}(\omega))d\mu$$

$$\underset{(4.7)}{\leqq} \liminf_k \int_\Omega f(\omega, u_{n_k}(\omega))d\mu + \varepsilon$$

$$\underset{(4.1)}{=} \liminf_n \int_\Omega f(\omega, u_n(\omega))d\mu + \varepsilon. \tag{4.8}$$

Passing $\varepsilon \downarrow 0$, we obtain (iii).

(ii) follows from (4.8) and the boundedness of $\{\omega \mapsto f(\omega, u_n(\omega))^-\}$. (i) is clear from (4.1). □

Theorem 4.2 *Let $(\Omega, \mathscr{E}, \mu)$ be a complete finite measure space, and X a metrizable Souslin space. Suppose that $\{u_n : \Omega \to X\}$ is a sequence of $(\mathscr{E}, \mathscr{B}(X))$-measurable functions and the corresponding sequence of Young measures*

$$\gamma_n = \int_\Omega \delta_\omega \otimes \delta_{u_n(\omega)} d\mu;\ n = 1, 2, \cdots$$

narrowly converges to $\gamma \in \mathfrak{Y}(\Omega, \mu; X)$. Furthermore $f : \Omega \times X \to \mathbb{R}$ is a Carathéodory function for which $\{\omega \mapsto f(\omega, u_n(\omega))\}$ is $\mathfrak{L}^1$-bounded and uniformly integrable.

Then f is γ-integrable and

$$\int_{\Omega\times X} f d\gamma = \lim_{n\to\infty} \int_\Omega f(\omega, u_n(\omega))d\mu.$$

Proof By Theorem 4.1, we have[94]

$$\int_{\Omega\times X} f(\omega, x)^+ d\gamma < \infty,$$

$$\int_{\Omega\times X} f(\omega, x)d\gamma \leqq \liminf_n \int_\Omega f(\omega, u_n(\omega))d\mu.$$

[94]We note that

$$\liminf_n \int_\Omega f(\omega, u_n(\omega))d\mu < \infty$$

since $\{\omega \mapsto f(\omega, u_n(\omega))\}$ is $\mathfrak{L}^1$-bounded.

Similarly it holds good that

$$\int_{\Omega\times X}(-f)(\omega,x)^{+}d\mu<\infty,$$

$$\int_{\Omega\times X}(-f)(\omega,x)d\gamma\leqq\liminf_{n}\int_{\Omega}(-f)(\omega,u_n(\omega))d\mu.$$

Combining these results, we can conclude that f is γ-integrable and

$$\begin{aligned}\limsup_{n}\int_{\Omega}f(\omega,u_n(\omega))d\mu&\leqq\int_{\Omega\times X}f(\omega,x)d\gamma\\&\leqq\liminf_{n}\int_{\Omega}f(\omega,u_n(\omega))d\mu.\end{aligned}$$

□

Corollary 4.1 *Let $(\Omega,\mathscr{E},\mu)$ be a complete finite measure space, and X a metrizable Souslin space. $\{u_n:\Omega\to X\}$ is a sequence of $(\mathscr{E},\mathscr{B}(X))$-measurable functions and $f:X\to\mathbb{R}$ is a continuous function such that $\{f\circ u_n\}$ is $\mathfrak{L}^1$-bounded and uniformly integrable. Furthermore suppose that a sequence*

$$\gamma_n=\int_{\Omega}\delta_\omega\otimes\delta_{u_n(\omega)}d\mu;\ n=1,2,\cdots$$

in $\mathfrak{Y}(\Omega,\mu;X)$ narrowly converges to $\gamma=\int_{\Omega}\delta_\omega\otimes\nu_\omega d\mu\in\mathfrak{Y}(\Omega,\mu;X)$. Then the following statements hold good.

(i) f is ν_ω-integrable for a.e. ω, and

$$\int_{\Omega\times X}|f(x)|d\gamma<\infty.$$

(ii) $\{f\circ u_n\}$ weakly converges to the function

$$\omega\mapsto\int_X f(x)d\nu_\omega$$

in $\mathfrak{L}^1(\Omega,\mathbb{R})$.

Proof

(i) By Theorem 4.2, the function $f_0(\omega,x)=f(x)$ is γ-integrable; i.e.

$$\int_{\Omega}\{\int_X|f(x)|d\nu_\omega\}d\mu<\infty.$$

(i) immediately follows.
(ii) If we define a function f_0 by

$$f_0(\omega, x) = g(\omega)f(x)$$

for $g \in \mathfrak{L}^\infty(\Omega, \mathbb{R})$, it satisfies the conditions in Theorem 4.2. In fact, the uniform integrability is verified in view of the following evaluation.

$$\begin{aligned}
&\sup_n \int_E |g(\omega)f(u_n(\omega))|d\mu \\
&\leqq \|g\|_\infty \sup_n \int_E |f(u_n(\omega))|d\mu \\
&\to 0 \quad \text{as} \quad \mu E \to 0.
\end{aligned}$$

(We here made use of the uniform integrability of $\{f \circ u_n\}$.) We also observe the $\mathfrak{L}^1$-boundedness of $\{f_0 \circ u_n\}$ by a similar evaluation as above and the assumption that $\{f \circ u_n\}$ is $\mathfrak{L}^1$-bounded.

Consequently,

$$\begin{aligned}
\int_{\Omega \times X} f_0 d\gamma &= \lim_{n\to\infty} \int_\Omega f_0(\omega, u_n(\omega))d\mu \qquad (4.9) \\
&= \lim_{n\to\infty} \int_\Omega g(\omega)f(u_n(\omega))d\mu.
\end{aligned}$$

However

$$\begin{aligned}
\int_{\Omega \times X} f_0 d\gamma &= \int_\Omega \{\int_X f_0(\omega, x)d\nu_\omega\}d\mu \qquad (4.10) \\
&= \int_\Omega g(\omega)\{\int_X f(x)d\nu_\omega\}d\mu.
\end{aligned}$$

Combining (4.9) and (4.10), we obtain

$$\lim_{n\to\infty} \int_\Omega g(\omega)f(u_n(\omega))d\mu = \int_\Omega g(\omega)\{\int_X f(\omega)d\nu_\omega\}d\mu$$

for any $g \in \mathfrak{L}^\infty(\Omega, \mathbb{R})$. □

4.2 Castaing–Clauzure Theorem

Let $(\Omega, \mathscr{E}, \mu)$ be a measurable space and a function $f : \Omega \times \mathbb{R}^l \times \mathbb{R}^k \to \mathbb{R}$ given. A functional $F : \mathfrak{L}^p(\Omega, \mathbb{R}^l) \times \mathfrak{L}^q(\Omega, \mathbb{R}^k) \to \mathbb{R}$ is defined by

$$F(u(\cdot), v(\cdot)) = \int_\Omega f(\omega, u(\omega), v(\omega)) d\mu.$$

Ioffe [25] provided a basic result concerning *l.s.c.* of F *w.r.t.* the combination of the strong topology on $\mathfrak{L}^p$ and the weak one on $\mathfrak{L}^q$. Ioffe's theorem paved the way to examining systematically the continuity property of nonlinear integral functionals appearing in calculus of variations. Castaing–Clauzure [12] established various similar results for an integral functional defined on the spaces of Bochner-integrable functions. Furthermore Valadier [47, 48, pp. 170–172] developed a route to these results via theory of Young measures. In this subsection, I follow Valadier's route.

Theorem 4.3 *Let $(\Omega, \mathscr{E}, \mu)$ be a complete finite measure space, and $(\mathfrak{X}, \|\cdot\|)$ a separable reflexive Banach space. Suppose that $\{u_n\}$ is a sequence in $\mathfrak{L}^1(\Omega, \mathfrak{X})$ which is $\mathfrak{L}^1$-bounded and uniformly integrable.*[95]

(i) There exists a metric ρ on $\mathfrak{X}$ which satisfies the following two conditions.

a. ρ defines a Lusin topology[96] *weaker than the weak topology $\sigma(\mathfrak{X}, \mathfrak{X}')$.*
b. The sequence

$$\gamma_n = \int_\Omega \delta_\omega \otimes \delta_{u_n(\omega)} d\mu; \quad n = 1, 2, \cdots$$

in $\mathfrak{Y}(\Omega, \mu; (\mathfrak{X}, \rho))$ is uniformly tight.

(ii) Assume that a sequence $\{\gamma_n\}$ in $\mathfrak{Y}(\Omega, \mu; (\mathfrak{X}, \rho))$ has a subsequence $\{\gamma_{n_k}\}$ which narrowly converges to

$$\gamma = \int_\Omega \delta_\omega \otimes \nu_\omega d\mu.$$

---

[95] Let $(\Omega, \mathscr{E}, \mu)$ be a measure space. A subset $\mathscr{F}$ of $\mathfrak{L}^1(\Omega, \mathfrak{X})$ is said to be **uniformly integrable** if there exists some $\delta > 0$, for each $\varepsilon > 0$, such that

$$\mu E \leqq \delta, \ E \in \mathscr{E} \Rightarrow \sup_{f \in \mathscr{F}} \int_E \|f(\omega)\| d\mu \leqq \varepsilon.$$

[96] A metrizable topological space X is called a **Lusin space** if there exist a Polish space P and a continuous bijection $f : P \to X$.

Any Borel set of a Lusin space is also a Lusin space. If a metrizable space X is decomposed into countable sets $A_n (n = 1, 2, \cdots)$, each of which is a Lusin space, then X itself is also a Lusin space. cf. Bourbaki [9, Part 2, pp. 200–206].

Then the following two statements hold good.

a. $\int_{\mathfrak{X}} \|x\| d\nu_\omega < \infty$ *a.e.*

b. *If we define a function* $u : \Omega \to \mathfrak{X}$ *by*

$$u(\omega) = \int_{\mathfrak{X}} x d\nu_\omega, \tag{4.11}$$

then $\{u_{n_k}\}$ *weakly converges to* u.[97] *(The right-hand side of (4.11) is called the* ***barycenter*** *of* ν_ω *on* $\mathfrak{X}$. *)*

Proof

(i) Since $\mathfrak{X}$ is separable and reflexive, $\mathfrak{X}'$ is also separable.[98] Let $\{\Lambda_n; n = 1, 2, \cdots\}$ be a dense subset of $\mathfrak{X}'$. If we define

$$\rho(x, y) = \sum_{n=1}^{\infty} \frac{1}{2^n} \cdot \frac{|\langle \Lambda_n, x - y \rangle|}{1 + |\langle \Lambda_n, x - y \rangle|}, \quad x, y \in \mathfrak{X},$$

then ρ is a metric on $\mathfrak{X}$, which defines the topology generated by $\{\Lambda_n\}$.

The set

$$K_n = \{x \in \mathfrak{X} | \|x\| \leqq n\}; \quad n = 1, 2, \cdots$$

is weakly compact (because of the reflexivity of $\mathfrak{X}$ and Alaoglu's theorem). Hence K_n is compact *w.r.t.* the topology defined by ρ. It follows that K_n is a Lusin space and its Borel subset $K_n \setminus K_{n-1}$ ($n \geqq 2$) is also Lusin. Since $\mathfrak{X}$ can be expressed as

$$\mathfrak{X} = K_1 \cup \{\bigcup_{n=1}^{\infty} (K_{n+1} \setminus K_n)\},$$

$\mathfrak{X}$ is Lusin (*w.r.t.* ρ). (cf. footnote 96.)

By the $\mathfrak{L}^1$-boundedness of $\{u_n\}$,

$$\sup_n \int_{\Omega} \|u_n(\omega)\| d\mu < \infty. \tag{4.12}$$

Taking account of the inf-compactness property of the function $x \mapsto \|x\|$, we know that $\{u_n\}$ is uniformly tight (Theorem 3.6). Hence $\{\gamma_n\}$ has a subsequence $\{\gamma_{n_k}\}$

[97] Let $(\Omega, \mathscr{E}, \mu)$ be a finite measure space, and $\mathfrak{X}$ a Banach space. The duality $\mathfrak{L}^p(\Omega, \mathfrak{X})' \cong \mathfrak{L}^q(\Omega, \mathfrak{X}')$ $(1 \leqq p, q < \infty, 1/p + 1/q = 1)$ holds good if and only if $\mathfrak{X}'$ has the Radon–Nikodým property *w.r.t*, μ. cf. Diestel–Uhl [20, pp. 98–100].

[98] A normed vector space $\mathfrak{V}$ is separable if $\mathfrak{V}'$ is separable (*w.r.t.* the strong topology). cf. Yosida [52, p. 126].

which narrowly converges to some Young measure

$$\gamma = \int_{\Omega} \delta_\omega \otimes \nu_\omega d\mu,$$

by Theorem 3.5.
(ii) If we define a function $f : \Omega \times \mathfrak{X} \to \mathbb{R}$ by $f(\omega, x) = \|x\|$, f is a Carathéodory function and

$$\{\omega \mapsto f(\omega, u_n(\omega)) = \|u_n(\omega)\|\} \tag{4.13}$$

is $\mathfrak{L}^1$-bounded and uniformly integrable. Hence it follows from Theorem 4.2 that

$$\int_{\Omega \times \mathfrak{X}} f d\gamma = \lim_{k\to\infty} \int_{\Omega} f(\omega, u_{n_k}(\omega)) d\mu,$$

i.e.,

$$\int_{\Omega \times \mathfrak{X}} \|x\| d\gamma = \lim_{k\to\infty} \int_{\Omega} \|u_{n_k}(\omega)\| d\mu < \infty.$$

Writing $M = \sup_n \|u_n\|_1$, we have

$$\int_{\Omega \times \mathfrak{X}} \|x\| d\gamma \leqq M. \tag{4.14}$$

So we can define the Bochner integral

$$u(\omega) = \int_{\mathfrak{X}} x d\nu_\omega. \tag{4.15}$$

Since

$$\| \int_{\mathfrak{X}} x d\nu_\omega \| \leqq \int_{\mathfrak{X}} \|x\| d\nu_\omega,$$

we obtain

$$\int_{\Omega} \|u(\omega)\| d\mu \leqq M,$$

that is $u \in \mathfrak{L}^1(\Omega, \mathfrak{X})$.

Finally we show that $\{u_{n_k}\}$ weakly converges to u. Suppose that D is any strongly dense subset of $\mathfrak{L}^\infty(\Omega, \mathfrak{X}')$. Since u_{n_k}'s and u are contained in an equi-continuous family $\{v \in \mathfrak{L}^1(\Omega, \mathfrak{X}) | \|v\|_1 \leqq M\} \subset \mathfrak{L}^\infty(\Omega, \mathfrak{X}')'$, it is enough to show that

$$\int_\Omega \langle g(\omega), u_{n_k}(\omega)\rangle d\mu \to \int_\Omega \langle g(\omega), u(\omega)\rangle d\mu \qquad (4.16)$$

$$\text{as } k \to \infty \text{ for all } g \in D.$$

It can be verified as follows. Since u_{n_k}'s and u are equi-continuous, there exists a small $\delta > 0$, for each $\varepsilon > 0$, such that

$$\|g - h\|_\infty \leqq \delta, \ g, h \in \mathfrak{L}^\infty(\Omega, \mathfrak{X}')$$
$$\Rightarrow \int_\Omega \langle (g - h)(\omega),\ u_{n_k}(\omega)\rangle d\mu \leqq \frac{\varepsilon}{2}, \qquad (4.17)$$
$$\int_\Omega \langle (g - h)(\omega),\ u(\omega)\rangle d\mu \leqq \frac{\varepsilon}{2}.$$

Assume that (4.16) holds good. For any $g_0 \in \mathfrak{L}^\infty(\Omega, \mathfrak{X}')$, there exists some $g \in D$ such that $\|g - g_0\| \leqq \delta$.

By (4.16), there is some large k_0 such that

$$\left| \int_\Omega \langle g(\omega),\ (u_{n_k} - u)(\omega)\rangle d\mu \right| \leqq \varepsilon \qquad (4.18)$$

$$\text{for all } \ k \geqq k_0.$$

It follows from (4.17) and (4.18) that

$$\left| \int_\Omega \langle g_0(\omega),\ (u_{n_k} - u)(\omega)\rangle d\mu \right|$$
$$= \left| \int_\Omega \langle (g_0 - g)(\omega), (u_{n_k} - u)(\omega)\rangle d\mu + \int_\Omega \langle g(\omega), (u_{n_k} - u)(\omega)\rangle d\mu \right|$$
$$\leqq \varepsilon + \left| \int_\Omega \langle g(\omega), (u_{n_k} - u)(\omega\rangle d\mu \right|$$
$$\leqq 2\varepsilon \quad \text{for all } k \geqq k_0.$$

Since this holds good for any $g_0 \in \mathfrak{L}^\infty(\Omega, \mathfrak{X}')$, $\{u_{n_k}\}$ weakly converges to u.

Thus we have only to show (4.16).

Let $\{\Lambda_n;\ n = 1, 2, \cdots\}$ be a countable dense set in $\mathfrak{X}'$. We denote by D the set of all the elements of $\mathfrak{L}^\infty(\Omega, \mathfrak{X}')$ which take their values only in $\{\Lambda_n;\ n = 1, 2, \cdots\}$. Then D is dense in $\mathfrak{L}^\infty(\Omega, \mathfrak{X}')$.[99]

Define a function $f : \Omega \times \mathfrak{X} \to \mathbb{R}$ by

$$f(\omega, x) = \langle g(\omega), x \rangle$$

for any $g \in D$. Then $x \mapsto f(\omega, x)$ is continuous on $(\mathfrak{X}, \rho)$ for each fixed $\omega \in \Omega$. Hence f is a Carathéodory function. Furthermore $\{f(\omega, u_n(\omega))\}$is uniformly integrable since

$$\sup_n \int_E |\langle g(\omega), u_n(\omega)\rangle| d\mu \leqq \|g\|_\infty \sup_n \int_E \|u_n(\omega)\| d\mu \to 0 \quad \text{as} \quad \mu E \to 0$$

by the uniform integrability of $\{u_n\}$, It is also easy to see the $\mathfrak{L}^1$-boundedness of $\{u_n\}$. By Theorem 4.2, we obtain

$$\begin{aligned}
\int_\Omega \langle g(\omega), u(\omega)\rangle d\mu &= \int_\Omega \langle g(\omega), \int_{\mathfrak{X}} x d\nu_\omega \rangle d\mu \\
&= \int_\Omega \{\int_{\mathfrak{X}} \langle g(\omega), x\rangle d\nu_\omega\} d\mu \\
&= \int_{\Omega\times\mathfrak{X}} f(\omega, x) d\gamma \\
&= \lim_{k\to\infty} \int_{\Omega\times X} f(\omega, x) d\gamma_{n_k} \\
&= \lim_{k\to\infty} \int_\Omega \langle g(\omega), u_{n_k}(\omega)\rangle d\mu.
\end{aligned}$$

□

We are now approaching a destination of our long journey. The following theorem due to Castaing–Clauzure [12] is one of the most general result concerning the continuity of nonlinear integral functionals.

Theorem 4.4 (Castaing–Clauzure) *Let $(\Omega, \mathscr{E}, \mu)$ be a complete finite measure space, S a metrizable Souslin space, and $\mathfrak{X}$ a separable reflexive Banach space.*

[99] For any $h \in \mathfrak{L}^\infty(\Omega, \mathfrak{X}')$ and $\varepsilon > 0$, the correspondence $\Gamma : \Omega \twoheadrightarrow \mathfrak{X}'$ defined by $\Gamma : \omega \mapsto\!\!\!\rightarrow B_\varepsilon(h(\omega))$ is measurable. Let $\Theta : \Omega \twoheadrightarrow \mathfrak{X}'$ be the constant correspondence $\Theta(\omega) = \{\Lambda_1, \Lambda_2, \cdots\}$. Then $\Gamma(\omega) \cap \Theta(\omega) \neq \emptyset$. Hence there exists a measurable selection $\varphi : \Omega \to \mathfrak{X}'$ of $\Gamma(\omega) \cap \Theta(\omega)$; i.e.

$$\|h(\omega) - \varphi(\omega)\| \leqq \varepsilon,\ \varphi(\omega) \in \{\Lambda_1, \Lambda_2, \cdots\}.$$

$\{u_n : \Omega \to S\}$ is a sequence of $(\mathscr{E}, \mathscr{B}(S))$-measurable functions which converges to a $(\mathscr{E}, \mathscr{B}(S))$-measurable function u_∞ in measure. $\{v_n : \Omega \to \mathfrak{X}\}$ is a sequence in $\mathfrak{L}^1(\Omega, \mathfrak{X})$ which weakly converges to $v_\infty \in \mathfrak{L}^1(\Omega, \mathfrak{X})$. $\mathfrak{X}$ is endowed with a Lusin topology defined by a metric ρ introduced in Theorem 4.3. A function $f : \Omega \times S \times \mathfrak{X} \to \mathbb{R}$ is a normal integrand (i.e. an element of $\mathfrak{G}(\Omega, \mathscr{E} \otimes \mathscr{B}(S) \times \mathscr{B}(\mathfrak{X});\ S \times \mathfrak{X})$) which satisfies the following two conditions.

a. $x \mapsto f(\omega, u_\infty(\omega), x)$ is convex.
b. $\{\omega \mapsto f(\omega, u_n(\omega), v_n(\omega))^-\}$ is $\mathfrak{L}^1$-bounded and uniformly integrable.

Then the statements (i) and (ii) hold good.

(i) If

$$\liminf_n \int_\Omega f(\omega, u_n(\omega), v_n(\omega))d\mu < \infty,$$

then

$$\int_\Omega f(\omega, u_\infty(\omega), v_\infty(\omega))^+ d\mu < \infty$$

(ii)

$$\int_\Omega f(\omega, u_\infty(\omega), v_\infty(\omega))d\mu \leqq \liminf_n \int_\Omega f(\omega, u_n(\omega), v_n(\omega))d\mu.$$

Proof Choose a subsequence $\{u_{n_k}\}$ (resp. $\{v_{n_k}\}$) of $\{u_n\}$ (resp. $\{v_n\}$) such that

$$\int_\Omega f(\omega, u_{n_k}(\omega), v_{n_k}(\omega))d\mu \to \liminf_n \int_\Omega f(\omega, u_n(\omega), v_n(\omega))d\mu \quad \text{as } k \to \infty. \tag{4.19}$$

A sequence

$$\theta_n = \int_\Omega \delta_\omega \otimes \delta_{v_n(\omega)} d\mu; \quad n = 1, 2, \cdots$$

in $\mathfrak{Y}(\Omega, \mu; \mathfrak{X})$ is uniformly tight. This fact can be proved in the same manner as in the first part of the proof of Theorem 4.3. So there exists a subsequence $\{\theta_{n_k}\}$ of $\{\theta_n\}$ which narrowly converges to a Young measure

$$\theta_\infty = \int_\Omega \delta_\omega \otimes \nu_\omega^\infty d\mu,$$

by Theorem 3.5. According to Theorem 4.3,

$$\int_{\mathfrak{X}} \|x\| d\nu_\omega^\infty < \infty \quad a.e. \tag{4.20}$$

and $\{v_{n_k}\}$ weakly converges to

$$v_\infty(\omega) = \int_{\mathfrak{X}} x d\nu_\omega^\infty. \tag{4.21}$$

Define a sequence $\{\gamma_n\}$ in $\mathfrak{Y}(\Omega, \mu; S \times \mathfrak{X})$ by

$$\gamma_n = \int_\Omega \delta_\omega \otimes \delta_{u_n(\omega)} \otimes \delta_{v_n(\omega)} d\mu; \quad n = 1, 2, \cdots$$

Then, by Theorem 3.9.

$$\gamma_{n_k} \to \int_\Omega \delta_\omega \otimes \delta_{u_\infty(\omega)} \otimes \delta_{v_\infty(\omega)} d\mu \quad \text{(narrowly)} \quad \text{as} \quad k \to \infty.$$

By Theorem 4.1,

$$\liminf_n \int_\Omega f(\omega, u_n(\omega), v_n(\omega)) d\mu < \infty$$

implies

$$\int_\Omega \{\int_{\mathfrak{X}} f(\omega, u_\infty(\omega), x)^+ d\nu_\omega^\infty\} d\mu < \infty. \tag{4.22}$$

(Of course, this relation holds good when we change n to n_k.)

We also obtain that

$$\begin{aligned} \int_\Omega \{\int_{\mathfrak{X}} f(\omega, u_\infty(\omega), x) d\nu_\omega^\infty\} d\mu &\leqq \liminf_k \int_\Omega f(\omega, u_{n_k}(\omega), v_{n_k}(\omega)) d\mu \\ &= \lim_{k\to\infty} \int_\Omega f(\omega, u_{n_k}(\omega), v_{n_k}(\omega)) d\mu \qquad (4.23) \\ &= \liminf_n \int_\Omega f(\omega, u_n(\omega), v_n(\omega)) d\mu. \end{aligned}$$

Since $x \mapsto f(\omega, u_\infty(\omega), x)$ is convex and *l.s.c.*, it can be expressed as a supremum of affine functions.

$$x \mapsto \langle \Lambda_n, x \rangle + b_n, \quad \Lambda_n \in \mathfrak{X}', \quad b_n \in \mathbb{R}; \quad n = 1, 2, \cdots.$$

Keeping (4.21) in mind, we obtain[100]

$$
\begin{aligned}
f(\omega, u_\infty(\omega), v_\infty(\omega)) &= \sup_n\{\langle \Lambda_n, v_\infty(\omega)\rangle + b_n\} \\
&= \sup_n \int_{\mathfrak{X}} \{\langle \Lambda_n, x\rangle + b_n\} d\nu_\omega^\infty \\
&\leqq \int_{\mathfrak{X}} \sup_n\{\langle \Lambda_n, x\rangle + b_n\} d\nu_\omega^\infty \quad \text{(Fatou's lemma)} \qquad (4.24)\\
&= \int_{\mathfrak{X}} f(\omega, u_\infty(\omega), x) d\nu_\omega^\infty.
\end{aligned}
$$

Integrating the both hand sides of (4.24), we have

$$
\begin{aligned}
&\int_\Omega f(\omega, u_\infty(\omega), v_\infty(\omega)) d\mu \\
&\leqq \int_\Omega \{\int_{\mathfrak{X}} f(\omega, u_\infty(\omega), x) d\nu_\omega\} d\mu \\
&\underset{(4.23)}{\leqq} \liminf_n \int_\Omega f(\omega, u_n(\omega), v_n(\omega)) d\mu.
\end{aligned}
$$

This proves the theorem. □

Remark 9 The condition b which the function f has to satisfy can be re-expressed as:

b' $\{\omega \mapsto f(\omega, u_n(\omega), v_n(\omega))^-\}$ is weakly relatively compact.

Furthermore $\mathfrak{L}^1(\Omega, \mathfrak{X})$ can be generalized to any $\mathfrak{L}^p(\Omega, \mathfrak{X}),\ p \geqq 1$.

4.3 Towards Calculus of Variations

In the classical calculus of variations, we always have to go along with a nonlinear integral functional of the type

$$x(\cdot) \mapsto \int_0^T f(t, x(t), \dot{x}(t)) dt$$

---

[100] The inequality in (4.24) is justified as follows.

$$\langle \Lambda_n, x\rangle + b_n \leqq f(\omega, u_\infty(\omega), x) \leqq f(\omega, u_\infty(\omega), x)^+.$$

By (4.22),

$$\int_{\mathfrak{X}} f(\omega, u_\infty(\omega), x)^+ d\nu_\omega^\infty < \infty \quad a.e.$$

Thus, Fatou's lemma applies.

defined on some space of functions. Of course, certain continuity is required in order to assure the existence of a solution of a problem to minimize this functional.

We now present a *l.s.c.* result of such a functional defined on the Sobolev space $\mathfrak{W}^{1,p}(\Omega, \mathfrak{X})$.[101] In my several works preceding Maruyama [35], I examined the existence of solutions for a variational problem governed by a differential inclusion in the case $\mathfrak{X} = \mathbb{R}^l$, by making use of the convenient properties of the weak convergence in the Sobolev space $\mathfrak{W}^{1,2}([0,T], \mathbb{R}^l)$; i.e.. if a sequence $\{x_n\}$ in $\mathfrak{W}^{1,2}([0,T], \mathbb{R}^l)$ weakly converges to some $x^* \in \mathfrak{W}^{1,2}([0,T], \mathbb{R}^l)$, then there exists a subsequence $\{z_n\}$ of $\{x_n\}$ such that

$$\text{(W)} \qquad \begin{cases} z_n \to x^* & \text{uniformly on } [0,T], \textit{ and} \\ \dot{z}_n \to \dot{x}^* & \text{weakly in } \mathfrak{L}^2([0,T], \mathbb{R}^l). \end{cases}$$

However it deserves a special notice that this property does not hold in the space $\mathfrak{W}^{1,2}([0,T], \mathfrak{X})$ if $\dim \mathfrak{X} = \infty$.

On the other hand, Let $\mathfrak{X}$ be a real Banach space with the Radon–Nikodým property (RNP). Then any absolutely continuous function $f : [0,T] \to \mathfrak{X}$ is Fréchet-differentiable *a.e.* (If the Banach space $\mathfrak{X}$ does not have RNP, this property does not hold.) Let $\{x_n\}$ be a sequence in $\mathfrak{W}^{1,p}([0,T], \mathfrak{X})$ which weakly converges to some $x^* \in \mathfrak{W}^{1,p}([0,T], \mathfrak{X})$. We should keep in mind that it is not necessarily true that the sequence $\{x_n\}$ has a subsequence $\{z_n\}$ which satisfies the property (W) if $\dim \mathfrak{X} = \infty$ even in the case $p = 2$. (See Cecconi [15] for a counter-example.)

The following theorem cultivated to overcome this difficulty is a generalization of the above result. Henceforth we denote by $\mathfrak{X}_s$ (resp. $\mathfrak{X}_w$) a Banach space $\mathfrak{X}$ endowed with the strong (resp. weak) topology.

Theorem 4.5 *Let $\mathfrak{X}$ be a real separable reflexive Banach space. And consider a sequence $\{x_n\}$ in the Sobolev space $\mathfrak{W}^{1,p}([0,T], \mathfrak{X})(p \geqq 1)$. Assume that*

(i) the set $\{x_n(t)\}_{n=1}^{\infty}$ is bounded (and hence relatively compact) in $\mathfrak{X}_w$ for each $t \in [0,T]$, and
(ii) there exists some function $\psi \in \mathfrak{L}^p([0,T], (0,\infty))$ such that

$$\|\dot{x}_n(t)\| \leqq \psi(t) \quad a.e.$$

Then there exist a subsequence $\{z_n\}$ of $\{x_n\}$ and some function $x^* \in \mathfrak{W}^{1,p}([0,T], \mathfrak{X})$ such that

(a) $z_n \to x^*$ uniformly in $\mathfrak{X}_w$ on $[0,T]$, and
(b) $\dot{z}_n \to \dot{x}^*$ weakly in $\mathfrak{L}^p([0,T], \mathfrak{X})$.

Remark 10 Since $\mathfrak{X}$ is separable and reflexive, the following results hold true. Assume that $p \geqq 1$.

[101] The following discussion is based upon Maruyama [35].

1° $\mathfrak{L}^p([0, T], \mathfrak{X})$ is separable.
2° $\mathfrak{L}^p([0, T], \mathfrak{X})'$ is isomorphic to $\mathfrak{L}^q([0, T], \mathfrak{X}')$, where $1/p + 1/q = 1$.
3° Any absolutely continuous function $f : [0, T] \to \mathfrak{X}$ is Fréchet-differentiable *a.e.* and the "fundamental theorem of calculus", i.e.

$$f(t) = f(0) + \int_0^t \dot{f}(\tau)d\tau; \ t \in [0, T]$$

is valid.

Proof of Theorem 4.5

(a) To start with, we shall show the equicontinuity of $\{x_n\}$. Since ψ is integrable, there exists some $\delta > 0$, for each $\varepsilon > 0$, such that

$$\|x_n(t) - x_n(s)\| \leqq \int_s^t \|\dot{x}_n(\tau)\|d\tau \leqq \int_s^t \psi(\tau)d\tau \leqq \varepsilon \quad \text{for all } n$$

provided that $|t - s| \leqq \delta$. This proves the equicontinuity of $\{x_n\}$ in the strong topology for $\mathfrak{X}$. Hence $\{x_n\}$ is also equicontinuous in the weak topology.

Taking account of this fact as well as the assumption (i), we can claim, thanks to the Ascoli–Arzelà theorem, that $\{x_n\}$ is relatively compact in $\mathfrak{C}([0, T], \mathfrak{X}_w)$ (the set of continuous functions of $[0, T]$ into $\mathfrak{X}_w$) with respect to the topology of uniform convergence.

By the assumption (i), $\{x_n(0)\}$ is bounded in $\mathfrak{X}$, say $\sup_n \|x_n(0)\| \leqq C < \infty$. and the assumption (ii) implies that

$$\left\| \int_0^t \dot{x}_n(\tau)d\tau \right\| \leqq \|\psi\|_1 \quad \text{for all} \quad t \in [0, T].$$

Hence

$$\begin{aligned} \sup_n \|x_n(t)\| &= \sup_n \left\| x_n(0) + \int_0^t \dot{x}_n(\tau)d\tau \right\| \\ &\leqq C + \|\psi\|_1 \quad \text{for all} \quad t \in [0, T]. \end{aligned}$$

Thus each x_n can be regarded as a mapping of $[0, T]$ into the set

$$M = \{w \in \mathfrak{X} |\ \|w\| \leqq C + \|\psi\|_1\}.$$

The weak topology on M is metrizable because M is bounded and $\mathfrak{X}$ is a separable reflexive Banach space. Hence if we denote by M_w the space M endowed with the weak topology, then the uniform convergence topology on $\mathfrak{C}([0, T], M_w)$ is metrizable.

Since we can regard $\{x_n\}$ as a relatively compact subset of $\mathfrak{C}([0, T], M_w)$, there exists a subsequence $\{y_n\}$ of $\{x_n\}$ which uniformly converges to some $x^* \in \mathfrak{C}([0, T], \mathfrak{X}_w)$.
(b) Since

$$\|\dot{y}_n(t)\| \leqq \psi(t) \quad a.e.$$

the sequence $\{w_n : [0, T] \to \mathfrak{X}\}$ defined by

$$w_n(t) = \frac{\dot{y}_n(t)}{\psi(t)}; \ n = 1, 2, \cdots$$

is contained in the unit ball of $\mathfrak{L}^\infty([0, T], \mathfrak{X})$ which is weak*-compact (as the dual space of $\mathfrak{L}^1([0, T], \mathfrak{X}'))$ by Alaoglu's theorem. Note that the weak* topology on the unit ball of $\mathfrak{L}^\infty([0, T], \mathfrak{X})$ is metrizable since $\mathfrak{L}^1([0, T], \mathfrak{X}')$ is separable. Hence $\{w_n\}$ has a subsequence $\{w_{n'}\}$ which converges to some $w^* \in \mathfrak{L}^\infty([0, T], \mathfrak{X})$ in the weak* topology. We shall write $\dot{z}_n = \dot{y}_{n'} = \psi \cdot w_{n'}$.

If we define an operator $A : \mathfrak{L}^\infty([0, T], \mathfrak{X}) \to \mathfrak{L}^p([0, T], \mathfrak{X})$ by

$$A : g \mapsto \psi \cdot g,$$

then A is continuous in the weak* topology for $\mathfrak{L}^\infty$ and the weak topology for $\mathfrak{L}^p$. In order to see this, let $\{g_\lambda\}$ be a net in $\mathfrak{L}^\infty([0, T], \mathfrak{X})$ such that $w^*\text{-}\lim_\lambda g_\lambda = g^* \in \mathfrak{L}^\infty([0, T], \mathfrak{X})$: i.e.

$$\int_0^T \langle \alpha(t), g_\lambda(t) \rangle dt \to \int_0^T \langle \alpha(t), g^*(t) \rangle dt \quad \text{for all} \quad \alpha \in \mathfrak{L}^1([0, T], \mathfrak{X}').$$

Then it is quite easy to verify that

$$\begin{aligned} \int_0^T \langle \beta(t), \psi(t) g_\lambda(t) \rangle &= \int_0^T \langle \psi(t)\beta(t), g_\lambda(t) \rangle dt \\ &\to \int_0^T \langle \psi(t)\beta(t), g^*(t) \rangle dt \\ &\text{for all} \quad \beta \in \mathfrak{L}^q([0, T], \mathfrak{X}'), \ 1/p + 1/q = 1 \end{aligned}$$

since $\psi \cdot \beta \in \mathfrak{L}^1([0, T], \mathfrak{X}')$. This proves the continuity of A.

Hence

$$\dot{z}_n = \psi \cdot w_{n'} \to \psi \cdot w^* \quad \text{weakly in} \quad \mathfrak{L}^p([0, T], \mathfrak{X}), \tag{4.25}$$

which implies

$$\left\langle \theta, \int_s^t \dot{z}_n(\tau)d\tau \right\rangle = \int_s^t \langle \theta, \dot{z}_n(\tau)\rangle d\tau \tag{4.26}$$

$$\to \int_s^t \langle \theta, \psi(\tau)\cdot w^*(\tau)\rangle d\tau \quad \text{for all} \quad \theta \in \mathfrak{X}'.$$

On the other hand, since

$$z_n(t) - z_n(s) = \int_s^t \dot{z}_n(\tau)d\tau \quad \text{for all} \quad n,$$

and $z_n(t) - z_n(s) \to x^*(t) - x^*(s)$ in $\mathfrak{X}_w$, we get

$$\left\langle \theta, \int_s^t \dot{z}_n(\tau)d\tau \right\rangle = \langle \theta, z_n(t) - z_n(s)\rangle \tag{4.27}$$

$$\to \langle \theta, x^*(t) - x^*(s)\rangle \quad \text{for all} \quad \theta \in \mathfrak{X}'.$$

(4.26) and (4.27) imply the relation

$$\langle \theta, x^*(t) - x^*(s)\rangle = \left\langle \theta, \int_s^t \psi(\tau)\cdot w^*(\tau)d\tau \right\rangle$$

$$\text{for all} \quad \theta \in \mathfrak{X}',$$

from which we can deduce the equality

$$x^*(t) - x^*(s) = \int_s^t \psi(\tau)\cdot w^*(\tau)d\tau. \tag{4.28}$$

By (4.25) and (4.28), we get the desired result:

$$\dot{z}_n \to \dot{x}^* = \psi \cdot w^* \quad \text{weakly in} \quad \mathfrak{L}^p([0, T], \mathfrak{X}).$$

□

In the proof of our Theorem 4.5, we made use of some ideas of Aubin and Cellina [3, pp.13–14].

Theorem 4.6 *Let $\{x_n\}$ be a sequence in $\mathfrak{W}^{1,p}([0, T], \mathfrak{X})$ $(p > 1)$, such that all the images of x_n's are contained in some closed ball $\bar{B}$ with the center 0 : i.e.*

$$x_n(t) \in \bar{B} \quad \textit{for all} \quad t \in [0, T] \quad \textit{and} \quad n.$$

Let $f : [0, T] \times \mathfrak{X}_w \times \mathfrak{X}_s \to \bar{\mathbb{R}}$ *be a proper convex normal integrand with the lower compactness property. Then there exist a subsequence* $\{z_n\}$ *of* $\{x_n\}$ *and* $x^* \in \mathfrak{W}^{1,p}([0, T], \mathfrak{X})$ *such that*

$$J(x^*) \leqq \liminf_n J(z_n), \tag{4.29}$$

where

$$J(x) = \int_0^T f(t, x(t), \dot{x}(t))dt.$$

Proof By assumption, we may restrict the domain of f to $[0, T] \times \bar{B}_w \times \mathfrak{X}_s$ provided that the sequence $\{x_n\}$ is concerned. Denoting $\bar{f} = f|_{[0,T]\times\bar{B}\times\mathfrak{X}}$, (restriction of f to $[0, T] \times \bar{B} \times \mathfrak{X}$), we have to show that there exist a subsequence $\{z_n\}$ of $\{x_n\}$ and some $x^* \in \mathfrak{W}^{1,p}([0, T], \mathfrak{X})$ such that

$$\int_0^T \bar{f}(t, x^*(t), \dot{x}^*(t))dt \leqq \liminf_n \int_0^T \bar{f}(t, z_n(t), \dot{z}_n(t))dt. \tag{4.30}$$

which is equivalent to (4.29).

The set $\bar{B}$ endowed with the weak topology is metrizable and compact. Hence it is a Polish space. According to Theorem 4.3, there exist a subsequence $\{z_n\}$ of $\{x_n\}$ and $x^* \in \mathfrak{W}^{1,p}([0, T], \mathfrak{X})$ such that

(a) $z_n \to x^*$ uniformly in $\bar{B}_w$, and
(b) $\dot{z}_n \to \dot{x}^*$ weakly in $\mathfrak{L}^p([0, T], \mathfrak{X})$.

(a) implies, of course, that $z_n \to x^*$ in measure. Thus applying Theorem 4.5, we obtain the desired relation (4.30). □

See Berkovitz [5, 6], Cesari [16], Ioffe [25] and Olech [40] for basic related results concerning the *l.s.c.* of integral functionals.

Acknowledgements Almost all the materials of this article are based upon my lectures delivered at Keio University and University of Tokyo. Some specific topics were also discussed in my lectures at University of California, Berkeley and Ohio State University. I would like to devote my sincere thanks to the audience for critical responses. The Japanese version (with some difference) is to appear in *Mita Journal of Economics* edited by Keio Economic Society, a generous permission by which must be cordially appreciated.

References

1. Arrow K, Hahn F (1971) General competitive analysis. Holden Day, San Francisco
2. Artstein Z (1980) Discrete and continuous bang-bang and facial spaces, or: look for the extreme points. SIAM Rev 22:172–185
3. Aubin JP, Cellina A (1984) Differential inclusions. Springer, Berlin

4. Aumann R, Katznelson A, Radner R, Rosenthal R, Weiss B (1983) Approximate purification of mixed strategies. Math Oper Res 8:327–341
5. Berkovitz LD (1974) Existence and lower closure theorems for abstract control problems. SIAM J Control Optim 12:27–42
6. Berkovitz LD (1983) Lower semi-continuity of integral functionals. Trans Amer Math Soc 192:51–57
7. Berliocchi H, Lasry JM (1973) Intégrandes normales et mesures paramétrées en calcul des variations. Bull Soc Math France 101:129–184
8. Billingsley P (1968) Convergence of probability measures. Wiley, New York
9. Bourbaki N (1966) Elements of mathematics; general topology, part 1,2. Addison-Wesley, Reading
10. Bourbaki N (2004) Elements of mathematics; integration. I. Springer, Berlin
11. Bourbaki N (2004) Elements of mathematics; topological vector spaces. Springer, Berlin
12. Castaing C, Clauzure P (1982) Semi-continuité des fonctionelles intégrales. Acta Math Vietnam 7:139–170
13. Castaing C, Valadier M (1977) Convex analysis and measureable multifunctions. Springer, Berlin
14. Castaing C, de Fitte RR, Valadier M (2004) Young measures on topological spaces. Kluwer Academic Publishers, Dordrecht
15. Cecconi JP (1987) Problems of the calculus of variations in Banach spaces and classes BV. In: Cesari L (ed) Contributions to modern calculus of variations. Longman Scientific and Technical, Harlow
16. Cesari L (1974) Lower semi-continuity and lower closure theorems without semi-normality conditions. Anal Math Pure Appl 98:381–397
17. Chatterji SD (1973) Disintegration of measures and lifting. In: Tucker DH, Maynard HB (eds) Vector and operator valued measures and applications. Academic Press, New York, pp 69–83
18. Choquet G (1969) Lectures on analysis, vol I. Benjamin, London
19. Dellacherie C, Meyer P-A (1978) Probabilities and potential. Hermann, Paris; North-Holland, Amsterdam
20. Diestel N, Uhl JJ Jr (1977) Vector measures. Amer Math Soc, Providence
21. Dunford N, Schwartz JT (1958) Linear operators, part 1. Interscience, New York
22. Evans L (1990) Weak convergence methods for nonlinear partial differential equations. Amer Math Soc, Providence
23. Heyer H (1977) Probability measures on locally compact groups. Springer, Berlin
24. Hoffmann-Jørgensen J (1971) Existence of conditional probabilities. Math Scand 28:257–264
25. Ioffe AD (1977) On lower semicontinuity of integral functionals I. SIAM J Control Optim 15:521–538
26. Ionescu Tulcea A, Ionescu Tulcea C (1970) Topics in the theory of liftings. Springer, Berlin
27. Itô K (1991) Kakuritsu-ron (probability theory) (in Japanese). Iwanami, Tokyo
28. Jirina M (1959) On regular conditional probabilities. Czech Math J 9:445–450
29. Kuratowski K, Ryll-Nardzewski C (1965) A general theorem on selectors. Bull Ac Pol Sc 13:397–403
30. Loève M (1963) Probability theory, 3rd ed. van Nostrand, Princeton
31. Maharam D (1958) On a theorem of von Neumann. Proc Amer Math Soc 9:987–994
32. Malliavin P (1995) Integration and probability. Springer, New York
33. Maruyama T (1979) An extension of Aumann-Perles' variational problem. Proc Japan Acad 55:348–352
34. Maruyama T (1995) Suri-keizaigaku no hoho (Methods in mathematical economics) (in Japanese). Sōbunsha, Tokyo
35. Maruyama T (2001) A generalization of the weak convergence theorem in Sobolev spaces with application to differential inclusions in a Banach space. Proc Japan Acad 77:5–10
36. Maruyama T (2006) Sekibun to kansu-kaiseki (Integration and functional analysis) (in Japanese). Springer, Tokyo

37. Maruyama T (2017) A note on variational problems via theory of Young measures. Pure Appl Funct Anal 2:357–367
38. Maruyama T (2018) Disintegration of Young measures and conditional probabilities. J Nonlinear Var Anal 2:229–232
39. Nagumo M (1929) Über die Gleichmassige Summierberkeit und ihre Anwendung auf ein variations-problem. Japanese J Math 6:178–182
40. Olech C (1975) Existence theory in optimal control problems – the underlying ideas. In: International conference on diffential equation. Academic Press, New York, pp. 612–629
41. Piccinini L, Valadier M (1995) Uniform integrability and Young measures. J Math Anal Appl 195:428–439
42. Sainte-Beuve MF (1974) On the extension of von Neumann-Aumann's theorem. J Funct Anal 17:112–129
43. Schwartz L (1973) Radon measures on arbitrary topological spaces and cylindrical measures. Oxford, London
44. Schwartz L (1975) Lectures on disintegration of measures. Tata Institute of Fundamental Research, Bombay (1975)
45. Valadier M (1972) Comparaison de troi théorèmes de désintégration. Séminaire d'analyse convexe exposé n°10. 10.1–10.21
46. Valadier M (1973) Désintegration d'une mesure sur un produit. C R Acad Sci Paris Sér 1 276:A33–A35
47. Valadier M (1990) Application des mesures de Young aux suites uniformément intégrables dans un Banach séparable. Séminaire d'analyse convexe, exposé n°3. 3.1–3.14
48. Valadier M (1990) Young measures. In: Cellina A (ed) Methods of nonconvex analysis. Springer, Berlin, pp 152–188
49. Valadier M (1994) A course on Young measures. Rendiconti dell'istituto di matematica dell' Università di Trieste. 26(suppl):349–394
50. von Neumann J (1931) Algebraische Repräsentaten der Funktionen: bis auf eine Menge von Masse Null. J Reine Angew Math 165:109–115
51. Warga J (1972) Optimal control of differential and functional equations. Academic Press, New York
52. Yosida K (1971) Functional analysis, 3rd ed. Springer, Berlin (1971)
53. Zaanen AC (1997) Introduction to operator theory in Riesz spaces. Springer, Berlin (1997)

Complete List of Works Published in Advances in Mathematical Economics Vol. 1–23

Vol. 1

Foreword

Gérard Debreu
On the use in economic theory of some central results of mathematical analysis

Research Articles

Laurent Calvet, Jean-Michel Grandmont, Isabelle Lemaire
Heterogenous probabilities in complete asset markets

Charles Castaing, Mohamed Guessous
Convergences in $L^1_X(\mu)$

Egbert Dierker, Hildegard Dierker
Product differentiation and market power

Shigeo Kusuoka
A Remark on default risk models

Hiroshi Shirakawa
Evaluation of yield spread for credit risk

Michel Valadier
Analysis of the asymptotic distance between oscillating functions and their weak limit in L^2

Survey

Kazuo Nishimura, Makoto Yano
Chaotic solutions in infinite-time horizon linear programming and economic dynamics

T. Maruyama (ed.), *Advances in Mathematical Economics*, Advances in Mathematical Economics 23, https://doi.org/10.1007/978-981-15-0713-7

Note
Shinichi Suda
Determinacy of monetary equilibria in an economy with no real risk

Vol. 2
Research Articles
Svetlana V. Anoulova, Igor V. Evstigneev, Volker M. Gundlach
Turnpike theorems for positive multivalued stochastic operators

Charles Castaing, Ahmed Gamal Ibrahim
Functional differential inclusion on closed sets in Banach spaces

Tatsuro Ichiishi, Semih Koray
Job matching: a multi-principal, multi-agent model

Shigeo Kusuoka
Term structure and SPDE

Ken Urai
Fixed point theorems and the existence of economic equilibria based on conditions for local directions of mappings

Akira Yamazaki
Efficiency of stochastic transfers in a directed graph

Alexander J. Zaslavski
Allocations of labour resources on trajectories for the model with discrete innovations

Vol. 3
Research Articles
Houcine Benabdellah, Charles Castaing
Weak compactness and convergences in $L^1_{E'}[E]$

Alexander D. Ioffe
Abstract convexity and non-smooth analysis

Seiichi Iwamoto
Recursive method in stochastic optimization under compound criteria

Shigeo Kusuoka
On law invariant coherent risk measures

Vladimir L. Levin
The Monge-Kantorovich problems and stochastic preference relations

Vol. 4

Research Articles

Alexander D. Ioffe, Alexander M. Rubinov
Abstract convexity and nonsmooth analysis: global aspects

Hiroyuki Ozaki
Dynamic programming with upper semi-continuous stochastic aggregator

Hiroshi Tateishi
On the existence of equilibria of equicontinuous games with incomplete information

Nicholas C. Yannelis
A Bayesian equilibrium existence theorem

Vol. 5

Research Articles

Guillaume Carlier
Duality and existence for a class of mass transportation problems and economic applications

Charles Castaing, Ahmed Gamal Ibrahim
Functional evolution equations governed by m-accretive operators

Takao Fujimoto, José A. Silva, Antonio Villar
Nonlinear generalizations of theorems on inverse-positive matrices

Leonid Hurwicz, Marcel K. Richter
Implicit functions and diffeomorphisms without C^1

Leonid Hurwicz, Marcel K. Richter
Optimization and Lagrange multipliers: non-C^1 constraints and "minimal" constraint qualifications

Shigeo Kusuoka
Monte Carlo method for pricing of Bermuda type derivatives

Historical Perspective

Isao Mutoh
Mathematical economics in Vienna between the Wars

Vol. 6

Research Articles

Charles Castaing, Paul Raynaud de Fitte
On the fiber product of Young measures with application to a control problem with measures

Dionysius Glycopantis, Allan Muir
The compactness of $Pr(K)$

Seiichi Iwamoto
Recursive methods in probability control

Shigeo Kusuoka
Approximation of expectation of diffusion processes based on Lie algebra and Malliavin calculus

Vladimir L. Levin
Optimal solutions of the Monge problem

Hidetoshi Nakagawa, Tomoaki Shouda
Valuation of mortgage-backed securities based on unobservable prepayment costs

Ken Urai, Akihiko Yoshimachi
Fixed point theorems in Hausdorff topological vector spaces and economic equilibrium theory

Akira Yamazaki
Monetary equilibrium with buying and selling price spread without transactions costs

Vol. 7

Research Articles

Chales Castaing, Paul Raynaud de Fitte, Anna Salvadori
Some variational convergence results for a class of evolution inclusions of second order using Young measures

Marco Frittelli, Emanuela Rosazza Gianin
Law invariant convex risk measures

Vladimir L. Levin
A method in demand analysis connected with the Monge-Kantorovich problem

Ryo Nagata
Real indeterminacy of equilibria with real and nominal assets

Jean-Paul Penot
The bearing of duality on microeconomics

Vol. 8
Research Articles
Laura Angeloni, Bernard Cornet
Existence of financial equilibria in a multi-period stochastic economy

Charles Castaing, Paul Raynaud de Fitte, Anna Salvadori
Some variational convergence results with applications to evolution inclusions

Freddy Delbaen
Hedging bounded claims with bounded outcomes

Antoni Bosch-Domènech, Joaquim Silvestre
The gain-loss asymmetry and single-self preferences

Prajit K. Dutta, Roy Radner
A game-theoretic approach to global warming

Gilles Evéquoz, Charles Alexander Stuart
On differentiability and bifurcation

Dionysius Glycopantis, Allan Muir, Nicholas C. Yannelis
On extensive form implementation of equilibria in differential information economies

Jean-Michel Grandmont
Fiscally stable income distributions under majority voting, Lorenz curves and bargaining sets

Hiroaki Hata, Jun Sekine
Solving long term optimal investment problems with Cox-Ingersoll-Ross interest rates

Leonid Hurwicz
Implementation with unknown endowments in a two-trader pure exchange economy

Tatsuro Ichiishi, Akira Yamazaki
Cooperative extensions of the Bayesian game

Alexander Ioffe
Variational problem associated with a model of welfare economics with a measure space of agents

Kunio Kawamata, Yasunari Tamada
Direct and indirect connections, the Shapley value, and network formation

M. Ali Khan, Tapan Mitra
Discounted optimal growth in the two-sector RSS model: a geometric investigation

Hidetoshi Nakagawa, Tomoaki Shouda
A prepayment model of mortgage-backed securities based on unobservable prepayment cost processes

Dov Monderer
Solution-based congestion games

Akihiko Takahashi, Yoshihiko Uchida
New acceleration schemes with the asymptotic expansion in Monte Carlo simulation

Naoki Watanabe, Shigeo Muto
Licensing agreements as bargaining outcomes: general results and two examples

Makoto Yano
The Bertrand equilibrium in a price competition game

Vol. 9
Research Articles
Takashi Adachi
Option on a unit-type closed-end investment fund

Takahiko Fujita, Ryozo Miura
The distribution of continuous time rank processes

Hirotaka Fushiya
Asymptotic expansion for a filtering problem and a short term rate model

Elyès Jouini, Walter Schachermayer, Nizar Touzi
Law invariant risk measures have the Fatou property

Mikio Nakayama
The dawn of modern theory of games

Manabu Toda
Approximation of excess demand on the boundary and euilibrium price set

Yuji Umezawa
The minimal risk of hedging with a convex risk measure

Na Zhang
The distribution of firm size

Vol. 10
Research Articles
Charles Castaing, Mohammed Saadoune
Komlós type convergence for random variables and random sets with applications to minimization problems

Jean-Philippe Garnier, Kazuo Nishimura, Alain Venditti
Capital-labor substitution and indeterminacy in continuous-time two-sector models

Takanori Ibaraki, Wataru Takahashi
Weak and strong convergence theorems for new resolvents of maximal monotone operators in Banach spaces

Seiichi Iwamoto
Golden optimal policy in calculus of variation and dynamic programming

Shigeo Kusuoka
A remark on law invariant convex risk measures

Anna Rubinchik, Shlomo Weber
Existence and uniqueness of an equilibrium in a model of spatial electoral competition with entry

Hiroaki Hata, Jun Sekine
Publisher's Errata: Solving long term optimal investment problems with Cox-Ingersoll-Ross interest rates

Vol. 11
Research Articles
Takuji Arai
Optimal hedging strategies on asymmetric functions

Charles Castaing, Christian Hess, Mohamed Saadoune
Tightness conditions and integrability of the sequential weak upper limit of a sequence of multifunctions

Chiaki Hara
Core convergence in economies with bads

Hidetoshi Komiya
A distance and a binary relation related to income comparisons

Vladimir L. Levin
On preference relations that admit smooth utility functions

Takashi Matsuhisa, Ryuichiro Ishikawa
Rational expectations can preclude trades

Robert J. Rossana
The Le Chatelier Principle in dynamic models of the firm

Tomoichi Shinotsuka
Interdependent utility functions in an intergenerational context

Vol. 12
Research Articles
Hideyuki Adachi
Unemployment and income distribution in the medium-run growth model

Charles Castaing, Mohammed Saadoune
Convergences in a dual space with applications to Fatou lemma

Alexander D. Ioffe
Variational analysis and mathematical economics 1: subdifferential calculus and the second theorem of welfare economics

Vladimir L. Levin
Smooth feasible solutions to a dual Monge–Kantorovich problem with applications to best approximation and utility theory in mathematical economics

Koichi Matsumoto
Optimal growth rate in random trade time

Hideatsu Tsukahara
Some properties of distortion risk measures

Vol. 13
Research Articles
Fettah Akhiat, Charles Castaing, Fatima Ezzaki
Some various convergence results for multivalued martingales

Jun Honda, Shin-Ichi Takekuma
A note on Aumann's core equivalence theorem without monotonicity

M. Ali Khan, Alexander J. Zaslavski
On two classical turnpike results for the Robinson–Solow–Srinivasan model

Shigeo Kusuoka
A certain limit of iterated conditional tail expectation

Truong Q. Bao, Boris S. Mordukhovich
Set-valued optimization in welfare economics

Nobusumi Sagara, Milan Vlach
Convexity of the lower partition range of a concave vector measure

Alexander J. Zaslavski
Good locally maximal programs for the Robinson–Solow–Srinivasan model

Historical Perspective
S. Todd Lowry
Pythagorean mathematical idealism and the framing of economic and political theory

Vol. 14
Research Articles
Takuji Arai, Takamasa Suzuki
How much can investors discount?

Alexander D. Ioffe
Variational analysis and mathematical economics 2: nonsmooth regular economies

M. Ali Khan, Adriana Piazza
An overview of turnpike theory: towards the discounted deterministic case

Koji Kuroda, Jun-ichi Maskawa, Joshin Murai
Stock price process and long memory in trade signs

Aychiluhim Habte, Boris S. Mordukhovich
Extended second welfare theorem for nonconvex economies with infinite commodities and public goods

Yumiharu Nakano
Partial hedging for defaultable claims

Keita Owari
Robust utility maximization with unbounded random endowment

Alejandro Jofré, R. Tyrrell Rockafellar, Roger J.-B. Wets
A time-embedded approach to economic equilibrium with incomplete financial markets

Wataru Takahashi, Jen-Chih Yao
Strong convergence theorems by hybrid methods for nonexpansive mappings with equilibrium problems in Banach spaces

Vol. 15

Research Articles

Charles Castaing
Some various convergence results for normal integrands

Samuel Drapeau, Michael Kupper, Ranja Reda
A note on robust representations of law-invariant quasiconvex functions

Toru Maruyama
On the Fourier analysis approach to the Hopf bifurcation theorem

Pavel Kocourek, Wataru Takahashi, Jen-Chih Yao
Fixed point theorems and ergodic theorems for nonlinear mappings in Banach spaces

Historical Perspective

Akira Yamazaki
On the perception and representation of economic quantity in the history of economic analysis in view of the Debreu conjecture

Note

Yuhki Hosoya
An existence result and a characterization of the least concave utility of homothetic preferences

Vol. 16

Research Articles

Charles Castaing, Marc Lavie
Some applications of Birkhoff-Kingman ergodic theorem

Yuhki Hosoya
Elementary form and proof of the Frobenius theorem for economists

Shigeo Kusuoka, Takenobu Nakashima
A remark on credit risk models and copula

Naoki Yoshihara, Roberto Veneziani
Profits and exploitation: a reappraisal

Notes

Kentaro Miyazaki, Shin-Ichi Takekuma
On the equivalence between the rejective core and the dividend equilibrium: a note

Takahiko Fujita, Naoyuki Ishimura, Norihisa Kawai
Discrete stochastic calculus and its applications: an expository note

Vol. 17

Research Articles

Charles Castaing, Paul Raynaud de Fitte
Law of large numbers and Ergodic Theorem for convex weak star compact valued Gelfand-integrable mappings

M. Ali Khan, Tapan Mitra
Discounted optimal growth in a two-sector RSS model: a further geometric investigation

Shigeo Kusuoka
Gaussian K-scheme: justification for KLNV method

Takashi Suzuki
Competitive equilibria of a large exchange economy on the commodity space ℓ^∞

Hisatoshi Tanaka
Local consistency of the iterative least-squares estimator for the semiparametric binary choice model

Vol. 18

Research Articles

Charles Castaing, Christiane Godet-Thobie, Le Xuan Truong, Bianca Satco
Optimal control problems governed by a second order ordinary differential equation with m-point boundary condition

Shigeo Kusuoka, Yusuke Morimoto
Stochastic mesh methods for Hörmander type diffusion processes

Survey Article

Alexander J. Zaslavski
Turnpike properties for nonconcave problems

Note

Yuhki Hosoya
A characterization of quasi-concave function in view of the integrability theory

Vol. 19
Charles Castaing, Christiane Godet-Thobie, Thi Duyen Hoang, Paul Raynaud de Fitte
On the integration of fuzzy level sets

Yuhki Hosoya
A theory for estimating consumer's preference from demand

Shigeo Kusuoka, Yusuke Morimoto
Least square regression methods for Bermudan derivatives and systems of functions

Alexander J. Zaslavski
Discrete time optimal control problems on large intervals

Vol. 20
Part I Research Articles
Takuji Arai
Local risk-minimization for Barndorff-Nielsen and Shephard models with volatility risk premium

Chales Castaing, Christaine Godet-Thobie, Le Xuan Truong, Fatima-Zohra Mostefai
On a fractional differential inclusion in Banach space under weak compactness condition

Yuhki Hosoya
On first-order partial differential equations: an existence theorem and its applications

Yoshiyuki Sekiguchi
Real radicals and finite convergence of polynomial optimization problems

Part II Expository Review
M. Ali Khan, Takashi Suzuki
On differentiated and indivisible commodities: an expository re-framing of Mas-Colell's 1975 Model

Part III Mini Courses
Vladimir Tikhomirov
Survey of the theory of extremal problems

Toru Maruyama
Fourier analysis of periodic weakly stationary processes: a note on Slutsky's observation

Vol. 21
Charles Castaing, Truong Le Xuan, Paul Raynaud de Fitte, Anna Salvadori
Some problems in second order evolution inclusions with boundary condition: a variational approach

M. Ali Khan and Yongchao Zhang
On sufficiently-diffused information in Bayesian games: a dialectical formalization

Shigeo Kusuoka
On supermartingale problems

Alexander J. Zaslavski
Bolza optimal control problems with linear equations and periodic convex integrands on large intervals

Vol. 22
Takuji Arai, Yuto Imai, and Ryo Nakashima
Numerical analysis on quadratic hedging strategies for normal inverse Gaussian models

Chales Castaing, Manuel D. P. Monteiro Marques, and Paul Raynaud de Fitte
Second-order evolution problems with time-dependent maximal monotone operator and applications

Yuhki Hosoya
Plausible equilibria and backward payoff-keeping behavior

Jun Kawabe
A unified approach to convergence theorems of nonlinear integrals

Takuma Kunieda, Kazuo Nishimura
A two-sector growth model with credit market imperfections and production externalities

Vol. 23
Charles Castaing, Manuel D.P. Monteiro Marques, Soumia Saïdi
Evolution problems with time-dependent subdifferential operators

Hélène Frankowska
Infinite horizon optimal control of non-convex problems under state constraints

Tomoki Inoue
The nonemptiness of the inner core

M. Ali Khan, Tapan Mitra
Complicated dynamics and parametric restrictions in the Robinson-Solow-Srinivasan (RSS) model

Shigeo Kusuoka
Some regularity estimates for diffusion semigroups with Dirichlet boundary conditions

Toru Maruyama
Disintegration of Young measures and nonlinear analysis

Index

T. Maruyama (ed.), *Advances in Mathematical Economics*, Advances in Mathematical Economics 23, https://doi.org/10.1007/978-981-15-0713-7

Printed in the United States
By Bookmasters